Environmental Technologies

New Developments

Edited by

E. Burcu Özkaraova Güngör

Environmental Technologies

Edited by E. Burcu Özkaraova Güngör

Republished by InTech

Janeza Trdine 9, 51000 Rijeka, Croatia

Technical Editor Teodora Smiljanic

Cover Designer Dino Smrekar

Additional hard copies can be obtained from orders@intechopen.com

Environmental Technologies, Edited by E. Burcu Özkaraova Güngör
p. cm.
ISBN 978-3-902613-10-3

For EU product safety concerns:
IN TECH d.o.o., Prolaz Marije Krucifikse Kozulić 3, 51000 Rijeka, Croatia,
info@intechopen.com or visit our website at intechopen.com.

Meet the editor

Dr Özkaraova Güngör works as an Assistant Professor at the Ondokuz Mayıs University since 2002. She graduated from the Environmental Engineering Department of Dokuz Eylül University, Izmir, in 1993. She was employed as a research assistant at the Boğaziçi University between 1996 and 2002. She holds a PhD in Environmental Technology on Soil Contaminant Interaction and Pollution Remediation and a MSc on Advanced Oxidation Processes. Over the past fifteen years, her primary research focus has been in the area of soil contaminant interaction, soil and ground water pollution, pollution control and remediation, and solid waste management. She worked as a project leader and researcher in projects like "Evaluation of the Soil Pollution Potential in the Middle Black Sea Region," "A Comparative Investigation of Heavy Metal Removal from Aegean, Mediterranean and Black Sea Region Soils," "Soil Pesticide Interactions and Investigation of Alternative Removal Methods." Regularly, she publishes academic papers, serves as peer-reviewer for various international academic journals, and for scientific activities. She is a member of the Chamber of Environmental Engineers.

Preface

There is no doubt that most of the environmental problems, which we are dealing with, are a result of improper management of industrial activities. Available techniques are used to reduce the emissions and the impact on the ecosystem, but stresses on the ecosystem continue. On the other side, the desire for a healthy environment increases, which produces the need for better tools to improve pollution prevention, control and cleanup capabilities. Besides the development of environmentally sound technologies based on waste minimization, energy efficiency and better use of resources, environmental technology research and development continues to lower future expenditures, to address specialized problems more efficiently and to achieve the required industrial and environmental standards. Special emphasis should be given to the reduction of risks to the ecosystem, which requires improved detection, monitoring and characterization of environmental changes. These practices will provide better information on exposure, enabling more precise environmental health risk assessments, which should be considered during the re-evaluation of new technology developments. Following these statements it should be realized that the performance of technology is related to well performed environmental management systems relying on collaborative or multi – stakeholder approaches.

This book on Environmental Technology takes a look at issues such as air, soil and noise pollution problems, environmental quality assessment, monitoring, modelling and risk assessment, environmental health impact assessment, environmental management and environmental technology development. It represents institutional arrangements, financial mechanisms and some sustainable technologies. The user can always count on finding both introductory material and more specific material based on national interests and problems. The user will also find ample references at the end of each chapter, if additional information is required. For additional questions or comments the user is encouraged to contact the author.

The book was a result of efforts by many experts from different professionals. I would like to acknowledge the authors, who are from different countries, for their contribution to the book. I wish to offer special thanks to Aleksandar Lazincia for his exceptional assistance and to the individuals and organizations, who either directly or indirectly contributed to this work.

E. Burcu Özkaraova Güngör
Ondokuz Mayıs University
Turkey

Contents

1

Biosensors for Life Sciences

Mihaela Badea, Liliana Rogozea, Mihaela Idomir and Nicoleta Taus
Transilvania University of Brasov
Romania

Doina Paula Balaban
Ovidius University from Constanta
Romania

Jean-Louis Marty, Thierry Noguer
University of Perpignan via Domitia
France

Gilvanda Silva Nunes
Universidade Federal do Maranhao
Brasil

1. Introduction into Research Problems

In the last period of time, the micro and nanotechnologies changed important fields of the molecular biology, in order to diagnose and treat at cellular and molecular level. In this area may be included also biosensors that use nanoparticles as immobilisation support (colloidal particles, carbon nanotubes, optic fibre) of the biological components that are used for cellular analysis *in vivo.*

The subject of this chapter propose as research the field of scientific investigation using biosensors based on the synergism of the knowledges from biophysics, biochemistry, electronics, biology, medicine, informatics and mathematic. Some of the classical and modern methods used in order to detect organophosphorus and carbamates pesticides, mycotoxins in the frame of EU regulations are presented, as partial results of two research programs for young researchers supported by MEdC – UEFISCSU Romania and a Balkan Environmental Association (B.EN.A.) fellowship supported by TUBORG-B.EN.A.

The new point of the work was to obtain and to optimize some new biosensors in order to be used for pesticide (organophosphorus and carbamates) and mycotoxins detection. The experimental part is still in work, so in this chapter will be presented some of the selected partial results.

A biosensor is a measurement system based on a combination of biochemical and electronic elements, which are in close contact each other and are incorporated in a single unit. A biochemical component (enzyme or biological material such as micro-organisms, plant or animal tissues and cells) is chosen for its selectivity toward the substrate or the inhibitor to be determined (Andreescu & Marty, 2006). The electronic signal-transducing element (electrochemical, optical detector, gravimetric detector) converts the biochemical response into electric and optic signals, which are amplified, measured and decoded by an appropriate electronic unit.

For enzyme based biosensor, the enzyme is situated inside of insoluble support and so it obtains a biphasic system. The enzyme can be reused after catalyse. An advantage of this technique is that final product is without enzyme. Other advantage of immobilizing the enzyme is the higher stability and activity (Badea&Coman, 2007; Coman et al. 2005).

The pollutants detection using biosensors offered viable alternative for usual chromatographic methods, the sensibilities for these two methods have been comparable after the immobilisation processes. Biosensors present many advantages: easy handling, compatibility with standard commercial equipment's, miniaturized possibility, and automatic measurement.

2. Generalities about Toxicity of Pesticides and Mycotoxins and their Detection Methods

The use of acutely toxic pesticides and mycotoxins associated with a weak or absent legislative framework regulating pesticide and mycotoxins use is one of the major reasons for the high incidence of poisoning in some developing countries (Kondardsen et al. 2003). Additional factors such as lack of information, low literacy, and education levels of the rural population, poor and inadequate working conditions, inadequate protection during pesticide application, and inappropriate spraying technology have also been shown to play important roles in the intoxication scenario (Hurtig et al., 2003; Karlsson, 2004).

2.1 Pesticides

A pesticide is any substance or mixture of substances intended for preventing, destroying, repelling, or mitigating any pest. Though often misunderstood to refer only to insecticides, the term pesticide also applies to herbicides, fungicides, and various other substances used to control pests.

A pesticides may be a chemical substance or biological agent used against pests including insects, plant pathogens, weeds, mollusks, bird, mammals, fish, nematodes (roundworms) and microbes that compete with humans for food, destroy property, spread disease or are a nuisance. Many pesticides are poisonous to humans (Coman et al., 2000)..

Organophosphorus and carbamate compounds are rapidly absorbed through the respiratory tract and through the digestive route, and to a lesser extent through the skin. After absorption, these compounds act by inhibiting the action of esterases, especially of acetylcholineesterases, following the interaction with the hydroxyl group of serine, which may determine: accumulation of acetylcholine which stimulates muscarinic and nicotinic receptors, increase cholinergic activity, and induce paralysis and death (Mijanovic & Zaciragic, 2006).

Organochlorine pesticides act primarily by altering the movement of ions across the nerve cell membranes, thus changing the ability of the nerve to fire. Organophosphate and carbamate pesticides act primarily at the synapses, altering the regulation of the transmission of the signal from one cell to the next (Hink et al., 2007).

A third, newer class of insecticides are the synthetic pyrethroids. These were developed because of their lower toxicity than OP and carbamates. These chemicals alter normal neuronal function by inhibiting ion movements across the nerve cell membrane, alterations in intracellular calcium ion concentrations and possibly by binding to GABA receptors.

Organophosphates are some of the most widely used pesticides in the world. They are used in agriculture, homes, gardens and veterinary practices, replacing the same uses as the organochlorines, many of which have been banned for years. In general, they are not persistent in the environment as they break down quickly. Because of their relatively fast rate of degradation, they have been a suitable replacement for the more persistent organochlorines.

Some of the early organophosphates were developed as nerve poisons for human warfare. The organophosphates recommended for non-residential uses are relatively toxic to vertebrate organisms. Their primary mode of action on insects and other animals is by phosphorylation of the acetylcholinesterase enzyme. This enzyme is necessary for controlling nerve impulse transmission between nerve fibres. A loss of this enzyme function results in an accumulation of acetylcholine, which causes unregulated nervous impulses. Higher levels of acetylcholine result in sensory and behavioural disturbances, incoordination and depressed motor function. Symptoms of acute poisoning develop during or after exposure, within minutes to hours, depending on method of contact. (Moser, 2007).

Carbamate pesticides are derived from carbamic acid and kill insects in a similar fashion as organophosphate insecticides. They are widely used in homes, gardens and agriculture. Like the organophosphates, their mode of action is inhibition of cholinesterase enzymes, affecting nerve impulse transmission. Because of carbaryl's relatively low mammalian oral and dermal toxicity and broad control spectrum, it has had wide use in lawn and garden settings.

In the literature it were performed different kind of analytical methods in order to detect organophosphorus and carbamates pesticides: liquid chromatography (Badea et al., 2004), immunoassay (Badea et al., 2004, Brun et al., 2004), biosensors (Schulze et al., 2003; Mulchandani et al., 2001; Pemberton et al., 2005, Badea et al, 2005; Badea et al., 2006; Ghosh et al., 2006).

In most countries, in order to sell or use a pesticide, it must be approved by a government agency. For example, in the United States, the EPA does so. Complex and costly studies must be conducted to indicate whether the material is effective against the intended pest and safe to use (Blasco et al., 2005; Neisheim 2002).

2.2 Mycotoxins

The ingestion of food containing mycotoxins, the toxic products of microscopic fungi (moulds), may have serious adverse health effects in humans and animals. Occasionally, occupational exposure to airborne mycotoxins may also occur. The mycotoxin contamination of foodstuffs may vary with geographical conditions, production and storage methods, and also with the type of food, since some food products are more suitable substrates for fungal growth than others (Pfohl-Leszkowicz & Manderville, 2007).

Ochratoxins are produced by several species of the fungal genera *Aspergillus* and *Penicillium*. These fungi are ubiquitous and the potential for contamination of foodstuffs and animal feed is widespread. Ochratoxin A, the major compound, has been found in more than 10 countries in Europe and the USA.

Ochratoxin A has been found in maize, barley, wheat, and oats, as well as in many other food products, but the occurrence of ochratoxin B is rare. Residues of ochratoxin A have been identified in the tissues of pigs in slaughterhouses, and it has been shown, under experimental conditions, that residues can still be detected in pig tissues one month after the termination of exposure.

Field cases of ochratoxicosis in farm animals (pigs, poultry) have been reported from several areas of the world, the primary manifestation being chronic nephropathy. The lesions include tubular atrophy, interstitial fibrosis, and, at later stages, hyalinized glomeruli. Ochratoxin A has been found to be nephrotoxic in all species of animals studied so far, even at the lowest level tested (200 µg/kg feed in rats and pigs). It has also been reported to produce teratogenic effects in mice, rats, and hamsters (Gresham et al, 2006; Pfohl-Leszkowicz & Manderville, 2007).

Ochratoxin A is a nephrotoxic mycotoxin which is carcinogenic to rodents and possesses teratogenic, immunotoxic and possibly neurotoxic properties. Further, it may be implicated as a factor in the human disease Balkan Endemic Nephropathy and the development of urinary tract tumours in humans. Human endemic nephropathy is a kidney disease of unknown etiology that has so far only been encountered in some areas of the Balkan Peninsula. The renal changes observed with this disease are comparable to those seen in ochratoxin A-associated nephropathy in pigs. Also, recent data from France and North Africa point towards a correlation between chronic interstitial nephritis and high exposure to ochratoxin A.

The effects of superoxide dismutase and catalase on ochratoxin A-induced nephrotoxicity were studied. Superoxide removes oxygen by converting it to hydrogen peroxide; this enzyme works in conjunction with catalase, which removes hydrogen peroxide within cells. Superoxide dismutase and catalase prevented most of the nephrotoxic effects induced by ochratoxin A, observed as enzymuria, proteinuria, and creatinaemia, and increased the urinary excretion of ochratoxin A (Sovoz et al., 2004).

Analytical techniques have been developed for the identification and quantitative determination of ochratoxin levels in the µg/kg range.

Ochratoxin A has been found in many commodities, including cereals, cereal products, coffee, grapes, grape juice, wine, cocoa and chocolate, beer, meat, pork products, pulses, milk and milk products, and spices. Several published analytical methods for the determination of ochratoxin A in maize, barley, wheat, wheat bran, wheat wholemeal, rye, wine, beer, and roasted coffee have been formally validated in collaborative studies. The methods are based on liquid chromatography (LC) with fluorescence detection, include a solid-phase extraction clean-up step with reversed-phase C18, silica gel 60, or immunoaffinity columns, and can guarantee detection of < 0.5 µg/kg (Pussemier et al, 2006).

The first LC method for determining ochratoxin A in maize and barley was validated in a collaborative study with materials spiked with ochratoxin A in the range of 10–50 ng/g. Ochratoxin A was extracted from grains with chloroform:aqueous phosphoric acid and isolated by liquid–liquid partitioning into aqueous bicarbonate solution that had been cleaned-up on a C18 (solid-phase extraction) cartridge. Identification and quantification were performed by reversed-phase LC with fluorescence detection.

The use of antibody-based immunoaffinity columns in the clean-up step has improved the analysis of ochratoxin A. Two methods based on immunoaffinity clean-up for determination of ochratoxin A in barley and roasted coffee have been developed and validated in collaborative studies under the auspices of the European Commission, Standard and Measurement Testing programme (Entwisle et al, 2000).

Screening methods based on TLC are also available. These methods are used in only a few laboratories since they do not provide an adequate limit of quantification (LOQ). Enzyme-linked immunoabsorbent assays (ELISAs) have been developed for the detection of ochratoxin A in pig kidney, animal and human sera, cereals, and mixed feed. The results obtained with these methods require confirmation since the antibodies produced often show cross-reactivity to compounds similar to ochratoxin A.

The new elements will be comparations of different chromatographic, spectral and enzymatic methods, trying to detect ochratoxin A also using biosensors, with the help of our partners from Romania, France and Brasil.

Aflatoxins are a family of fungal toxins produced mainly by two *Aspergillus* species which are especially abundant in areas of the world with hot, humid climates. *Aspergillus flavus,* which is ubiquitous, produces B aflatoxins. *A. parasiticus,* which produces both B and G aflatoxins, has more limited distribution. Major crops in which aflatoxins are produced are peanuts, maize and cottonseed, crops with which *A. flavus* has a close association. Human exposure to aflatoxins at levels of nanograms to micrograms per day occurs mainly through consumption of maize and peanuts, which are dietary staples in some tropical countries.

Aflatoxin M1 is a metabolite of aflatoxin B1 in humans and animals. Human exposure to aflatoxin M1 at levels of nanograms per day occurs mainly through consumption of aflatoxin-contaminated milk, including mothers' milk. Measurement of biomarkers is being used increasingly to confirm and quantify exposure to aflatoxins. In large studies realized in China, it was observed that risk for hepatocellular carcinoma was elevated among people with aflatoxin metabolites in urine, after adjustment for cigarette smoking and hepatitis B surface antigen positivity (Huang et al, 2003; Yu et al, 2002).

Extensive experimental studies on the carcinogenicity of aflatoxins led to a evaluation of the evidence as follows: sufficient evidence for carcinogenicity of naturally occurring mixtures of aflatoxins and of aflatoxins B1, G1 and M1, limited evidence for aflatoxin B2 and inadequate evidence for aflatoxin G2. The principal tumours induced were liver tumours.

The use of resistant varieties of seed and of pesticides, and careful drying and storing procedures can reduce fungal infestation and thus diminish food contamination by aflatoxins. The toxin is not eliminated from foodstuffs or animal feeds by ordinary cooking or processing practices and, since pre-and post-harvest procedures do not ensure total protection from aflatoxin contamination, techniques for decontamination have been developed. The toxin is generally concentrated in a small proportion of seeds that are often different in colour.

Biological and chemical procedures have been developed for the detection and determination of aflatoxins and other mycotoxins. The bioassay techniques that are currently available are not suitable for routine screening purposes, their detection levels being not low enough. The chemical assay techniques, although more accurate and faster, are not always specific. The presence of a certain toxin is usually confirmed by derivative formation and its toxicity verified by bioassay.

The aflatoxins are concentrated by evaporation of the chloroform, and then separated by thin-layer chromatography (TLC). Aflatoxins are intensely fluorescent when exposed to long-wave ultraviolet radiation, which makes it possible to determine these compounds at extremely low levels. An analyst experienced in this field can detect 0.5 ng aflatoxin B1 on a TLC plate. In most methods, the intensity of fluorescence of the sample is compared with that of a standard (Stroka & Anklam, 2000).

The methods using high-pressure liquid chromatography become the methods of choice for mycotoxin analyses because of their sensitivity and improved accuracy, and because they can be applied to a number of mycotoxins including aflatoxins B1, B2, G1, and G2 (Stroka et al., 2000; Castegnaro et al., 2006).
Immunoassays are also important in the qualitative and quantitative detection steps of aflatoxins (Badea & Coman, 2004; Coman & Badea, 2004; Sapsford et al, 2006).
It were performed and are in progress studies in the frame of these research projects, by collaboration between representatives from Transilvania University of Brasov, (Romania) from Sanitary Veterinary Direction and for Food Safety, Brasov (Romania), from BIOMEM, University of Perpignan via Domitia (France) and .Universidade Federal do Maranhão, Sao Louis (Brasil).

3. Importance and Relevance of the Scientific Content

The research projects propose the analysis and the optimisation of some detection possibilities of several bioactive compounds with toxic potential (mycotoxins, organophosphorus and cabamates pesticides) from water samples, foods and from biological samples, using enzymatic, chromatographic and spectral methods. Analysis using the biosensor technology is part of this area of research and offers the advantages such as miniaturization, easy sample manipulation, and the possibility of *in-situ* determination which further substantially diminishes the errors resulted from sample processing operations, with simple and low-cost instrumentation, fast response times, minimum sample pre-treatment, and high sample throughput. Biosensors are devices consisting of biological active protein species immobilized on the surface of physical transducers.
In the last period of time, there were reported also several enzymatic methods that may possible the detection of pollutants (pesticides, mycotoxins) from different samples using oxidoreductase and hydrolase. The use of enzyme-based biosensors is presented also for other fields as: medicine, agriculture, food industry, biotechnology.
Our research group intend to add some original contributions to the developing of this kind of methods, using free or immobilised enzymes (biosensors).
The existence of an experimental nucleus also in the frame of Transilvania University of Brasov, Romania, makes possible a large use of this technique and also the implementing of some subjects in courses and laboratory practice in the curricula of the students from specialisation general medicine, medical college, physics-chemistry and not only their theoretical discussions of this processes.
All these reference elements will constitute the base for the theoretical and experimental research, the young team of specialists being eager to bring new contributions to the knowledge level in the field of life and earth, by studies of some toxic compounds and analysis of food hygiene with impact over human and animal public health.
It is important to dedicate considerable time and energy to planning of the activities for detection of some toxic compounds (pesticides, mycotoxins) from water, foods, and biological samples, using enzymatic, chromatographic and spectral methods, as good planning makes work much easier in the long run and helps to avoid problems and misinterpretations. After the optimization studies using references samples, it will be tested the presence of the bioactive compounds with toxic effects from real samples (waters, foods, biological samples) in order to report the exceeded the maximum limits admitted by European Union environmental regulations.

The objectives of the theoretical and experimental research will be attended by well-established activities that will be performed during the project financing.

4. Experimental Procedures

4.1 Principle of the Experimental Method

There were obtained enzyme-based biosensors that were tested for detection of some pollutants compounds from reference and real samples, using amperometric detection. There were compared the experimental results for different commercial and mutants acetylcholinesterase and different pollutants compound.

4.2 Reagents and Equipments

Reagents

- Acetylcholinesterase (AChE) *Electric eel* – commercial enzyme Sigma Aldrich Co (St. Louis, MO, USA).
- Acetylcholinesterase (AChE) - *Drosophila melanogaster* wild type and genetic modified (E107W, E107Y, G406, I199V), obtained by genetic engineering using recombinant DNA - PBS Company (Toulouse, France)
- Electrochemical Mediator 7,7,8,8-tetracyanoquinodimethane (TCNQ), hidroxyethyl-cellulose (HEC) - Sigma Aldrich Chemie GmbH, (Steinheim, Germania)
- Substrate acetylthiocholine chloride (ATCh), pyridin-2-aldoxyme metachloride (2-PAM) - Sigma Aldrich Co (St. Louis, MO, USA).
- Polyvynil alcohol with stirylpiridinium groups SPP-S-13(bio) (PVA-SbQ), polymerization degrees 1700 and 2300 bio were provided by Toyo Gosei Kogoyo Co., Ltd. (Tokyo, Japan).
- Graphite - TIMREX TAS Graphite, M-058 - from TIMCAC LTD., Graphites and Technologies (Bodio, Switzerland).
- The plastic bed used for transducer obtaining – Electrodag PF-410, 423SS, 6037SS-Acherson (Plymounth, UK)
- Chlorpyriphos methyl oxon, Diazinon - CHEM SERVICE, West Chester, PA (USA) 99% purity. Pesticide stock solution was prepared in acetonitrile.
- Methyl paraoxon (98% purity) – Dr Ehrenstorfen GmbH, D86199 (Augsburg, Germania)
- The precursors that were used in sol-gel immobilisation: TMOS (tetrametoxysilane) (99% purity) and MTMOS (methyl thiometoxisilane) (98% purity) - Sigma Aldrich Chemie GmbH, (Steinheim, Germania). The precursors hydrolysis was realized in acid medium (HCl 1mM), and for immobilization was used also PEG600
- All other reagents used have had analytical purity

Equipments and Consumables

- Equipment for screen printed transducer obtaining – DEK 248, UK
- System with 3 screen printed electrodes, obtained in University of Perpignan via Domitia France
- Amperometric measurements were realized using a potentiostat METROHM 641 VA DETECTOR (Metrohm, Sweden), working potential being 100mV
- The signal was measured using BD40 (Kipp & Zonen, Flatbed Recorder, Olanda) equipment

- pH measurement were performed using PHM 220 MeterLab, Radiometer Copenhagen.
- Neon lamp for photopolymerization
- Waterbath –Buchi Waterbath B-480

4.3 Working Procedure

The transducer was realised in University of Perpignan via Domitia, BIOMEM, France, using screen-printed procedure. The reference electrode is considered Ag/AgCl, and auxiliary electrode –graphite. Working electrode contained mediator layer deposed to a graphite layer. The enzyme could be immobilised using different procedures.

There were tested two immobilisation methods frequently used in enzymatic biosensors field research: the method that use PVA-SbQ and sol-gel methods. The immobilised enzymes were *Electric eel* AChE (Sigma) and wild-type and genetic modified *Drosophila melanogaster* AChE.

Amperometric determinations are based on the measurement of electric current intensity generated in redox processes of an electrochemical species, working at a constant potential. An example is the transformation of acetylthiocholine in acetic acid and thiocholine, in presence of acetylcholinesterase. Thiocholine forms dithiocholine in presence of TCNQ mediator, liberating two protons and two electrons.

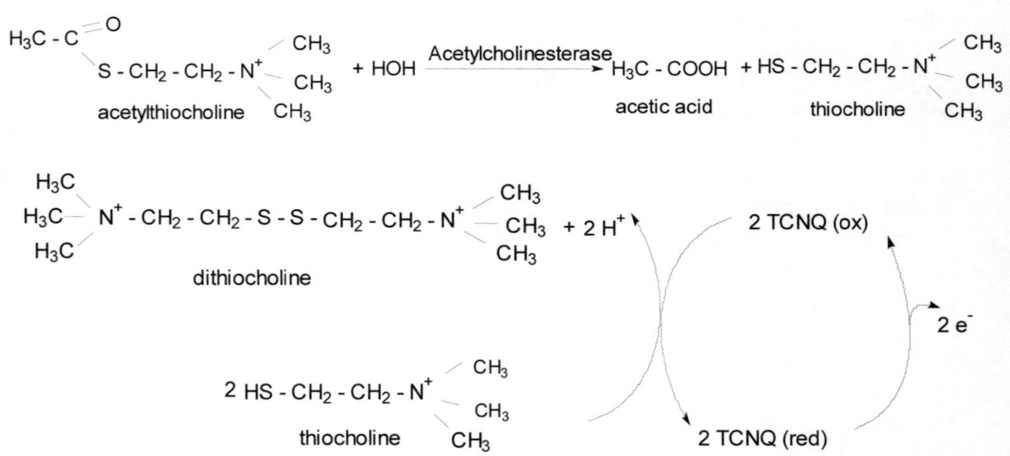

After the biosensor optimisation (stability, reproducibility, calibration), there were tested the influence of different pollutants (organophosphorus and carbamates pesticides, aflatoxins) (Badea et al, 2005; Gurban et al, 2005; Sikora et al, 2005). The obtained results were presented as degree of inhibition or residual enzymatic activity for each experimental condition.

4.4 Results and Discussions

(a) Enzyme immobilisation using PVA-SbQ

The PVA-SbQ enzyme immobilisation method presents the advantage that doesn't involve covalent binding, which determine the variation in enzyme conformation. It doesn't appear intermediate product, which may determine the enzyme denaturation. Around the enzyme it is formed a polymer network, the enzyme being included in the polymer cavities (Fig.1.)

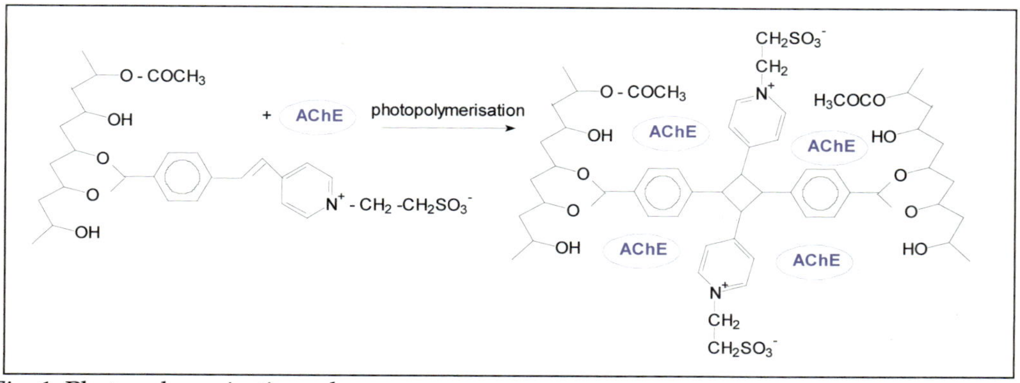

Fig. 1. Photopolymerisation schema

The values of electric signals for immobilisation of E107W and *Electric eel* are bigger than in case of AChE Dm wild type using. For E107W and Electric eel immobilised enzyme it was observed a signal decrease after the firsts assays (Fig. 2.).

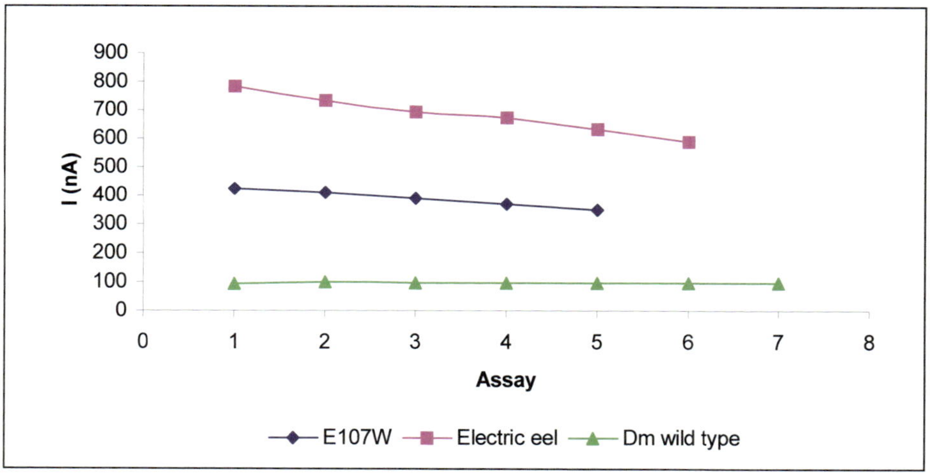

Fig. 2. The operational stability after 2 days drying of biosensors which use immobilised AChE Dm wild type, E107W and *Electric eel* enzymes using PVA-SbQ method; work conditions: buffer solution pH 7; 33% PVA (type PVA-SbQ 2300) in enzymatic mixture; work potential 100mV vs Ag/AgCl; [ATCh]=1mM

For the biosensors containing AChE Dm wild type it was obtained the best stability, the standard deviation representing 1,4% from the average of the electric signal. When it was used enzyme from *Electric eel*, it was immobilised a higher enzyme quantity and that determine the leaking the enzyme from the PVA network.

Two days after the immobilisation of 2mU AChE Dm wild-type and respectively PVA-SbQ 2300 (2:1), the sensor presented 85.5 % stability after 10 min incubation in buffer. The change

of the mixture ratio (1:2) indicated a standard deviation of the experimental values representing 23,21% from the mean of all values obtained for 150 min analysis (double percentage then accepted value 10%).
For these experimental conditions the electric signals versus the enzyme activity (EA) have been presented in Table 1.

Enzyme	EA / electrode (mU/electrode)	I ± STDEV (nA)
AChE Dm wild type	2.24	97.13 ± 1.36
Electric eel	6;75	685.83 ± 69.60
E107W	0.74	389.00 ± 30.08

Table 1. Electric signal for 2 days storing of the biosensors containing AChE Dm wild type, E107W and *Electric eel*; immobilization using PVA –SbQ method; working conditions: buffer solution pH 7; 33% PVA (type PVA-SbQ 2300) in enzymatic mixture; working potential 100mV vs Ag/AgCl; [ATCh]=1mM

For studying the influence of PVA percent from mixture over the enzymatic biosensor stability, there were utilised in the experimental studies electrodes that contain 50% PVA-SbQ (type PVA-SbQ 1700), too. Analysing the biosensor answer after successively preservation in buffer system (20 min each), followed by washing steps with distilled water, the experimental values were plotted.
It was observed that the electric signal decrease in time for the same acetylthiocholine substrate concentration (1mM) in the reaction mixture. The standard deviation calculated for the experimental values represent 20,27% from the average value obtained in the experimental case previously analysed in 160 minutes, double percent over the literature accepted value for the relative variations of biosensor signals.
The failure of a good operational stability observed for using of PVA method for enzyme immobilisation enforce the testing and optimisation of other possibilities to obtain enzymatic biosensors that will be used in organophosphorus pesticide monitoring.

(b) Enzyme immobilisation using sol-gel method
The enzymes immobilisation on the measuring electrode using the sol-gel method involves the obtaining of a SiO_2 network and the precursor's polymerisation, resulting a network that includes the enzyme. The method involves two steps: precursor's hydrolysis and the condensation in the presence of enzyme. The sol-gel material which is obtained gives important properties to the biosensor: rigidity, thermal and photochemical stability, chemical inertia, functionality in water and organic environment. It presents the advantage of a single immobilisation stage. Processes are achieved at low temperatures comparable with optimum temperature for enzyme action. It still presents the disadvantage of a diffusing barrier, as the toxicity of intermediary reaction products.
In our experiments it were analysed the properties of biosensors obtained using different hydrolysis time of the precursors, mixture method of precursors with enzyme solution, drying time before their utilisation, type of acetylcholinesterase, to find an optimum method for pesticide analysis from different environmental matrices (water, food).
The precursors solutions were prepared (Table 2.).

Precursors	TMOS (μL)	MTMOS (μL)	Deionised water (μL)	HCl 1mM (μL)	PEG600 (μL)
1	5	15	44	40	4
2	10	10	44	40	4
3	15	5	44	40	4
4	20	10	44	40	4

Table 2. Precursors volumes used to test sol-gel method for enzyme immobilisation

It was deposed a final mixtures that have 1mU enzymatic activity / electrode before immobilisation.

The ratio between the precursors was presented in Table 3. There were tested different immobilisation methods, following the signal stability (repeatability) and reproducibility for the same experimental conditions.

Metho d	Precursors	Mixture 1 (HEC+graphite +TCNQ) (μL)	Mixture 2 (precursors) (μL)	Ratio Enzyme: Mixture 1 : Mixture 2
M.1.1.	1	50	50	1:1:1
M.1.2.		25	50	1:1:2
M.2.1.	2	50	50	1:1:1
M.2.2.		25	50	1:1:2
M.3.1.	3	50	50	1:1:1
M.3.2.		25	50	1:1:2
M.4.1.	4	50	50	1:1:1
M.4.2.		25	50	1:1:2

Table 3. Reagents volumes used to test sol-gel immobilisation method for E107Y

It was tested ***biosensor stability***. For mutant E107Y, there were tested the previous methods, for different drying times. Using statistics, the medium values and their corresponding standard deviations were presented in Fig. 3.

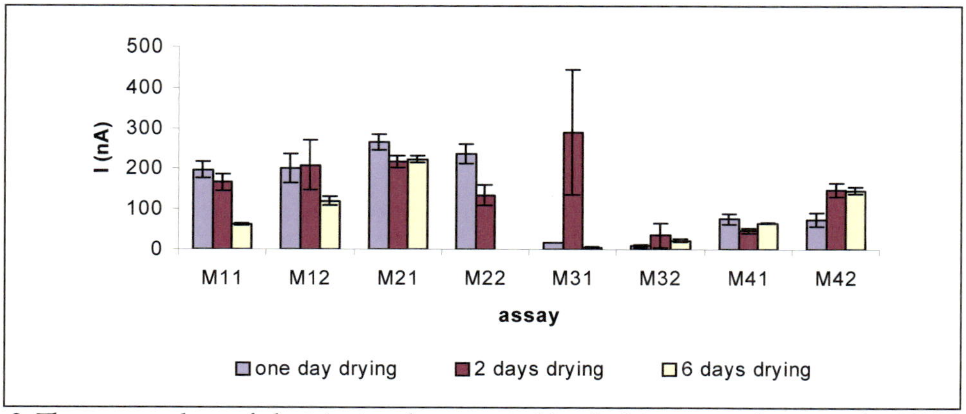

Fig. 3. The mean values of electric signals generated by the biosensors that contains E107Y immobilised using different sol-gel methods (12h hydrolysis time of the precursors)

For some experiments, the standard deviations of the signals represent values higher than 10% from the mean of these determinations (Table 4.).

%	M11	M12	M21	M22	M31	M32	M41	M42
one day drying	10,65	17,91	7,13	10,10	3,44	38,41	18,25	23,35
2 days drying	12,36	28,64	6,45	19,25	53,27	85,24	13,46	11,28
6 days drying	3,78	8,30	3,54	nd	36,47	14,86	1,55	6,12

where nd – undetermined

Table 4. The percentile values of the standard deviations corresponding to the mean values of electric signals generated by the biosensors containing immobilised E107Y using different sol-gel methods (12h-hydrolysis time of the precursors)

It was observed an increase of the signals stability during the increase of drying time, even that the mean values are lower than in the first analysis. The transducers obtained using M.3.1. and M.3.2. methods present, also for 6 days drying time, big values of the experimental values (bigger than 10%), so they cannot be used in the next experiments. It may be recommended the use of biosensors obtained by M.4.1 and M.4.2. methods, these presenting the smallest values.
The similar studies were realised for 6h-hydrolysis time of the precursors. It is not recommended in any case the use of transducers that contains the mutant E107Y immobilised using M.3.1 and M.3.2 methods in the next experimental studies, also because of the low signals, but especially because of the big variations of the currents obtained for the same experimental conditions.
The ***electrode calibration*** was realised following the current intensity variations from analysis system. The current intensities resulted from successively injection, in the same quantity of buffer system, of known amounts of acetylthiocholine were plotted versus the final concentration of enzymatic substrate from reaction mixture.
The calibration of the biosensors containing cholinesterase from Dm E107Y is presented in Fig.4.

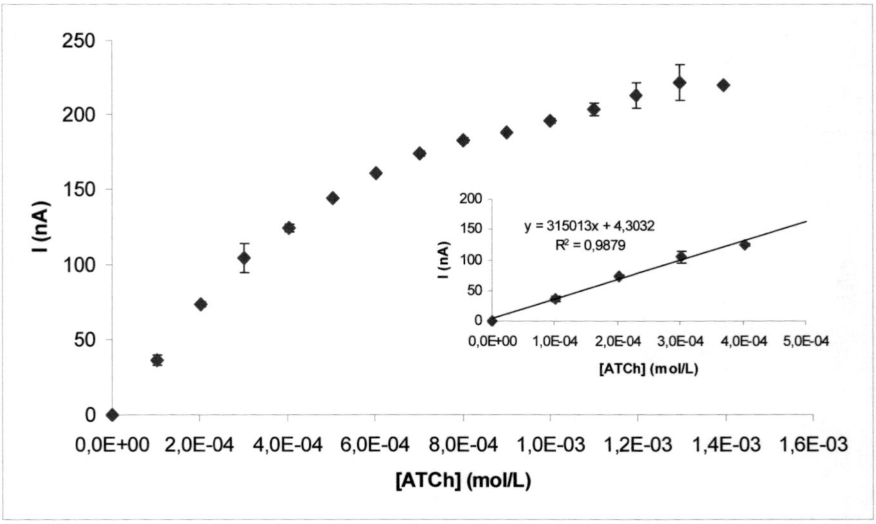

Fig. 4. Biosensors calibration; Dm E107Y mutant AChE immobilised using M.4.1. sol-gel method; working potential 100mV vs Ag/AgCl

Michaelis-Menten allure can be seen, with a linear dependence for concentration of substratum smaller then $4 \cdot 10^{-4}$M, characterised by a correlation coefficient close to the unitary value. It was characterized the kinetic of the reaction, corresponding to immobilised acetylcholinesterase. For this, it was realised the Lineweaver-Burk representation (Fig. 5.).

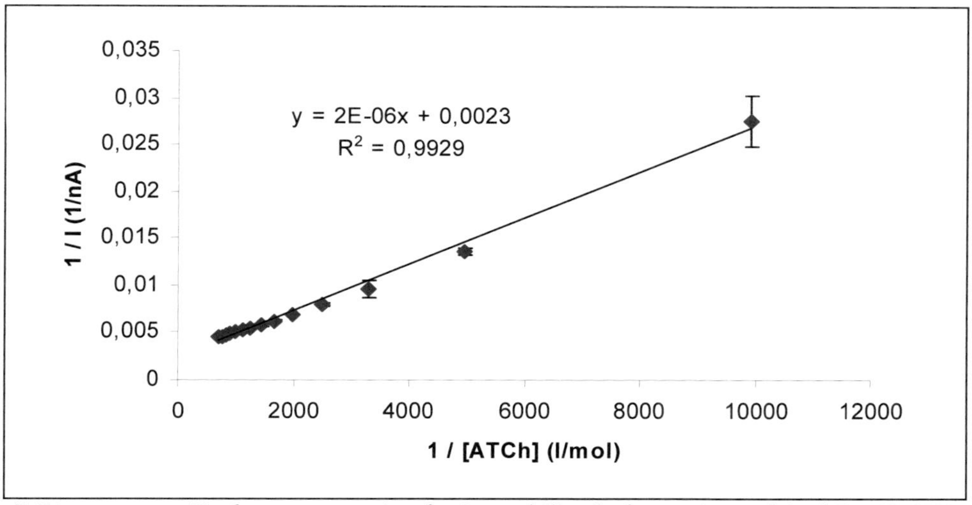

Fig. 5. Lineaweaver-Burk representation for immobilised of genetic modified Dm E107Y acetylcholinesterase using M.4.1. sol-gel method; working potential 100mV vs Ag/AgCl

The same steps were followed for mutant of AChE (*Drosophila melanogaster*) E107W and AChE from *Electric eel*. Using the information from calibration curves and from Lineweaver-Burk equations, there were determinate biosensors sensitivities and apparent Michaelis-Menten constants (Table 5.).

Parameter	AChE Electric eel	Mutant AChE *Drosophila melanogaster*	
		E107W	E107Y
K_M (mM) - for free enzyme	0,21	0,55	0,51
K_M (mM) - for immobilised enzyme	0,29	0,66	0,87
Biosensor sensitivity (mA·L/mol)	245,75	80,53	315,01
I max (nA)	144,92	82,64	434,78

Table 5. Michaelis-Menten constants for AChE from Electric eel and mutants *Drosophila melanogaster* E107W and E107Y and the slopes of the linear dependencies from the calibration curves

A lower Michaelis-Menten constant for *Electric eel* AChE indicates a higher affinity of enzyme for their substrate (Coman et al, 2003) and the slope indicates a higher transformation rate of the substrate in reaction product, comparing with mutant *Drosophila melanogaster* AChE (E107W and E107Y).

The influence of organic solvents over the enzyme biosensors answers was also tested. Depending by the solvent used and their quantity used in the experimental studies, it was necessary to study the enzyme behaviour in these organic media .
For the inhibition tests in presence of organophosphorus pesticides there were used small volumes of pesticide solutions in acetonitrile (2-20 μL) in 5 mL buffer system pH 7.
Biosensors response was tested adding controlled volumes of acetonitrile in the reaction mixture, studying the influence of this solvent over the enzymatic activity.
Over a certain value of the percent of the organic solvent in the system, it was observed a decrease of the electric signal. This fact may be explained by the enzyme inhibition, because of the changes from reaction media, in presence of the tested solvent.
The experimental results were statistically analysed and the residual enzymatic activity was plotted for each situation (Fig. 6.).

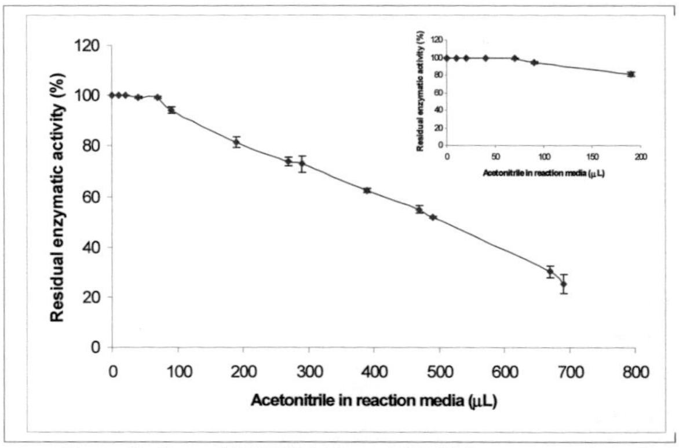

Fig. 6. Residual enzymatic activity of E107Y Dm AChE in presence of acetonitrile; work conditions: sol-gel immobilisation method M.2.1.; 12h precursors hydrolysis 6 days drying

For the values that there will be used in the following studies, it wasn't present a significant inhibition due by the presence of this organic solvent. So, in the following inhibition studies it may be assumed that only the influence of the organophosphorus pesticide from the synthetic samples or the organophosphorus and carbamates pesticides are important. The similar results were also obtained by other researchers for other immobilised enzymes, using different immobilisation method and different working procedure (Avramescu et al., 2002; Montesinos et al., 2001).
The ***inhibition tests*** regard the organophosphorus pesticides action over different origin acetylcholinesterase. It was analysed the biosensor signals obtained before inhibition and compared them to those obtained after a certain inhibition period with different pesticide solutions (after inhibition).
The pesticide selection for the experimental studies was realised term by their preponderant use in Romania (CODEX, 1996). The mutant enzyme systems selection followed to use those AChE that have inhibition constants different by the commercial AChE (Sigma), for the studied pesticides.
The degree of inhibition increases with the increase of the concentration of organophosphorus pesticide because of the binding of the pesticides to the serine hydroxyl function, which inhibits the enzyme. For concentration lower then 10^{-7}M methyl paraoxon,

the possibility of linearisation the dependence of inhibition degree versus the pesticide concentrations is observed for Electric eel AChE.
This dependency allows the estimating of the detection limits of the biosensor, pesticides concentration which cause inhibits levels of 10%, 20%, respectively 50%.
It was studied also the influence of methyl paraoxon over the genetic modified acetylcholinesterase activity. So, different pesticide concentrations were added in analysis system, maintaining the same sol-gel immobilisation method (M.4.1.) (Fig. 7, Fig. 8).

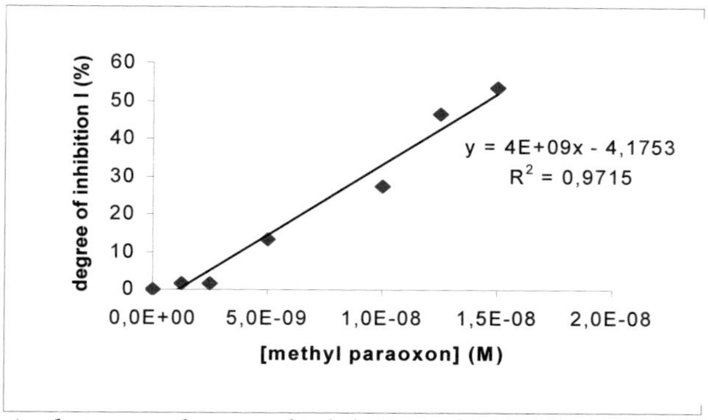

Fig. 7. The linearity between degree of inhibition of E107W acetylcholinesterase and different methyl paraoxon concentrations

Using this dependency there might be obtained I20 and respectively I50 with the values $6.6{\cdot}10^{-9}$M, respective $1.4{\cdot}10^{-8}$M methyl paraoxon, versus E107W, the results being in concordance with literature data (Andreescu et al., 2002) The detection limit for organophosphorus pesticides analysis is around 10^{-9} - 5 10^{-7} M, depending by the pesticide and by biosensors obtaining protocol (Evtugyn et al., 1996; Ivanov et all, 2003).
The same increase of inhibition degree was also observed for immobilised mutant Dm acetylcholinesterase E107Y, using the same procedure sol-gel M.4.1. (Fig. 8.).

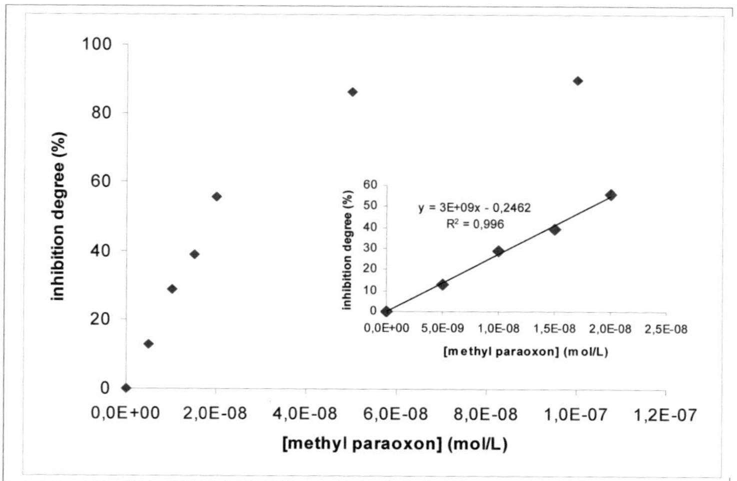

Fig. 8. The dependence of inhibition degree of E107Y acetylcholinesterase in presence of different methyl paraoxon concentrations

Using these dependence there were obtained I20 and I50 with values $6{,}75\cdot10^{-9}$M, and respectively $1.67\cdot10^{-8}$M methyl paraoxon, versus E107Y. The results are comparable with the data obtained for E107W enzyme.

But, comparing the slopes of the calibration lines for both enzyme it was observed a value higher for E107W AChE than E107Y enzyme (Table 6).

Enzyme	Inhibition constant for methyl paraoxon	Sensitivity (slope of dependency I% versus methyl paraoxon concentration)
E107W	3.52	$4\cdot10^{-9}$
E107Y	1.00	$3\cdot10^{-9}$

Table 6. The slopes of linear dependence between enzymatic inhibition degree and methyl paraoxon concentration

These observation resulted from experimental studies may be explained whereas by the inhibition constant. A high value of inhibition constant represents high enzyme sensitivity for the organophosphorus pesticide in the system, explaining the quick increase of the inhibition degree for the same pesticide concentration (the slope of the dependence is higher).

For E107W analysis, the sensitivity and the specificity for methyl paraoxon is more accentuated than other two studied acetylcholinesterase. It may be recommended E107W enzyme for methyl paraoxon analysis, being more sensitive for testing of low concentration of this pesticide in the analysed matrices.

The same experiments were done for aflatoxin detection using biosensors. For a concentration of aflatoxin 10^{-6}M the inhibition of acetylcholinesterase from *Electric eel* was higher than the inhibition in the presence of mutant AChE Dm I 199V.

For ***reactivation studies***, there were analysed the influence of methyl chlorpyrifos (MCP) samples over enzymatic biosensors containing E107Y immobilised using method sol-gel M.2.1. After incubation steps with pesticide solutions for 10 min, there were analysed the electric signals generated for adding of 1mM acetylthiocholine. After measurement of inhibition (inhibition degree, residual enzymatic activity), the transducers were incubated for different times with different concentration of 2-PAM reactivator. It was studied the possibilities of enzymatic biosensor reactivation.

The enzymatic inhibition in presence of methyl chlorpyrifos (MCP) was compared with the situation in which was used chlorpyrifos methyl oxon (MCPO), the product of the first pesticide metabolisation (obtained after water treatment with hypochloride solutions).

For a good appreciation of the experimental results, there were calculated and plotted the dependencies of inhibition degree for each experimental situations (Fig. 9.)

It was observed an increase of E107Y inhibition degree versus the chlorpyrifos methyl concentration increasing in analytical system. It was not observed an inhibition for pesticide concentration lower than 10^{-12}M. The reactivation increases versus incubation time in presence of 2-PAM increasing (5 or 10 min) and with reactivator concentration increasing (2mM or 10mM). Chlorinated sample presents a higher enzyme inhibition because of oxon derivative obtaining, with a pronounced toxic activity.

The tests were repeated with 3 different electrodes, the averages of inhibition degrees and their standard deviation being presented in Fig. 10.

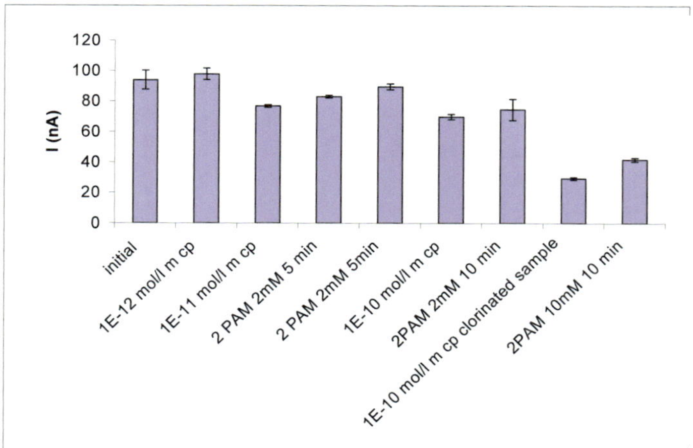

Fig. 9. Electric signals generated at the enzymatic electrodes (M.2.1.sol-gel method ; AChE type *Drosophila melanogaster* E107Y immersion in samples

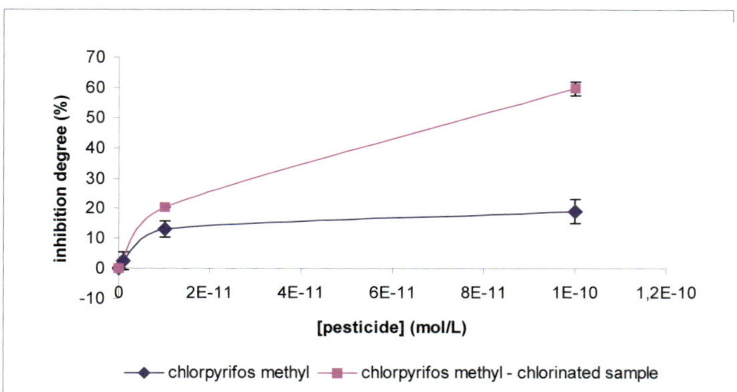

Fig. 10. Inhibition degree of enzymatic biosensors which contains immobilised AChE E107Y using sol-gel method M.2.1., in presence of chlorpyrifos methyl and chlorinated sample

It was observed a different inhibition for the same concentration of pesticide on the system for these two analysed samples. For 10^{-10}M chlorpyrifos methyl the difference between the two-inhibitors increase significantly, the chlorinated sample presenting a inhibition 3 times higher versus the sample untreated with hypochloride. Normally, chlorpyrifos methyl sample should not determine the acetylcholinesterase inhibition, the oxon form being the compound with affinity to the active site of AChE.

But, the existence of 99% purity of MCP may be a possible explanation for the inhibition existence, the remaining 1% probably being the active form MCPO, that modify the activity of AChE.

So, the "good intentions" of hypochlorides adding in water samples for their purification, also favourites the oxonic metabolites of organophosphorus thioderivatives obtaining, compound that has an inhibition activity over acetylcholinesterases, producing neurotoxic effects more important to the living organisms from this media, versus untreated samples.

The study of inhibition action of oxon was studied also for small concentration in the system. The increase of reactivator concentration and the increase of reactivation time determine an increase of biosensor reutilization possibilities, increasing their lifetime.
The representation of inhibition degree may me realised versus the pesticide concentration or versus decimal logarithm of these values Fig. 11.

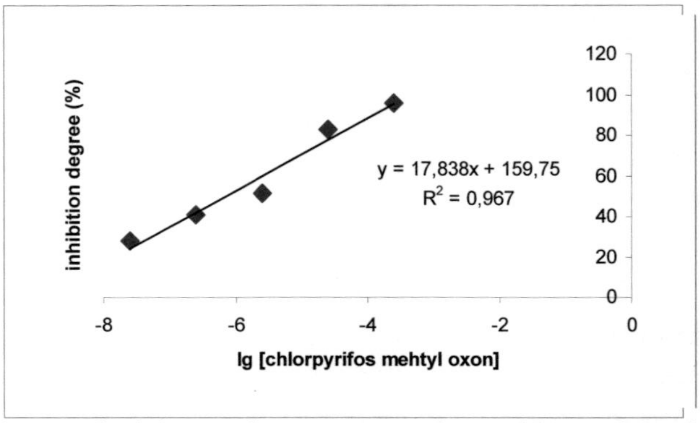

Fig. 11. The dependency of inhibition degree versus decimal logarithm of pesticide concentration

It was observed the possibility of linearization of this dependency, that permit the detection of the unknown concentration in the same experimental conditions.
Applications in real samples analysis was done for water sample testing: water from Villeneuve Lake and Rear River (Perpignan, France) and Timis River (Brasov, Romania). There were used for beginning the biosensors that used immobilised E107W and B107Y acetylcholinesterase using method M.4.1. The experimental data obtained for samples testing were statistically analysed. It was distinguished a more sensitive action of biosensor that use E107W enzyme versus the other that use E107Y enzyme, the values of inhibition degree for tested samples being higher in the first case. For E107W immobilised based biosensor, the inhibition effect is more accentuated, obtaining values higher than the similar situations that use E107Y cholinesterase. River waters, in both cases, presents compounds with inhibitory activity over the two-studied acetylcholinesterases (Badea et al., 2005).
Using the equation of regression lines obtained for analysis of methyl paraoxon standard solutions, it was calculated the equivalent concentration on this pesticide. The equivalent methyl paraoxon concentrations from these three water samples are in the admitted limit by the regulation of the European Economic Community (EEC) and U.S. Environmental Agency (EPA) (EPA Procedures, 1998).
There were tested also baby food products using biosensors for organophosphorus and carbamates detection (Nunes et al., 2005; Badea et al, 2005). These values are lower than the maximum admissible limits from agro-food, so the tested product respects the national legislation. The recovery degree was variable between 40-55%, the matrices being very complex and in the extraction steps it may be possible to loss part of the active compounds.
To increase the coefficient of efficiency of the organophosphorus pesticides (methyl paraoxon and diazinon), there were also tested other solvents (Badea et al., 2004): hexane, and dichloromethane for diazinon and methyl paraoxon extraction from water and fruit juice. (apple), using biosensors and HPLC-UV (Badea et al., 2005). The recovery degree for methyl paraoxon using biosensors was a little bit lower than the similar result obtained by

chromatographic method, maybe due by the fact that this new extracted compound from juice may have an activator action over the acetylcholinesterase Dm E107Y.
The methods that use biosensors with immobilised acetylcholinesterase may be used for pesticide screening, being rapid and low cost techniques. Methods were compared with the chromatographic techniques, the difference between the results being explained.

5. Conclusions and Future Work

The biosensors offer the possibility to toxicological investigations from different matrices: water, biological liquids, foods, detecting specific and with very good sensitivities low concentrations of organophosphorus and carbamates pesticides. The applicability of the method is also in environmental protection (Coman et al., 2000), but also in analysis with medical interest (Nivolini et al., 1995; Akyilmaz et al., 2006).
The biosensors obtained and characterised have been used PVA-SbQ and sol-gel immobilisation method. Sol-gel method was optimised for immobilisation of different kind of acetylcholinesterase to be able to be use in sensitive detection of some organophosphorus pesticides (methyl paraoxon, clorpyrifos methyl oxon, diazinon) from synthetic samples. The methods were also applied for real sample testing (water, baby food, juice), testing different extraction methods from initial matrices. The extracts were tested also using HPLC-UV, for method validation and for the explanation of the phenomenon that appear during the extraction step. It was concluded that biosensors might be used successfully for rapid, sensitive and specific determination of the organophosphorus and carbamates pesticides and some times mycotoxins (still in work for optimisation procedures) from different media.
An important area of research in ChE biosensors is directed toward the development of automated and continuous systems for measuring ChE inhibitors in flow conditions. The automation can be obtained by a computer controlled-programmable valve system which allows reproducible pumping of different reagents including buffers, substrate and inhibitor solutions, reactivating agents and real samples. Further improvement in sensitivity and selectivity can be obtained with the use of sensitive multienzymes which also allow discrimination between the insecticides and other interferences.
In the future, compact and portable devices specifically designed for in-field analysis and development of arrays of multiple sensors will constitute another area of intensive research for ChE biosensors. A ChE biosensor array could be designed with each sensor containing a different immobilized enzyme (wild type and mutants ChEs extracted from different sources). This could allow sensitive detection and differentiation of multianalyte mixtures. Considerable progress is expected in genetic engineering for the production of more selective and sensitive ChEs.
The future work would like to make the integration of a fully autonomous electrochemical biosensor with pattern recognition techniques for the detection and monitorize the pollutants level in different matrices. The system will provide a continuous, real-time monitoring of immobilized enzymes activities upon exposure to pollutants compounds (Karasinski et al, 2005).
Our research group have also in work some animal studies, testing the modification of some enzymes activity after the aflatoxin and ochratoxin innoculation, intending in future to use also cell models.

The authors would like to thank to the financial support of the projects for excellence for young researchers (CEEX code 163 no. 5898/2006 and CEEX cod 96 no. 1459/2006) that made possible documentation and the experimental studies
The results that characterise Sigma acetylcholinesterase-based biosensors and their applications behaviour were obtained during the Mihaela Badea's scholarship, which has been selected and funded by the Balkan Environmental Association (B.EN.A) in 2004, with the TUBORG-B.EN.A. sponsorship funds, which support a program of projects and fellowships for environmental in Romania.
The authors would like to thank to PBS Company (Toulouse, France) for the mutants enzymes that was provided during the experimental research

6. References

Andreescu S. & Marty J.L., (2006). Twenty years research in cholinesterase biosensors: from basic research to practical applications, *Biomol Eng*. Mar;23(1):1-15.

Avramescu A, Andreescu S, Noguer T, Bala C, Andreescu D.& Marty JL., (2002), Biosensors designed for environmental and food quality control based on screen-printed graphite electrodes with different configurations., *Anal Bioanal Chem.* 2002 Sep;374(1):25-32.

Akyilmaz E, Yasa I, & Dinckaya E. (2006), Whole cell immobilized amperometric biosensor based on Saccharomyces cerevisiae for selective determination of vitamin B1 (thiamine), *Anal Biochem.* 2006 Jul 1;354(1):78-84

Badea M. & Coman Gh., (2007). *Biochimie. Tehnici enzimatice de analiza*, Ed. Univ.Transilvania Brasov, ISBN 978-973-635-943-9, Brasov

Badea M., Romanca M., Drăghici C., Marty J. L., Marques C. V. V. C. O., Mendes D. R., Amarante Jr. O. P. & Nunes G. S., (2006), Multidisciplinary Collaboration for Environmental Protection using Biosensors. Detection of Organophosphate Insecticides in Aqueous Medium, *Journal of the Brazilian Chemical Society*, Vol. 17, No. 4, 807-811

Badea M., Noguer T., Drăghici C., Coman Gh., Idomir M., Gocan S. & Marty J. L., (2005), Laboratory Classes Regarding the Biosensors field in University of Perpignan and Transilvania University of Braşov, *Environmental Engineering and Management Journal*, IV(2), 2005, p. 205-212, ISSN: 1582-9596.

Badea M., Voina I., Coman Gh., Dumitrescu L., Drăghici C. & Idomir M., (2005), Pollutants detection using hapten derivatization, *Environmental Engineering and Management Journal*, IV(2), p. 223-228, ISSN 1582-9596

Badea M., Coman Gh., Drăghici C., Nunes G. S., & Marty J. L., (2005), Biosenzori enzimatici utilizați pentru testarea probelor de apă şi alimente, *A 7-a Conferință Națională pentru Protecția Mediului prin biotehnologii Agricole, Silvice, de Alimentație şi sănătate publică, BIOTASAS*, Braşov, 27 - 28 mai 2005, p. 457-462; ISBN 973-87505-0-4 / Fascicula 2 – ISSN-973-85051-8-6

Badea M., Bala C., Rotariu L., Gocan S., Cobzac S., Coman Gh., Drăghici C. & Marty J. L., (2005), Liquid Chromatography Used for Some Organophosphorus Pesticide Detection, *International Conference EnvEdu 2005*, Braşov, Sept. 8-10, 2005, poster PB30, pg. 69, ISBN 973-635-554-3

Badea M, Romanca M., Coman Gh., Idomir M., Nunes G. S., Mendes D. R., & Marty J. L., (2005), Collaboration Study for Environmental Control by using AChE-Based Biosensors for the Analysis of Organophosphate Insecticide Residues, *Bulletin of the*

Transilvania University of Brasov, Series D, Badea, M., Coman, Gh., Dima, L. (Editors), Special Editions for the 2nd International Conference on Trends in Environmental Education EnvEdu 2005, Transilvania University Press, Brasov, 2005, ISBN 973-635-555-1, pg. 36-42

Badea M., Coman Gh., Marginean A., Munteanu M., Radu G, Munteanu C., Haba B., Beke R., Bandac M., Sica M., Idomir M., Keresztes A., Toma S., & Marty J. L., (2005), The Influence of Some Usual Solvents on the Acetylcholinesterase Activity in Ellman Method and Biosensors, *International Conference EnvEdu 2005,* Braşov, Sept. 8-10, 2005, poster PB31, pg. 70, ISBN 973-635-554-3

Badea M., Plotogea A. C., Coman Gh., Drăghici C., Grusea M. D. A., Gocan, S., Nunes G. S., & Marty J. L., (2005), Caracterizarea unor biosenzori enzimatici cu acetilcolinesteraze de tip Electric eel, *Jurnal Medical Braşovean,* nr. 2, an 1, 2005, pg. 65-69, ISSN 1841-0782

Badea M.& Coman Gh., (2004) *Enzyme immunoassay* In: Analytical Biotechnology. Principles and Applications, Billiet, H., Drăghici, C., Coman, Gh. (Eds.), Ed. Universităţii Transilvania din Braşov, (2004), pg. 202-209

Badea M., Florescu M. & Coman Gh., (2004), *Immunofluorescence,* In: Analytical Biotechnology. Principles and Applications, Billiet, H., Drăghici, C., Coman, Gh. (Eds.), Ed. Universităţii Transilvania din Braşov, (2004), pg. 195-201

Badea, M.; Gocan, S. & Marty, J. L., (2004), Chromatographic and enzymatic methods for organophoshorus pesticides detection from water samples. *Bulletin of the Transilvania University of Brasov,* 11, 209-215. ISSN: 1582-9596.

Badea M. & Coman Gh., *Enzyme immunoassay,* In: Analytical Biotechnology. Principles and Applications, Billiet, H., Drăghici, C., Coman, Gh. (Eds.), Ed. Universităţii Transilvania din Braşov, (2004), pg. 202-209.

Brun EM, Garces-Garcia M, Banuls MJ, Gabaldon JA, Puchades R, & Maquieira A., (2005), Evaluation of a novel malathion immunoassay for groundwater and surface water analysis., *Environ Sci Technol.* 2005 Apr 15;39(8):2786-94

Blasco C., Font G. & Pico Y. (2005). Analysis of pesticides in fruits by pressurized liquid extraction and liquid chromatography–ion trap–triple stage mass spectrometry, *Journal of Chromatography A,* 1098, 37–43

Castegnaro M, Tozlovanu M, Wild C, Molinie A, Sylla A, & Pfohl-Leszkowicz A., (2006), Advantages and drawbacks of immunoaffinity columns in analysis of mycotoxins in food, *Mol Nutr Food Res.* 2006 May;50(6):480-7

CODEX- Codexul produselor de uz fitosanitar omologate pentru a fi utilizate în Romania, Ed. Iriana, Bucureşti, 1996

Coman Gh., Badea M., Drăghici C., Ciurea C. & Virtosu A. (2005) *Biochimie. Elemente de biochimie descriptiva si energetica biochimica,* Ed. Univ. Transilvania, ISBN 973-635-570-5 Braşov,

Coman Gh. & Badea M., (2004) *Immunochemical Methods,* In: Pollution and Environmental Monitoring, Ian Colbeck, Drăghici C., Dana Perniu (Eds), Ed. Academiei Romane, (2004), pg. 229-241,ISSN 1584-0506, ISBN 973-27-1156-6.

Coman Gh., Badea M., Drăghici, C., Dumitrescu L. & Tiut M., (2003), *Biochimie. Reacţii care se desfăşoară in organismele vii,* Ed. Ermetic, Brasov, ISBN 973-97733-1-1,

Coman, Gh., Pop, M., Taus, N., Brezina, A. & Florescu, M., (2000), Implicații biochimice şi clinice ale pesticidelor organofosforice, *Terapeutică, farmacologie şi toxicologie clinică*, 4(4), 48

Coman Gh., Pop M., Drăghici C., Greavu G., D. Panait, E. Trifu, P. Grigore & Tiut M., (2000) Pesticides. Possible Action Mechanism In Living Systems, *Bulletin of the Transilvania University of Brasov*, Series B: Mathematics, Physics, Chemistry, Medicine, Philology, 1999, 6 , 109-114.ISSN: 1582-9596.

Entwisle AC, Williams AC, Mann PJ, Slack PT & Gilbert J., (2000), Liquid chromatographic method with immunoaffinity column cleanup for determination of ochratoxin A in barley: collaborative study, *J AOAC Int*. 2000 Nov-Dec;83(6):1377-83).

EPA Procedures for Water Analysis (1998), In: *GC Bulletin* Supelco, 1998, 775 C

Evtugyn, G.A., Budnikov, H.C., & Nicolskaya EB, I (1996), Influence of surface-active compounds on the response and sensitivitybof cholinesterase biosensors for inhibition determination, *Analyst*, 121, 1911-1915.

Gresham A, Done S, Livesey C, MacDonald S, Chan D, Sayers R, Clark C. & Kemp P. (2006), Survey of pigs' kidneys with lesions consistent with PMWS and PDNS and ochratoxicosis. Part 2: pathological and histological findings, *Vet Rec*. 2006 Dec 2;159(23):761-8.

Ghosh D., Dutta K., Bhattacharvay D. & Sarkar P., (2006) Amperometric detection of pesticides using polymer electrodes., *Environ Monit Assess*. 2006 Aug;119(1-3):481-9.

Gurban A. M., Grusea D., Noguer T., Badea M., Coman Gh. & Marty J. L., (2005) Optimization of Prussian Blue Modified Transducers for the Development of Oxido-reductase Biosensors, *Bulletin of the Transilvania University of Brasov*, Series D, Badea, M., Coman, Gh., Dima, L. (Editors), Special Editions for the 2nd International Conference on Trends in Environmental Education EnvEdu 2005, Transilvania University Press, Brasov, 2005, ISBN 973-635-555-1, pg. 92-97 ISSN: 1582-9596.

Hinck JE, Blazer VS, Denslow ND, Echols KR, Gross TS, May TW, Anderson PJ, Coyle JJ; & Tillitt DE. (2007), Chemical contaminants, health indicators, and reproductive biomarker responses in fish from the Colorado River and its tributaries, *Sci Total Environ*. 2007 Jun 1;378(3):376-402.

Huang XH, Sun LH, Lu DD, Sun Y, Ma LJ, Zhang XR, Huang J, & Yu L., (2003), Codon 249 mutation in exon 7 of p53 gene in plasma DNA: maybe a new early diagnostic marker of hepatocellular carcinoma in Qidong risk area, China., *World J Gastroenterol*. 2003 Apr;9(4):692-695.

Hurtig, A.K., San Sebastian, M., Soto, A., Shingre, A., Zambrano, D., Guerrero & W., (2003). Pesticide use among farmers in the Amazon basin of Ecuador., *Arch. Environ.* Health 58, 223–228.

Ivanov A, Evtugyn G, Budnikov H, Ricci F, Moscone D, & Palleschi G. (2003), Cholinesterase sensors based on screen-printed electrodes for detection of organophosphorus and carbamic pesticides., *Anal Bioanal Chem*. 2003 Oct;377(4):624-31.

Karasinski J, Andreescu S, Sadik OA, Lavine B & Vora MN., (2005), Multiarray sensors with pattern recognition for the detection, classification, and differentiation of bacteria at subspecies and strain levels, *Anal Chem*. 2005 Dec 15;77(24):7941-9

Karlsson, S.I., (2004), Agricultural pesticides in developing countries – a multilevel governance challenge, *Environment* , 46, 22–41

Konradsen, F., van der Hoek, W., Cole, D.C., Hutchinson, G., Daisley, H., Singh, S., & Eddleston, M. (2003) Reducing acute poisoning in developing countries—options for restricting the availability of pesticides. *Toxicology* 192, 249-261

Mijanovic M. & Zaciragic A., (2006), Comparative determination of xenobiotics adsorption., *Bosn J Basic Med Sci.* 2006 Aug;6(3):54-6.

Montesinos, T., Perez-Munguia, S., Valdez, F., & Marty, J.L., *Anal.Chim. Acta.*, 431 (2001), 231.

Moser VC., (2007) Animal models of chronic pesticide neurotoxicity, *Hum Exp Toxicol.* Apr;26(4):321-31.

Mulchandani A, Chen W, Mulchandani P, Wang J, & Rogers KR., (2001), Biosensors for direct determination of organophosphate pesticides, *Biosens Bioelectron.,* 2001 Jun;16(4-5):225-230

Nesheim, O.N. (2002). Toxicity of pesticides. UF/IFAS EDIS Document PI-13. http://edis.ifas.ufl.edu/PI008.

Nunes G. S.; Marques C V V C O; Castillo A; Badea M, & Marty J L. (2005), Determinação de resíduos de inseticidas carbamatos em alimentos infantis utilizando biossensores amperométricos à base de enzimas acetilcolinesterases mutantes., *Acta Toxicológica Argentina*, Buenos Aires, Argentina, 13(2), 2-8

Pfohl-Leszkowicz A & Manderville RA., (2007), Ochratoxin A: An overview on toxicity and carcinogenicity in animals and humans., Mol Nutr Food Res. 2007 Jan;51(1):61-99

Pemberton RM, Mottram TT, & Hart JP., (2005), Development of a screen-printed carbon electrochemical immunosensor for picomolar concentrations of estradiol in human serum extracts., J *Biochem Biophys Methods.* 2005 Jun 30;63(3):201-12.

Pussemier L, Pierard JY, Anselme M, Tangni EK, Motte JC, & Larondelle Y., Development and application of analytical methods for the determination of mycotoxins in organic and conventional wheat, Food Addit Contam. 2006 Nov;23(11):1208-18

Sapsford KE, Taitt CR, Fertig S, Moore MH, Lassman ME, Maragos CM,& Shriver-Lake LC., Indirect competitive immunoassay for detection of aflatoxin B1 in corn and nut products using the array biosensor, Biosens Bioelectron. 2006 Jun 15;21(12):2298-305

Schulze H, Vorlova S, Villatte F, Bachmann TT, & Schmid RD., (2003), Design of acetylcholinesterases for biosensor applications, *Biosens Bioelectron.* 2003 Mar;18(2-3):201-9.

Sikora T., Eremia S., Plotogea A., Noguer T., Badea M, Coman Gh., Trojanowicz M & Marty J. L., Detection of Chlorpyrifos Oxon Using Screen Printed Electrodes with Different Acetylcholinesterases, *Bulletin of theTransilvania University of Brasov*, Series D, Badea, M., Coman, Gh., Dima, L. (Editors), Special Editions for the 2nd International Conference on Trends in Environmental Education EnvEdu 2005, Transilvania University Press, Brasov, 2005, ISBN 973-635-555-1, pg. 168-173 ISSN: 1582-9596.

Soyoz M, Ozcelik N, Kilinc I, & Altuntas I., The effects of ochratoxin A on lipid peroxidation and antioxidant enzymes: a protective role of melatonin, *Cell Biol Toxicol.* 2004 Jul;20(4):213-9.

Stroka J, & Anklam E., Development of a simplified densitometer for the determination of aflatoxins by thin-layer chromatography, *J Chromatogr A.* 2000 Dec 29;904(2):263-8.

Stroka J, & Van Otterdijk R, Anklam E., Immunoaffinity column clean-up prior to thin-layer chromatography for the determination of aflatoxins in various food matrices, *J Chromatogr A*. 2000 Dec 29;904(2):251-6.

Yu SZ, Huang XE, Koide T, Cheng G, Chen GC, Harada K, Ueno Y, Sueoka E, Oda H, Tashiro F, Mizokami M, Ohno T, Xiang J. & Tokudome S., (2002), Hepatitis B and C viruses infection, lifestyle and genetic polymorphisms as risk factors for hepatocellular carcinoma in Haimen, China., *Jpn J Cancer Res*. 2002 Dec;93(12):1287-92.

Ecological, Economic and Marketing Aspects of the Application of Biofertilisers in the Production of Organic Food

Drago Cvijanović[1], Gorica Cvijanović[2] and Jonel Subić[3]
[3]Institute of Agricultural Economics, Belgrad
[2]Maize Research Institute, Zemun Polje, Belgrade-Zemun
[3]Institute of Agricultural Economics, Belgrad
Serbia

1. Introduction

Having ability to produce on his own necessary products, man has become independent of nature and its generosity. By this ability man provided for instance his independent development within nature. Due to the further development of his knowledge he managed to elevate the production to the level of the science and therefore to aim many processes towards the desirable direction. Today, there are no more or less important plant species, but only more or less important technology of their production. Crop yield expressed through the gain/profit, which is anyway an ultimate objective of the production, depends, first of all, on the investments into the production. Fertilisation is one of very important cultivation measures. The application rate of mineral fertilisers can be distributed in the soil in the determined concentrations, which is their advantage over biological nitrogen. Nitrogen mineral fertilisers are more soluble and due to their greater solubility they are more accessible to plants as plants consume much less energy when they absorb easy available nitrogen forms. However, in order to obtain high yields, mineral fertilisers, especially of nitrogen ones, are often applied in the agricultural production. It is considered that the application of mineral fertilisers is the most advantageous and the fastest way to increase crop yields. The significance of the application of mineral fertilisers in the crop production in the European Union (EU-15) is confirmed by the records made in 2001: out of the totally applied amount of 15,286,903 t, 4,178,000 t were applied in France; 589,717 t in Ireland; 1,405,913 t in Italy; 416,000 t in the Netherlands; 2,613,413 t in Germany, etc. (www.fao.org). On the other hand, only 272,000 t of mineral fertilisers were applied in Serbia (www.poljoprivreda.info). The structure of the mineral fertiliser application in Serbia is unsatisfactory and presents a problem as complex and more expensive NPK fertilisers are used in smaller amounts, while the use of nitrogen fertilisers is much greater. Namely, 63%, i.e. 37% of the total use of mineral fertilisers were related to nitrogen, i.e. NPK fertilisers, respectively.

Plants use 40-50% of nitrogen incorporated by mineral fertilisers, then, soil microorganisms transform 15-20 % of nitrogen fertilisers into gaseous compound, while 20-30% are integrated into the soil organic matter. The remaining amounts of nitrogen are leached or can be involved into metabolism of effective microorganisms and in such a way they can inhibit the development of beneficial soil microorganisms.

It must not be forgotten that some adverse effects of high application rates of nitrogen fertilisers can occur in the nutrition chain of man. Young plants have no ability to transfer great amounts of incorporated nitrogen into proteins and other organic forms. The nitrogen surplus can be accumulated in green parts of plants, which can be reduced in digestive organs into nitrites and carcinogenic nitrosoamines that can induce general toxicosis. Moreover, a share of nitrates in the surplus is leached into deeper layers and causes pollution of aquatic basins. The application of high rates of nitrogen mineral fertilisers results in the disturbance of the natural equilibrium in grasslands by the alteration of their floristic composition. Results obtained in many studies show that under influence of these fertilisers a participation of grasses has been increasing while the participation of legumes has been decreasing. The unilateral use of higher rates of nitrogen fertilisers decreases the contents of sugar and starch in sugar beet, the oil content in oil crops, then it prolongs the growing season, affects wheat lodging and reduces grain yield in maize, leads to the modification of the floristic composition of weed associations. Nitrogen fertilisers are the most dangerous mineral fertilisers from the aspect of pollution of the environment (Marchner & Römheld, 1992). In addition, mineral fertilisers are relatively expensive - they are approximately five-fold more expensive than biological nitrogen. High rates of nitrogen mineral fertilisers can alter agrochemical properties of the soil. Their application during a longer period of time can lead to the increase of salts in the soil, can change pH of the soil solution, can lead to a change of soil biogeny and of those microbes that are important for soil fertility. Besides, growers seek any possible way to reduce the production costs, i.e. cost prices of all agricultural products so that their products can be competitive in local and global food markets (Cvijanović, D. et al. 2001).

Microbes have the most important role in the matter and energy cycle in the soil and take significant part in biodiversity conservation. The change in biodiversity, defined as a change within the living world of a certain ecosystem, is a good indicator in protection. The soil is a complex system in which the majority of biological processes occur due to microorganisms. Control of soil microorganisms with the aim to provide optimum conditions for the crop production means the following:

a. monitoring of microbiological activity (abundance, enzymic activity and biodiversity);
b. the relation between beneficial and pathogenic microorganisms;
c. the application of beneficial and effective microorganisms as inoculates in order to direct certain processes.

The intensive use of chemicals in the conventional agricultural production is one of the factors that affect the reduction in biodiversity, nitrogen leaching, soil and water contamination with pesticides, nitrites and heavy metals. Soil, as a very important agricultural resource for any country, is not any more of a non-limiting area as it used to be, hence a great attention has to be paid to the soil management and utilisation. National strategies of conservation and protection of nature and biodiversity have been implemented in many European countries (Nielsen & Winding, 2002). Soil, i.e. ecosystem control action guidelines were presented in The Sixth Environmental Action Programme „Environment 2010: Our Future, Our Choice" (Hubert et al., 2001).

On the other hand, growers have the following very positive effects and multiple benefits from the application of microbiological fertilisers in the crop production (Subić et al. 2006):

- yield increase;
- possibility to reduce the application of mineral fertilisers;
- profit increase;
- conservation of the soil as a limiting resource necessary for food production.

The aspect of how and in which way we should protect not only regions in which the food is produced, but the complete environment that is necessary for plants, animals and man is especially important (Katić B. et al. 2005). It can be achieved by reducing all inputs that lead to the endangerment of soil, water and air, and whereby of plants, animals and man. Microbiological fertilisers are the appropriate approach and a mode of maintaining and increasing of obtained yields of all crops with the reduction of the mineral fertilisers application.

The significant aspects of this issue are contemporary market trends that change marketing activities (which are essentially economic ones) towards the ethic, human and **ecological aspect**. As a rule, developed markets are in developed countries with customers who have become choosy due to their incomes, i.e. with consumers who have become aware of the fact that they use ecological products and healthy food. There are significant natural resources (first of all soil and water) in our country that are not polluted and therefore provide real possibilities for the production of healthy and safe, i.e. organic food (Cvijanović, D. et. al. 2001b).

2. A Possible Application of Biofertilisers as Microbiological Fertilisers

Soil microorganisms are very heterogeneous and the most abundant group that makes the soil a very complex and dynamic system. The abundance, activity and diversity of microorganisms are considered a significant indicator of a potential and effective soil fertility. Based on the knowledge of functions of certain groups of microorganisms, microbiological processes that are significant for the plant growth and the soil health can be directed towards a desirable direction (Milošević et al. 2004, 2006, Cvijanović G. et al. 2005). Studies and utilisation of microorganisms and their metabolic processes for the crop production optimisation provide a long-term conservation of soil qualities. Therefore, the application of microbial inoculates as biofertilisers, stimulators or biocontrols of phytopathogens in the food production, satisfies the concept within the system of sustainable agriculture: stability and quality of yield, maintenance of the ecological equilibrium that is reflecting on both, health safety of food and economic effect. The application of the information gained within the filed of biological nitrogen fixation means the application of effective microorganisms (that fix atmospheric nitrogen *Rhizobium/ Bradyrhirizobium, Azotobacter, Azospirillum,* bacteria of the genus *Bacillus, Pseudomonas,* blue-green algae *Nostoc, Anabaena* and mycorrhizal fungi) as inoculates that increase the soil biological activity and quality of field and vegetable crops (Milošević & Jarak, 2005). Genetically modified species do not encompass the mixture made of effective microorganisms. The soil species, that are an integral part of microbial niches, encompass this mixture. Biological fixation is a significant and interesting process not only in a biological, but also in a practical sense. According to studies carried out by Babeva & Zenova (1989) the amount of 160-190 kgN.ha^{-1} is annually fixed in the biosphere in the process of biological nitrogen fixation. A total of 80% of nitrogen is fixed within the microbial system, while 7% within these 80% are fixed in the association of diazotrophs and non-leguminous plant species. Nitrogen, accumulated in the soil in such a way, is not only cheap, but it is also harmless, as it is in a form of organic compounds and it is accumulated during the growing season.

Pure cultures of microorganisms are applied in the plant production in the form of microbiological fertilisers. The microbiological fertiliser for soybean encompassing the highly-effective strains *Rhizobium/Bradyrhizobium japonicum* has the greatest application. The initial attempts of the application of the symbiotic bacteria as a microbiological fertiliser in Germany dates as far back 1896 (more concrete: legumes were treated; non-legumes were treated much later, i.e. in 1923). The application of symbiotic bacteria has started spreading in other regions and today it is a compulsory measure, especially for soils on which soybean has not been previously grown. Moreover, the application of selected active strains of symbiotic nitrogen fixing bacteria resulted in higher yields (3.662 $kg.ha^{-1}$ JUS - macro trial, Agroinstitut, Sombor, 2005), higher protein content (34.55% - macro trial, Agroinstitut, Sombor, 2005), reduction of nitrogen mineral fertilisers utilisation (60 $kgN.ha^{-1}$ replacement for approximately of 130kg UREA - macro trial, Agroinstitut, Sombor, 2005). By such an application of this fertiliser, soybean should be treated as stubble with 30 $kgN.ha^{-1}$, as a low level of nitrogen allows recognition and infection of roots. In soils well supplied with nitrogen, nitrogen mineral fertilisers are not applied in the production. Over 200 $kgN.ha^{-1}$ could be fixed in a symbiotic association with other plant species (other legumes, clover, alfalfa). Plants in the association with symbiotic bacteria can satisfy up to 80% of their nitrogen requirements. The effect of plant inoculation depends on a plant species, applied agro-amelioration treatments, soil types and bacterial species (Bashan and Levanony, 1990).

Studies on and the application of non-symbiotic, associative biofertilisers were increasingly distributed in the practice during the last decades of the previous century. The application of associative nitrogen fixing bacteria (*Azotobacter, Azospirillum, Derxia,* etc.) in the production of wheat, maize, sugar beet, sunflower and some vegetable crops, indicates to the possibility of replacement of up to 60 $kgN.ha^{-1}$ (Govedarica et al. 1997), or even up to do 150 $kgN.ha^{-1}$ (Dőberainer, et al. 1972) depending on a strain. Particularly good results of the application of associative biofertilisers were obtained in the production of nursery plants of vegetable crops, which matured faster and more uniformly after replanting (tomato leaf area - approximately three times greater; pepper leaf area - approximately two times greater; tomato stalk length - greater by approximately 14 cm; pepper root length - greater by about 40%; tests carried out by Jugoinspekt, Belgrade, 2006).

In the countries of eastern Asia, a microbiological fertiliser with blue-green algae *Nostoc* and *Anabena*, that are able to bind atmospheric nitrogen in the amount of up to 20-30 $kgN.ha^{-1}$ annually, is used in the production of rice and cotton. If a microbiological fertiliser without supplements of mineral nitrogen is applied, the rice yield can be increased by 33% (Venkataraman, 1981), as it produces many biologically active growth substances. Great amounts of biomass remain after harvest; hence by the application of this fertiliser the content of a organic part of the soil is increased. Bacteria and fungi capable to supply plants with phosphorus are also very important as biofertilisers. Bacterium *Bacillus megaretium* var. *phosphaticum* is able to supply plants with phosphorus as it is capable to mineralise phosphorus in the organically bound form, and therefore it is recommendable for the use in soils rich in phosphorus. Previous results show that this fertiliser cannot replace mineral phosphorus fertilisers, but it should be a growth stimulator. Mišustin & Emcev (1987) assumed that a stimulatory effect of *Bac. megaterium* could be explained by the ability of the bacterium to extract active growth substances (thiamine, pyrodoxine, biotin, nicotinic acid, B_{12}), hence it can be recommended as a growth and development stimulator of field crops.

The benefit of the application of microbiological fertilisers can be of a great importance for both, growers and processors (Subić et al. 2006). If we take sunflower as an example (macro trial, Institute „Tamiš" Pančevo, 2005):

1. grower's benefit:
   -the average yield increase (+35 €.ha^{-1});
   -the average reduction of mineral fertilisers (+6.5€.ha^{-1});
   -the average profit increase (+38 €.ha^{-1});
   -plus effect on the soil;
2. processor's benefit:
   -the average increase of crude oil (+5 %);
   -the average profit increase (+12 €.t^{-1} processed raw material);
   -plus transport and storage savings, as well as, a greater amount of raw material.

Hence, the application of associative nitrogen fixing bacteria has a multiple significance for the yield increase, lowering of mineral fertiliser rates, profit increase, as well as, a positive effect on the soil.

The application of biofertilisers is the unavoidable future and therefore a rational and efficient utilisation of mineral fertilizers can be provided by combination with microbiological preparations.

3. Significance of Biological Nitrogen Fixation in Agriculture

Experimental and theoretical studies on the effects of plant inoculation with nitrogen fixing bacteria are economically and ecologically justified for their broad application and commercialisation as microbiological fertilisers. These fertilisers can be liquid, solid, wet and powdered after lyophilisation of strains. The effect of an inoculate depends on the applied agro-amelioration treatments, soil types, plant species, species and strains of nitrogen fixing bacteria, cell concentration in the preparation, the mode of preparation (Wani et al. 1994). Preparations encompass effective microorganisms that can be applied as single or a mixture of strains of one or more species. According to Higa (1991), effective microorganisms contain selected species of microorganisms including dominant populations of bacteria, actinomycetes, fungi and others. All applied species or strains have to be compatible in a liquid culture that is used as an inoculate. The effect of certain strains or species can hardly be determined by the application of inoculates with mixed species or strains of the same species. If such inoculates are based on the principles of a natural ecosystem, or if, due to their abundance and efficiency, their starting position is better than the position of the autochthons population, then their effectiveness will be great.

According to results gained by Burns, 1995, Bachan & Levanony, 1990, biofertilisation with associative nitrogen fixing bacteria showed a positive effect in the production of non-leguminous plants:

- maize;
- wheat;

- sugar beet;
- potato;
- tobacco, as well as, other vegetable crops.

Studies on the application of *Azotobacter, Azospirillum, Derxia, Klebsiella, Pseudomonas* and others as biofertilisers in the production of the most important field crops (maize, wheat, sugar beet and sunflower) show that they can, depending on the strain, mode and the form of the application, replace from 20 to 60 $kgN.ha^{-1}$. Plant inoculation with associative nitrogen fixing bacteria and phosphorus as mineralisers significantly increases yields and biomass of field crops (Govedarica et al., 1997a) and productive and technological quality of sugar beet (Mrkovački et al. 2007) and resistance to phytopathogens (Burns, 1995). The application of biofertilisers provide plants with an easier intake of phosphorus and potassium, absorption of active growth substances and vitamins, auxins, gibberellins produced by biofertilisers, hence it is their advantage over chemical fertilisers. Professor Dr. Mitar Govedarica initiated studies on the application of associative nitrogen fixing bacteria in field and vegetable crops in Serbia. This author also gave the greatest contribution to studying of their effects on yield and quality of filed crops, as well as, on elements of soil biogeny. Inoculation of wheat seed shows a positive effect on a spike length (7-30%) and a number of fertile spikelets (12-25%) (Govedarica, et al. 2001). Inoculation with *Azotobacter chroococcum* led to the increase of maize yield, while the effect depended on the strain, hybrids and the amount of applied NPK fertilisers. Long-term studies point out to a positive effect of different stains of *Azotobacter chroococcum* on grain yield of maize and applied rates of 60 $kgN.ha^{-1}$, indicating to a possibility of a replacement of a certain amount of nitrogen mineral fertilisers (Cvijanović, G. et al. 2005., 2006.) According to Doberenier (1997), there is a possibility to replace 30% of mineral nitrogen with biofertilisers and certain amounts of phosphorus and potash fertilisers.

The profitability of the application of microbiological fertilisers could be best observed in the model with maize and sunflower (studies were carried out during 2005):

1. macro trial, „Agroinstitut", Sombor:
   -hybrid (ZP 680);
   -number of plants (54,400);
   -moisture (32.4%);
   -content of total proteins (8.46%);
   -grain yield at 14% moisture (12,648 $kg.ha^{-1}$);
   -average yield increase (+288 $kg.ha^{-1}$);
   -average profit increase (+216 $€.ha^{-1}$);
2. macro trial, Institute „Tamiš", Pančevo:
   -hybrid (Velja);
   -yield (3,654 $kg.ha^{-1}$);
   -moisture (13.3%);
   -admixtures (1.8%);
   -oil content (43.86%);
   -yield JUS (3,567 $kg.ha^{-1}$);
   -average yield increase (+221 $kg.ha^{-1}$);
   -average profit increase (+414.3 $€.ha^{-1}$).

Long-term studies were performed on slightly calcareous chernozem in the experimental field of the Maize Research Institute, Zemun Polje at Zemun Polje in the vicinity of Belgrade. The effect of associative biofertilisers on quality and quantity of yields of maize hybrids of

various FAO maturity groups (FAO 600-700 and FAO 300-400), then on fundamental parameters of soil biogeny at different rates of nitrogen mineral fertilisers was observed. The seed was treated immediately prior to sowing with inoculates that encompassed the same amounts of different highly efficient strains of the following species *Azotobacter chroococcum, Azospirillum lipoferum, Brijerinckia Derx, Klebsiella planticola, Azotobacter vinelandi, Pseudomonas Bacillus magaterium* and *Bacillus subtillis* at the fertilisers rates of 60, 80, 90, 120, 150 and 160 kg N. ha^{-1}. All cropping practices were performed on optimum dates with high quality.
Results show a positive effect of biofertilisers on the level and stability of yield of maize of both maturity groups. The highest yield (9.84 t.ha^{-1} - 13.62 %) was recorded in hybrids of FAO 600-700 in the variant with bacterisation and fertilising with 80 kgN.ha^{-1}. Yield was increasing with the increase of mineral nitrogen rates over 80 kg.ha^{1}, while it was decreasing, i.e. increasing in the variant with, i.e. without bacterisation, respectively.

FAO 600-700 FAO 300-400

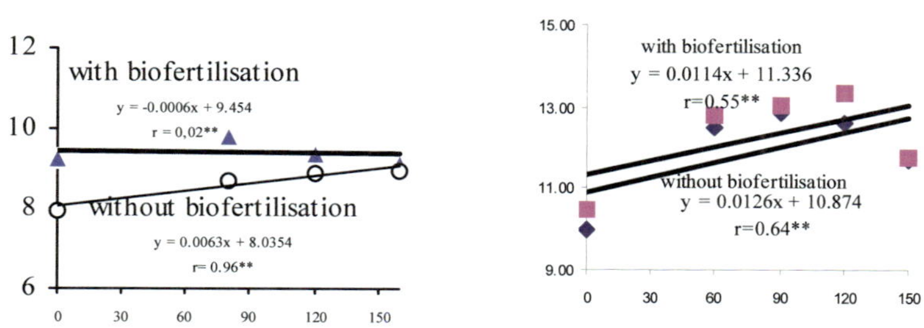

RATES OF MINERAL NITROGEN RATES OF MINERAL NITROGEN

Fig.1. Dependence of maize hybrid yields on the application of biofertilisers, bacterisation and rates of mineral nitrogen

The three-year average grain yield of hybrids of FAO 600-700 over all fertiliser rates amounted to 9.36 t.ha^{-1} at 14% moisture, which was higher by 8.58%, i.e. 740 kg.ha^{-1} than the yield recorded in the variant without bacterisation. The two-year average grain yield of hybrids of FAO 300-400 over all rates of fertilising amounted to 12.29 t.ha^{-1}, which was higher by 3.02%, i.e. 360 kg.ha^{-1} than the yield recorded in the variant without bacterisation. The increase of the growth and quality of plant production is a refection of supremacy of beneficial microorganisms. A regression line in obtained yields in both hybrids grown on chernozem (Figures 1 and 2) shows that the yield was significantly higher in the variant with bacterisation and fertiliser rates of 80 and 120 kgN.ha^{-1} and that the yield could be stabilised at a high level. According to the performed correlation analysis it can be stated that a correlation of maize yield and higher rates of mineral nitrogen including bacterisation was highly significant. The regression line in the obtained yield in the variant without bacterisation is in accordance with studies carried out by Vesković (1989) who did not record higher maize yields with rates of mineral nitrogen above 100 kg.ha^{-1}, and who determined that nitrogen rates higher than 135 kg.ha^{-1} adversely affected maize yields.

Mode of bacterisation	Proteins	Fertiliser rates kgN.ha^{-1}				Average
		0	80	120	160	
Inoculated	%	8.87	9.44	10.07	9.98	9.59
	Index level	110.32	108.00	109.10	109.19	109.22
Uninoculated	%	8.04	8.74	9.23	9.14	8.78
	Index level	100.00	100.00	100.00	100.00	100.00
	LSD 5%	LSD 1%				
Inoculation **	0.176	0.269				
Inoculation x Fertilisers**	0.188	0.274				

Table 1. Effects of bacterisations on total proteins in maize grain

Long-term studies (Cvijanović, G. 2002) on effects of inoculation with biofertilisers show the identical effect on the maize grain protein content and applied rates of mineral nitrogen. Due to its high energetic value, maize is all over the world, including Serbia, mainly used as feed. Although, maize grain proteins lack two amino acids (lysine and tryptophane), they do however provide approximately 20% of total proteins in mixtures, hence the mode of its increase is very important from the ecological aspect. Based on obtained results it is noticeable that maize seed bacterisation led to the increase of the content of total maize grain proteins. This increase, on the average, amounted to 9.22%, so the content of total proteins in the variant with bacterisation amounted to 9.59%, which is very important, as the maize grain protein content ranges from 6 to 12% (Table 1).

According to long-term results it was determined that mineral fertilisers had significantly affected the dynamics of abundance of systematic groups under conditions of bacterisation in the rhizosphere of both hybrids of different maturity groups (Tables 1 and 2). A total number of microorganisms is a good parameter for the soil evaluation. This number increased in the variant with bacterisation at each fertiliser rate in hybrids of both maturity groups. As for the abundance of *Azotobacter* a considerable decrease of their number was observed under high rates of mineral nitrogen. The decrease of the *Azotobacter* abundance in hybrids of FAO 600-700 in the variant with 160 kgN.ha^{-1} was lower by 10.81 % than in the control variant, which led to the decrease of the number on the average over all rates of fertilising by 0.33 %. However, their long-term application result in the decrease of the *Azotobacter* abundance. Due to its susceptibility, this important species of nitrogen fixing bacteria very vigorously responses to changes in the environmental conditions by lowering its number. High rates of mineral nitrogen cause disturbance of the equilibrium of microbiological processes in any soil, and the activity of the majority of beneficial microorganisms is reduced. The reduction of the enzymic activity of microorganisms occurs due to the fact that the level of mineral nitrogen determines the induction of the expression of the gene regulating the nutrient uptake. Furthermore, great amounts of mineral nitrogen inhibit the process of free nitrogen fixation, as they adversely affect the enzyme nitrogenase, which is also caused by lowering the number of this group of nitrogen fixing bacteria.

Fertilisers kgN.ha^{-1}	Total number of microorganisms		Number of *Azotobacter*-a		Dehydrogenase activity	
	10^7.g^{-1}soil	Index level	10^1.g^{-1}soil	Index level	µgTPF.g^{-1} soil	Index level
0	153.80	100.00	168.21	100.00	438	100
80	159.20	103.51	181.58	**107.94**	481	**109.81**
120	162.22	**105.47**	171.76	**102.11**	460	105.02
160	166.50	**108.25**	150.04	89.19	463	105.70
Average	162.60	105.72	167.66	99.67	468	106.84

Table 2. Effects of bacterisation and fertilisers on parameters of soil biogeny in rhizospheres of maize hybrids of FAO 600-700

The analysis of gained results indicates certain regularity in the increase of their number at low rates of mineral nitrogen. The highest number and the percentage of the increase were determined at fertiliser rates of 80-120 kgN.ha^{-1}. Beside the total number of microorganisms and *Azotobacter,* a dehydrogenase activity is also an important parameter of soil biogeny. The greater value of dehydrogenase points out to faster proceeding of oxidoreduction processes in the soil, that is to faster and greater mineralisation of fresh organic matter. The highest values of dehydrogenase and the percentage of increase were determined in the variant with bacterisation and a fertiliser rate of 80 kgN.ha^{-1}, which correlate to previous two parameters. Based on the statistical analysis of results gained over years it can be concluded that the total number of microorganisms was significantly increased in the variant with bacterisation and fertilising, as well as, in the interaction of these two factors.

Fertilisers kgN.ha^{-1}	Total number of microorganisms		Number of *Azotobacter*		Dehydrogenase activity	
	10^7.g^{-1}soil	Index level	10^1.g^{-1}soil	Index level	µgTPF.g^{-1} soil	Index level
0	245.50	100.00	47.08	100.00	407	100
60	355.08	144.63	174.18	369.46	526	129.17
90	412.05	**167.84**	180.30	**382.39**	438	107.58
120	302.03	123.03	98.12	208.40	440	**108.05**
150	158.63	64.61	57.49	122.18	113	27.64
Average	307.01	125.05	139.29	296.36	379	93.21

Table 3. Effects of bacterisation and fertilisers on parameters of soil biogeny in rhizospheres of maize hybrids of FAO 300-400

Obtained results point out to the compatibility of selected species of nitrogen fixing bacteria in the inoculum, as seed bacterisation favoured the growth and multiplication of introduced diazotrophs, and their enzymic activity which reflected upon the increase of the abundance and enzymic activity of the autochthons microbial community, which presents a good base for the evaluation of the soil productivity. Biofertilisers applied to the soil induce changes in microbiological communities among which competition for space and energy occur. These changes are more pronounced if hydrothermal conditions during the year are more extreme. Based on such results and under the assumption that agro-meteorological conditions match the average year for maize production, higher yields can be expected.

Bacteria synthesising phosphatase (*Bacillus, Pseudomonas, Azotobacter, Enterobacter, Serratia, Streptomyces*) dwell in plant roots and significantly affect mineralisation of phosphorus organic compounds an could be used in the agricultural production as a supplement of phosphorus mineral fertilisers.

According to results, seed inoculation with *Bacillus, Micrococcus, Enterobacter, Pseudomonas, Flavobacterium, Serratia* positively affects the length and dry matter weight of the root and the height of above ground plant parts. Also, maize seed bacterisation resulted in the increase of the phosphorus percentage in the root, while the phosphorus content in the above ground parts was at the level of control. All types of phosphorus as mineralisers caused the increase of the phosphorus content in maize plants in the variants with standard rates of nitrogen mineral fertilisers. The increase of over 100% on the average were caused by *Bacillus* (strain 26), *Pseudomonas* and *Flavobacterium*.

<table>
<tr><th>Variants</th><th>Root length (cm)</th><th>Weight (g)</th><th>Total P</th></tr>
<tr><td>Control</td><td>29.3</td><td>0.17</td><td>4.02</td></tr>
<tr><td rowspan="2">Bacillus megaterium B. megaterium + Azotobacter chroococcum</td><td>32.2-36.7</td><td>0.18-0.23</td><td>4.01-4.14</td></tr>
<tr><td>39.6-43.0</td><td>0.24-0.29</td><td>4.91-5.92</td></tr>
</table>

Table 4. Effects of *Azotobacter chroococcum* and *Bacillus megaterium* on maize

According to obtained results it can be concluded that free and associative microorganisms can be successfully used as biofertilisers in the form of microbiological fertilisers. Studies should be, first of all, aimed at the production methods providing high- quality microbiological fertilisers that should encompass effective microorganisms that initiate certain microbiological processes, then should be greatly competitive and supply plants with assimilatives and support their growth. In order to fulfil these criteria, microorganisms are selected, and then studies are aimed at the selection of microorganisams according to a plant genotype.

As selerogical tests showed that symbiotic and some species of free nitrogen fixing bacteria (*Pseudomonas, Arthobacter* and *Azotobacter*) were related, further studies should be aimed at inventing the best combination of symbiotic and associative mixtures in microbiological fertilisers for legumes. Moreover, new studies should be directed at solving problems of inoculation: a) mechanisms of recognition and binding microorganisms to hosts or soil particles; b) a role of plant genotypes and genetic engineering of microbial communities; c) selection of microorganisms; d) new technologies e) possibilities of mixed cultures.

4. Marketing in the Function of Organic Food

It is the indisputable fact that there are agricultural regions that are not at all contaminated or are contaminated at a very insignificant level, i.e. according to all elements of pollution the soils in such regions are significantly bellow a maximum level of pollution and are much less polluted than soils in European countries. According to certain data, 95% of agricultural areas in Europe are not suitable for the production of organic food. Fortunately, our country, except several industrial centres, is a very favourable region for such a production.

As already stated, massive pollution of environments, on one hand, and high standards on the other hand, that are, as a rule, implemented in highly developed countries, lead to a conclusion that the importance of organic food has been increasing and that it will be even more pronounced in the future.

In relation to the application of biofertilisers in the production of organic food, the issue of the marketing orientation of enterprises, farms and growers, i.e. the issue of Serbian organic food entering the "global market" has been becoming very important.

It is a well known fact that our agricultural enterprises, including farms, used to deal and exist (develop) under conditions of socialistic production relations. The business activities of our companies were adjusted to legal-governmental frames of that time.

Under such administrative-centralistic relationships, the aims of companies were to produce sufficient amounts of goods, which would find their way to the markets (Cvijanović D. & Milenković S. 1995). This used to be a principal characteristic of a business orientation of companies, hence these specific relations were the elements of the market, meaning that demands were greater than supplies. Engineers and other technical experts used to have a dominant place and a role within companies, while the production was the most importat function. Other company functions followed the production development, but the attention was not paid to costs and financial effects of such a production.

A marketing business orientation has established with the development and application of scientific and technological achievements within the field of agriculture. The market has become saturated and the importance of consumers has been observe, hence the need to stimulate the demand for the manufactured demands has arisen. In such an orientation, business leaders have become financiers, sales specialists, etc., while engineers and other technical experts lost their supremacy.

The marketing business orientation means that companies pay special attention to needs and wishes of consumers (buyers) and that they try to satisfy them with the lowest costs. As a matter of fact, the business orientation is leaned towards the market with a great competition, meaning that the main issue is to sell not to produce. Therefore, company leaders aim their activities towards buyers and towards forming a strong bond among scientists, financiers and marketing specialists.

It is not easy to introduce the marketing business orientation into an agro-industrial company, especially into our companies that have been operating under special conditions of inner and outer sanctions during the last ten years. In addition, the whole country including agriculture has been lagging behind the technically developed world. An especial problem of our agriculture is a problem of duality and implement of legal frames for the commodity production (Cvijanović, D. 1996).

The marketing business orientation has been introduced step by step. In short terms, it is necessary to maintain and increase the volume of production and sale, while in long terms it is necessary to develop operative marketing, marketing planning, performance and control. Managers should create a critical mass of collaborates and not only followers and those submissive to authorities, meaning the actions should be clear and decisive, while implementing of such an orientation should be flexible,

Regardless of a type of property and a size of a company, the special emphasise should be put on the position and the image of consistence of the marketing business orientation. In other words, it is necessary to determine the strategy of marketing activities so that the company could be competitive in the market (local and especially international one).

Making significant business decisions will be an objective only if they are based on the adequate marketing information. It is not possible to perform a proper marketing analysis, marketing planning & to make a correct business decisions without data on users, competitions, market conditions and all other factors related to business.
The special attention should be paid to the production of organic food, which is as a rule expensive, has a limited market and is produced in the rural regions of each country including Serbia (Cvijanović, D. and Trifunović, B., 1995a)
One of principal methods of gathering information on market business that can be used to plan current and alter the total production is the marketing analysis of each country, and in the case of our country is the analysis of markets in developed countries of Europe, Asia and America.

5. The image of a Company and its Organic Product (OP)

- facts and evaluation of image elements,
- evaluation of some scopes of business of a given company,
- experience gained with organic products and companies,
- information on OP supplied by a company,
- information channels on a company and OP,
- data on consumers of OP produced by a given company,
- evaluation of a company and OP in relation to the competition,
- differences related to a company comparing to other companies in the region,
- opinions of interviewees on the possibilities to improve OP, purchase, offer, advertising, information

6. The Competitor's Image

- evaluation of OP and some other scopes of business of main competitors,
- competitor's position,
- information channels on competitors

7. Evaluation of the Serbian Market for the Products Manufactured by a Given Company

- opinions and standpoints about supplies of OP produced by a given company in our country
- evaluation of the channels of purchasing of raw materials necessary for the OP production.

8. Habits and Needs to Use OP Produced by a Given Company

- needs to use certain types (modalities) of OP,
- habits, expectations and needs related to OP and the company that products such products,
- reasons to chose particular OPs produced by a given company

9. Standpoints Related to the OP Production of a Given Company

- significance of such products in relations to same or similar produces manufactured by a given company or competitive companies,
- standpoint related to the improvement of the production within the analysed field.

In order to actually recognise set goals it is necessary to perform quantitative studies, as parameters obtained in such studies are reliable and valid due to a greater number of interviewees belonging to the target groups relevant from the aspects of a company for which the studies were performed. The meaning of the working methods that are very important in such studies is as follows:

- making a questionnaire,
- training of an interviewer,
- field work,
- data coding and feeding into a computer,
- statistical processing of data,
- analysis of gained results,
- systematisation of results,
- preparation of results,
- oral presentation, and
- writing reports.

If a given company is not able to perform the stated study, then an authorised agency will do it on behalf of a company.

According to all data obtained by our own efforts or efforts of engaged authorised agency we have to have an answer to the question: "WHERE ARE WE NOW?" and due to it we can much clearer foresee directions of future activities which should enable us to determine "WHERE DO WE WANT TO GET TO?" and to set general guidelines and strategies in order to win wished positions.

The simplest auxiliary method in making decisions at the stage of planning is the S.W.O.T. analysis that makes differences between strengths, weaknesses, opportunities and threats. This analysis helps us to determine:

- a target group in the target market,
- desirable position,
- communication aims,
- unique message,
- message strategies,
- tactics of communication,
- platform of communication, and
- directions of further operations.

This is a method and a model of a future food production development in Serbia that fits into a new European concept that encompasses actions aimed not only at the increase of production and performing agriculture, but also at ecological functions - conservation of biodiversity, socio-economic function (conservation of traditional rural values, cultural inheritance). It means the production, marketing and advertising of microbiological inputs within the production of organic food, as well as, possibilities of export of food produced in such a way in Serbia into European countries.

10. Reference

Babeva, I.P., Zenova, A.R. (1989): Biologija počv. Izd. Moskovkogo univerziteta, Moskva

Bashan, Y., Levanony, H. (1990): Current status of Azospirillum inoculation technology: Azospirillum as a hallenge for agriculture. Can. J, Microbiolog., 36, 591-608

Bashan, Y., Levanony, H. (1990): Current status of Azospirillum inoculation techmology: Azospirillum as a challenge for agriculture. Can. J, Microbiolog., 36, 591-608

Burns, R. G. (1995): Enumeration, Survival, and Beneficial Activities of Microorganisms Introduced into Soil. 145-164. In: Huang, P. M., et al., (eds.) Environmental impact of soil component interaction, Metal, Other Inorganics, and Microbial Activities, CRC Pres. Inc. London

Cvijanović, D. i Milenković, S. (1995): Marketing poslovna orijentacija agro-industrijskog preduzeća, Zbornik radova sa simpozijuma Inovacije u agrobiznis menadžmentu, Aranđelovac 16-18. februar 1995. 332.02:65.012 (082) 338.434 (082). 129-135.

Cvijanović, D. i Trifunović, B. (1995a): Marketing semenskog kukuruza i ruralne sredine Srbije, Rad saopšten na Međunarodnom naučnom skupu Vlasinski susreti '95", 06-08. septembra 1995. Objavljen u časopisu za ruralni razvoj "Viđenja", Beograd. YUISSN 0354-6713, UDK 339.138:633.15 (497.11), 100-109.

Cvijanović, D. (1996): Savremeni marketing u funkciji ekologije, u "Proizvodnja zdravstveno-bezbedne hrane, ekonomsko-ekološki aspekt", Monogr, Novi Sad, mart 1996. CIP 633/635:152(497.11) (082), 103-111.

Cvijanović, D., Mirić, M., Jeremić, M., Kovačević, B., Mladenović, G., i Trifunović, B. (2001): Pariteti i dispariteti cena semena i drugih repromaterijala u našoj poljoprivredi, Časopis „Agroinovacije" YU ISSN 1450-9504, 213-218.

Cvijanović, D. i Mirić, M. (2001a): Odnosi cena semena i ostalih repromaterijala u jugoslovenskoj poljoprivredi, „Agroznanje„ UDK: 339.5.057.7; ISSN: 1512-6412; Tesliću-Republika Srpska (BiH), 13-16.mart 2001.

Cvijanović, D., Dojčinović Nataša i Radović Ljiljana (2001b): Marketing u funkciji proizvodnje zdravstveno bezbedne hrane, (2001): Rad saopšten 14-17. novembra 2001. godine u Subotici, na Međunarodnom simpozijumu: «HRANA U 21. VEKU»; Zbornik rezimea, str. 262-263.

Cvijanović, Gorica (2002): Uticaj diazotrofa na prinos i mikrobiločku aktivnost kod kukuruza, pšenice i soje, Doktorska disertacija; Poljoprivredni fakultet Univerziteta u Novom Sadu

Cvijanović, G., Lalević, B., Milošević, N. (2004): Microorganisms and their significance in the conversation of the agroecological environment, Proceedings Ecoist· 05, Bor 01-04.06.2005. 337-341

Cvijanović, G., Jovanović, Ž., Govedarica, M., Milošević, N., Cvijanović, D. (2005): Ecological and economic effects of the bacterisation application within system of sustainable agriculture, Savremena poljoprivreda Novi Sad Vol. 54, 3-4, UDC: 63(497.1)(051)-"540.2", YUISSN 0350-1205, 115-119,

Cvijanović, G., Milošević, N., Cvijanović, D. (2006): Biological Nitrogen as development function of new fertilizing technology in crop production , Economics of Agriculture Year, 53 Special edition, UDC 631.147:631.847, YU ISSN 0352-3462,39-44

Cvijanović, G., Cvijanović, D., Subić, J. (2006): *„Različiti sistemi đubrenja u proizvodnji kukuruza u funkciji održivosti poljoprivrede i ruralnih naselja".* Ekonomika poljoprivrede, Год./Vol. LIII, бр./N⁰ 3 (525-934). UDC 338.43:63, YU ISSN 0352-3462, UDK 631.895:633.15, str. 903-914.

Döberenier,J., Day, Y. M., Dart, P. J. (1972); Nitrogenase activity and oxygen sensivity of the Paspalium notatum-Azotobacter paspalli association J. Gen. Microbiol. 71., No 1, 103-106

Govedrica, M., Milošević, N., Jarak, M. (1997): Biločka azotofiksacija u poljoprivredi: Mogućnost primena i perspektiva, Zbornik radova Naučnog instituta za ratarstvo i povrtarstva Novi Sad, No 29. 27-35.

Govedarica, M., Milošević, N., Jarak, M., Jeličić, Z., Protić, R. (1997a): Diazotrophs and their Activity in Maize and Wheat, Biological Nitrogen Fixation for 21st Century, In: Elmerich et al (eds). Current Plant Science and Biotechnology in Agriculture, Vol. 31, 408 Academic Publish London

Govedarica, M.(2001): Mogućnost primene biofertilizatora u proizvodnji neleguminoznih biljaka Zbornik naučnih radova PKB INI Agroekonomik, 3, 1, 69-76

Higa T. (1991): Effective microorganisms: A biotechnology for mankind. Proceeding of the First International Conference on Kyusei Nature farming Research Center, Atami, Japan, 1-20

Hubert, S. Syed, B., Freudenschuss, A.,Emstsen,V., Loveland, P.- (2001):Propsal for a Europen soil monitoring and assessment framework. Tedhnical report, No 61, Europen Environment Agency, Copenhagen, Danmark

Katić, B., Nikolić Marija i Cvijanović, D. (2005): Zakonodavstvo Srbije u funkciji zaštite životne sredine, Zbornik radova I, ruralni razvoj i zaštita životne sredine, Vlasotince-Vlasinsko jezero, CIP 338.43 (082) 631.95 (082); ISBN 86-902797-2-5; 1-14

Marchner, H., Römheld, V. (1992): Optimieuring der Stickstoffdüngung bei gleichzeitiger Verringerung der Umwltbelastung. Inter.Tagung Optimierung der Stickstoffdungumg. Hohenheim, 187-192

Makro ogled, Agroinstitut Sombor, 2005.

Makro ogled, Institut „Tamiš" Pančevo, 2005.

Milošević,N., Govedarica,M., Cvijanović, G. (2004): Microorganisms as indicators of herbicide effect on biological activity of soil. Acta herbologica, 13, (1), 243-250.

Milošević, N., Jarak, M: (2005): Značaj azotofiksacije u snabdevanju biljaka azotom. U: R. Kastor (ed.) Azot-agrohemijski, agrotehnički, fiziološki i ekološki aspekti, Naučni institut za ratarstvo i povrtarstvo, Novi Sad, 305-352.

Milošević, N., Cvijanović, G., Tintar, B. (2006): Herbicides effects on microbial activity on agricultural soil, Herbologia, ISSN 1840- 0809 Vol. 7, No. 2, 57-71

Mrkovački, N., Mezei, S., Čačić, N., Kovačev, L. (2007): Efekat primene različitih tipova inokulacije šećerne repe, Zbornik radova Naučnog instituta za ratarstvo i povrtarstvo Novi Sad, ISSN 0354-7698, Sveska 43, 201-209

Nielsen, M. N., Winding, A. (2002): Microorganisms as indicators of soil health. National Environmental Research Institute. Technical report, 388, 1-86, Denmark

Primena asocijativnih biofertilizatora u proizvodnji rasada povrtarskih biljaka, Jugoinspekt, Beograd, 2006.

Subić, J., Arsenijević, Đ., Mihajlović, D. (2005): Metode za ocenu održivog razvoja na poljoprivrednim gazdiunstvima. Tematski zbornik „Multifunkcionalna poljoprivreda i ruralni razvoj", Institut za ekonomiku poljoprivrede - Beograd, Departman za ekonomiku poljoprivrede i sociologiju sela Poljoprivrednog fakulteta - Novi Sad, Ekonomski fakultet - Subotica, Društvo agrarnih ekonomista Srbije, Društvo agrarnih ekonomista Srbije i Crne Gore. str. 253-267, CIP 338.43.02(497.1*497.16)(082), COBISS.SR-ID 127317260, ISBN 86-82121-30-1.

Subić, J., Cvijanović, D., Marković, B. (2006): Ocena ekološke održivosti na poljoprivrednim gazdinstvima u opštini Mali Zvornik. Ekonomika poljoprivrede. Tematski broj. Međunarodni naučni skup: Multifunkcionalna poljoprivreda i ruralni razvoj I – razvoj lokalnih zajednica Год./Vol. LIII, бр./N⁰ TB (13-667). UDC 338.43:63, YU ISSN 0352-3462, УДК 631.95(497.11 Mali Zvornik), str. 57-63.

Tintar, B., Milošević, N., Marinković J., Cvijanović, G. (2007): Mikrobiološka svojstva černozema na različitim lokalitetima u okolini Novog Sada, Zbornik radova Instituta za ratarstvo i povrtarstvo Novi Sad, ISSN 035-7698 Vol. 43., 311-318.

Venkataraman, G.S. (1981): Blue-green algae for rice production-a manual foor its promotion. FAO Soils. Bulletin, 46, Food and agriculture organization on the united nations, Rome, 3-103.

Vesković, M. (1988): Bilans organske materije u zemljištu i prinos kukuruza na černozemun Zemun Polja pri različitim sistemima đubrenja, Doktorska disertacija Poljoprivredni fakultet , Beograd-Zemun

Wani, S.P., Rupela, O.P., Lee, K.K. (1994): BNF Technology for Sustainable Agriculture in the Semi-Arid Tropics. 15th World Congress of Soil Science, Acapulco, 4a, 245-262

www.fao.org

www.poljoprivreda.info

3

Environmental Problems Induced by Pollutants in Air, Soil and Water Resources

Murat Deveci & Fusun Ekmekyapar
Namık Kemal University
Turkey

1. Introduction

The rapid increase of population and intensive agriculture in our planet has resulted in large quantities of organic and inorganic wastes being discharged into environment, thus giving rise to serious environmental problems and deterioration of the agroecosystems. This process may also cause a risk in the human health. The potential problems in environment caused by pesticides, heavy metals, fertilizers, agricultural residues, wastewater, sewage sludge, solid wastes, atmospheric fallout and transgenic organisms. The results are an increase in toxic elements in air, soil and water resources. Once heavy metals enter the environment, they are very difficult to remove.

2. Important

The increase in animal and vegetable production obtained by using new technologies and methods has undoubtedly raised the productivity. However, it is not possible to provide an everlasting increase of product through the new methods and techniques implemented. Even though a quantitative product increase is provided by this way, some environmental problems also appear. Chemical products implemented in soil and plants with developed agricultural applications, various wastes pollute air, soil and water resources and make them uninhabitable for alive things living on them. Soils are systems having a high level of buffering power against external factors compare to water and air. However, the problems encountered when some deteriorations are created by some pollutants added in to the system become complex, difficult and costly to be corrected in the same degree. Some amount of these substances getting into bottom layers of the soil with rain and irrigation waters and then to underground waters deteriorate the quality of waters and make them impossible to drink.
In this chapter, discussion is focused on the environmental impact of agricultural and industrial practices in air, soil and water resources. Our work is an attempt at describing some environmental problems. Remediation options and strategies are considered in the following conclusions.

3. Information

3. 1 Pesticides

The fields convenient for agriculture is continuously decreasing due to many reasons like increase in residential areas in parallel to the continuous increase in world population, opening of new urban residential areas, establishment of factories, increase in the number of highways and vehicles. Since the area of the world is limited opening new fields for agriculture is not possible in order to meet the requirement of increasing population. Pesticides come first among the inputs used to increase product amount to be obtained from unit of area. Pesticides are chemical compounds used with the aim of removing micro and macro pests in the agriculture. Use of pesticides in agricultural struggle applications appears the easiest and the cheapest method. This situation increases use of these compounds for long years. Pesticides found a widespread area of use in the measures oriented to human health during and after First and Second World Wars.

3.1.1 Environmental Risks in Pesticide Using

Increasing amount of pesticide using also creates a general and potential danger like use of other toxic materials. Three main problems determine the limits in continuous use of pesticides:

a. Organisms become resistant against pesticides in time.
b. Some pesticides do not undergo biodegradation easily, but remains resisting in the environment they are implemented or carried.
c. They also harm some living things other than those targeted.

3.1. 2 Mobility of Pesticides in Soil

Pesticides are generally sprayed or applied on plants, soil surface and inside of soil. Pesticides applied may encounter one or more of following cases;

3.1.2.1 Evaporation

Atmospheric analyses indicated that some pesticides like DDT and dieldrin are mixed with the air. These chemical substances reaching to atmosphere from the soil can be mixed with soil or surface waters again with rain. Pesticides having the characteristic of mixing with the air by evaporating can be carried to very long distances with air flows from regions they are applied (Taylor & Spencer, 1990). Mixing of pesticides into the atmosphere through evaporation in the soil or their mobility in soil profile depend on vapor pressures of pesticides, adsorption characteristics of the soil, soil pH, soil temperature, texture of soils, and water content of the soil (Haktanır, 1983). Increase in temperature and soil moisture increase the evaporation speed of pesticides from the soil. Evaporation ratios of some pesticides are indicated in Table 1 which was put forward through researches made by Jiang and Cai in 1990.

3.1.2.2 Adsorption

Clay minerals and organic matter play a role in retention of pesticides in the soil. Adsorption occurs in oxides and hydroxides in sandy soil organic substance of which is low but containing Al and Fe. Pesticides like diquat, paraquat and dinoseb are in cationic form, and they are adsorbed in clay minerals. Metal ions like Cu, Fe, Mn, Co and Ni are effective in connection of pesticide molecules with clay minerals or soil organic matter.

Wang et al., (1989) and introduced that some characteristics of the soil like clay content of the soil, clay type, organic substance amount, soil structure, water content, temperature and pH affect adsorption processes. According to Shan et al., (1994), clay soil contains more pesticides compare to sandy soil. Adsorptions of pesticides according to the texture of soil occur in following sequences;

Sand>Sandy loam>Loam>Clay

On the other hand, dissolubility of pesticides also affects the adsorption. Pesticides having more dissolubility have a less adsorption.

Pesticide	Molecular Weight	Vapor Pressure (mm Hg)	Water Solubility (mg/l)	Glass Film	Water	Soil
Trifluralin	335	65.0	0.3	99.5	92.6	6.2
Lindane	290	9.4	10.0	88.0	89.4	8.6
Methhyl-parathion	263	9.7	60.0	23.8	15.8	14.4
Carbofuran	221	20	500.0	97.9	3.6	15.6

Table 1. Volatilization Rate of Pesticides From Glass Film, Water and Soil (%)[a]

3.1.2.3. Washing and diffusion

Washing of pesticides towards bottom layers of the soil occurs in the form of molecular diffusion and mass transport. Diffusion characteristics of pesticides, soil structure and humidity content of the soil are all effective in transport with diffusion. Movement of pesticides in form of mass transport is equivalent with washing. Increasing adsorption conditions decreases washing of pesticides. On the other hand, texture of the soil is also highly effective on washing. Sandy soil texture in the areas of intensive agriculture creates the risk of becoming polluted in groundwaters with pesticide residues. For this reason, residue amounts permitted in groundwaters have been determined by some international institutions like WHO and EPA. Some of these concentrations are indicated in Table 2. The EC Directive sets a maximum admissible concentration of 0.5 µg/l (0.0005 mg/l) for pesticides in total, and 0.1 µg/l for any individual pesticide.

Pesticides reach surface water resources with different ways. For example, they contaminate through their application in water to fight with water plants and water insects, through carriage of soils, plants and organisms containing pesticide residues to water resources with different ways, through discharge of pesticide production industry wastes into water resources, through washing of pesticide boxes and tools and equipments used in insecticide application, and through sedimentation of pesticide residues carried due to atmosphere pollution as a result of powder or liquid pesticide applications into water resources.

While some part of pesticide molecules reached to surface water resources through these ways dissolves in the water, other part remains suspended, and remaining part accumulates in the sediment. Then, pesticide is released from the sediment continuously.

[a] Jiang & Cai, (1990)

Pesticides	WHO G.V. µg/l	EPA MCL µg/l
Alachlor	20	2
Atrazin	2	3
Carbofuran	5	40
Chlordane	0.2	2
2,4 D	30	70
heptachlor/heptachlor epoxide	0.03	0.4/0.2
methoxychlor	20	40
silvex	9	50

Table 2. Drinking Water Standards[b]

3.1.2.4 Chemical decomposition

A large amount of pesticides can decompose with pure chemical events. Especially aluminum and iron oxides from soil compounds catalyzes the decomposition. Hydrolysis, oxidation, isomerization, ionization and salt formation among chemical decomposition reactions are not catalyzed. Clay content and pH of the soil are effective on chemical decomposition. Crushing especially in acid nature soils depends on the increase in hydrogen ion concentration close to clay mineral surfaces.

3.1.2.5 Biochemical decomposition

The most important part of decomposition of pesticides in the soil is composed of biochemical decompositions like many other toxic substances causing environmental pollution (petroleum and its derivatives, oils, detergents etc.). Microorganisms participating in such type of decomposition use -OH, -COO, NH_2, -NO_2 groups included in pesticide molecules as nutrients. Soil temperature, soil moisture, organic matter content of the soil and soil pH which affect the activities of microorganisms also affect biochemical decomposition processes. Most of pesticides are new compounds for soil microorganisms. For this reason, an inability can be seen in initial biological decomposition speed of due to adaptation absence of microflora. On the other hand, new compounds which emerge in different stages of biochemical decomposition may be sometimes more toxic than main compound. The list of microorganisms participating in biochemical decomposition processes of pesticides is given in Table 3.

3.1.2.6 Uptake by plants

Some pesticides do not harm plants since their chemical structure deteriorates after they are taken away within plants. By this way the amount of pesticides decreases. However, it is known that in agricultural fields where pesticide with content of dinitroanilin and metalaxyl is used for long years, seed germination decreases and so causes abnormal seed germinations. Cheng, (1985), PCBN, examined the effects of folpet, aliette and metalaxyl fungicides on beneficial endotrophic mycorrhiza (VAM) which leads a symbiotic life in roots of bean and clover plants for two years, and stated that fungicides can cause decreases in product amounts

---

[b] Twort, et al., (1994)

Microorganism	Pesticides That Can Be Degraded by the Microorganism
Achromobacter	DDT, Carbaryl, 2,4-D, MCPA
Agrobacterium	DDT, Dalapon
Arthrobacter	DDT, Malathion, Diazinon, 2,4-D, MCPA, Simazin, Propanil
Bacillus	DDT, EPN, Parathion, Methyl-Parathion, Fenitrothion, Toxaphene
	Dalapon, Linuron, Monuron, Lindane
Corynebacterium	DDT, 2,4-D, MCPA, Dalapon, Dinoseb, Paraquat, Diquat
Flavobacterium	Parathion, Methyl-Parathion, Malathion, Diazinon, 2,4-D, MCPA, Dalapon
	Chlorpyrifos
Pseudomonas	DDT, Toxaphene, Malathion, Parathion, Dichlorovos, PCP, Diazinon, Phorate
	Carbaryl, 2,4-D, MCPA, Dalapon, Dinoseb, Monuron, Simazine, Paraquat, Lindane
Xanthomonas	DDT, Parathion, Fenitrothion, Monuron
Aerobacter	DDT, Methoxychlor, Lindane, Toxaphene
Esherichia	DDT, Lindane, Prometryne, Amitrole
Streptococcus	DDT, Diazinon, Simazine, Dalapon
Nocardia	DDT, 2,4-D, 2,4-DB, Dalapon, Maleic hydrazide
Streptomyces	Diazinon, Dalapon, Simazine
Aspergillus	DDT, Trichlorphon, Linuron, Carbaryl, MCPA, Atrazine, 2,4-D, Dalapon, Monuron
	Simazine, Simetryne, Prometryne, Trifluralin
Cephalosporium	Atrazine, Prometryne, Simetryne
Cladosporium	Atrazine, Prometryne, Simetryne
Fusarium	DDT, Trichlorphon, Fenitrothion, Carbaryl, Simazine, Atrazine, Chlordimeforn, Lindane
Peniciolium	DDT, Carbary, Trichlorphon, Parathion, Atrazine, Prometryne, Simazine, Propanil,
Rhizopus	DDT, Fonofos, Carbaryl, Atrazine, Trichlorphon
Trichoderma	DDT, Lindane, Dalapon, Atrazine, Simazine, Dichlorovos, Parathion, Malathion, PCP
Chlamydomonas	Metobromuron, Atrazine
Chlorella	Phorate, Parathion

Table 3. Microorganisms and Degradability of Pesticides[c]

3.1.3 Persistences and Effects of Pesticides on the Living

Resistance of pesticide molecules against physical, chemical and biological decomposition displays their persistence feature. Pesticide residue amount in the soil is determined by

[c] Huang, P.M. & Iskandar, I.K., (2000)

physicochemical characteristics, soil factors, agricultural and environmental factors. The factors effective on pesticide remnants in the soil are stated in Table 4.

Properties of pesticide:	Water solubility, vapor pressure, pKa, pKb, stability, polarity, ionizabilty
Soil factors:	Soil texture and structure, content of organic matter, salinity, moisture content,
	porosity, temperature, pH, cation exchange capasity (CEC), permeability,
	kind and content of heavy metal ion, kind and population of microorganism,
	hydrolic conductivity
Agricultural factors:	Cropping pattern, cropping practices, crop type, pesticide formulation,
	application method, time and rate, frequence and times, irrigation time and volume
Environmental factors:	Rainfail, air temperature, evapotranspiration, illumination intensity and time, wind

Table 4. Factor That Influence the Pesticide Residues in the Soil[d]

Carbamate group pesticides are the group which should be preferred in terms of environmental pollution since they have a low level of persistence. The most resistant ones against decomposition processes and undesired ones in terms of environmental pollution are chlorine hydrocarbons and inorganic pesticides. On the other hand, chlorine hydrocarbon pesticides have the characteristic of accumulating in adipose tissue of mammalians. By this way, they may cause more toxic effects in receiving living group by accumulating from one living to another. Mercury included in the content of pesticides is an important environmental pollutant. It can reach high concentrations in food chain since it is accumulated in animals. Mercury reaching high concentrations in fish and mussels may cause human deaths due to eating of these livings.

Indicator species are selected as experimental animal in toxicity tests of pesticides and LD_{50} (Lethal Dose) parameter is based on while valuating results. LD_{50} is defined as concentration required for death of half among organisms being tested in a certain time period (24 hours, 48 hours etc.). The lesser the LD_{50} value of a pesticide through mouth, skin or respiration is, the higher the characteristic of making acute intoxication of the pesticide is in that ratio.

The residues of pesticides especially on vegetables and possible risks of them on human health has become the prior subject of pesticide researchers who evaluate vegetable quality recently. (Colume et al., 2001; Padron-Sanz et al., 2005).

Maximum Residue Levels (MRLs) are not exceeded if pesticides are applied according to appropriate agricultural techniques, but unconscious applications may lead to harmful remnants containing environmental pollution and possible health risks. Reductions frequently made in Maximum Remnant Levels (MRLs) accepted by the international institutions like EU and EPA and determination of levels by urgently creating purposive multi-residue methods are dramatical changes (Colume et al.,2001).

[d] Zhu, (1994)

World public opinion has reached a highly sensitive position against allergen, mutagen and cancerogenic effects created by pesticide residues on soil, water and foods depending on extinction events occurred in bird species feeding with accumulated pesticide residue. Forbiddance of production and consumption of pesticides causing cancer has been recommended by World Health Organization (WHO) and International Cancer Research Institutions, some has been forbidden and production of some other has been decreased. Some among them are DDT, endosulphan, fenitrothion, fenthion, malathion, parathion and trifluralin.

3.2 Heavy Metals

It is known that heavy metals forming an essential pollutant group even in trace amounts have harmful effects on human health. An important amount of heavy metals tends to accumulate in livings, and their dissolubility in water is so low. The processes which cause heavy metals reaching to toxic and cancerogenic amounts in soil and water resources following their increase in the atmosphere depending on climate conditions can be listed as follow:

- Industrial flue gases
- Local and intercity vehicle traffic
- Fossil fuels
- Mines

Other processes creating heavy metal pollution are as follow: Agricultural irrigation made with domestic and industrial wastewaters. Contamination with fertilizers and pesticides. Contamination with leaking waters from solid waste storing areas, forestry activities.

The most important ones among heavy metals are Pb, Cd, Hg, Cr, Fe, Cu, Mn, Zn, Ag, As, and Boron. According to Uslu & Turkman, (1987), the amount of these elements in water resources may exceed determined standards depending on the pollutant resource and hydrochemical atmosphere. All of them excluding iron exist in underground waters almost always below 1 mg/l concentrations. The reason why the concentration is such low is that they are adsorbed in clay minerals, iron and mangan hydroxides or soil inorganic substances as well as their low level of dissolubility.

According to the EPA (1985), atmospheric lead concentration range from 0.000076 $\mu g/m^3$ in remote areas to 10 $\mu g/m^3$ near point sources. Average annual lead concentration in air in most areas were reported to be below 1.0 $\mu g/m^3$. The EPA calculated the average intake of lead from respiration to be approximately 1 μg/day. This very low compared to the maximum drinking water intake, which would be 100 μg/day, assuming there are 50 μg/l of lead present and daily water intake of 2 liters.

Nriagu & Pacyna, (1988) stated that 38 thousand ton cadmium and 1 million ton lead are contaminated in the soil all over the world in every year through atmospheric fallout sweeping to the atmosphere, fly ashes, urban swinging, fertilizers and sewage sludge. The resources causing great anxiety both in public opinion and scientific world and creating heavy metal pollution are indicated in Table 5. Atmospheric fallout and coal ashes sweeping to the atmosphere constitutes the most important part among these resources.

Heavy metals can not only prevent waste waters from cleaning spontaneously but also restrict use of waste waters in agricultural irrigation in treated or untreated forms.

According to Ekmekyapar & Kaykıoglu, (2007), heavy metals also restrict the use of sludge for agricultural purposes. Characterization of sludge should be made with attention in these types of application and agricultural soils should not be permitted to be overloaded with heavy metals. The U.S. Environmental Protection Agency (US EPA, 1993) has established regulations for the disposal of sewage sludge on land. (Table 6).

Source	**Lead**	**Cadmium**	**Mercury**
		kt/yr	
Agricultural and food wastes	1.5-2.7	0-3.0	0-1.5
Animal wastes, manure	3.2-20	0.2-1.2	0-0.2
Logging and other wood wastes	6.6-8.2	0-2.2	0-2.2
Urban refuse	18-62	0.9-7.5	0-0.26
Municipal sewage sludge	2.8-9.7	0.02-0.34	0.01-0.8
Miscellaneous organic wastes, and excrata	0.02-1.6	0-0.01	-
Metal manufacturing awstes	4.1-11	0-0.08	0-0.08
Coal ash	45-242	1.5-13	0.4-4.8
Fertilizer	0.4-2.3	0.03-0.25	-
Peat (agricultural and fuel use)	0.4-2.6	0-0.11	0-0.02
Commercial product waste	195-390	0.8-1.6	0.6-0.8
Atmospheric fallout	202-262	2.2-8.4	0.6-4.3
TOTAL	479-1113	5.6-38	1.6-15

Tablo 5. Additions of Lead, Cadmium and Mercury to Soils[e]

	Limit Conc. for	**Cumulative Loading Limit**		**Limit Conc.for**	**Annual**
	Sludge	**Area Basis**	**Soil Basis**	**"Safe Sludge"**	**Loading Limit**
Pollutant	**mg/kg dry wt**	**kg/ha**	**mg/kg soil**	**(mg/kg d.wt)**	**kg/ha/y**
Arsenic	75	41	31	41	2.0
Cadmium	85	39	29	39	1.9
Chromium	3000	3000	2260	1200	150
Copper	4300	1500	1130	1500	75
Lead	840	300	226	300	15
Mercury	57	17	13	17	0.85
Molybdenum	75	18	14	18	0.90
Nickel	420	420	316	420	21
Selenium	100	100	75	36	5.0
Zinc	7500	2800	2100	2800	140

Tablo 6. U.S. Environmental Agency Limit Values for the Use of Sewage Sludge on Land[f]

[e] Nriagu & Pacyna, (1988)

According to Huang & Iskandar, (2000), soil loading capacity for heavy metals refer to the maximum load of heavy metals the soil is capable of holding within a given environmental unit and a given duration of time without the risk of exceeding the criteria for environmental quality, affecting the yield and biological quality of agricultural products, polluting the environment.

Arsenic compounds intake in body are connected in blood by hemoglobin protein and prevent the activity of many enzymes. The most harmful ones among mercury compounds are alkyl mercury compounds and they block oxygen supply mechanism of living tissues. Mercury is also used as a fungusite in agriculture. Mercury can also enter into the environment through mining and use of fossil fuels. Mercury is present in coal in the range of 10 to 46 000 mg/kg, though generally it is in the range of 200 to 400 mg/kg (U.S. EPA, 1985). Pb, Cd, Cr, Cu, Ni and Zn may be present in toxicological amounts in soils around highways with heavy traffic (Scanlon, 1991; Sezgin et al., 2003; Charlesworth et al., 2003). The important ones among negative effects created by lead are lead apoplexy, sense defectiveness, cerebral disorders and digestive system disorders. Traffic origin emissions stated by Novotny & Olem, (1994) are indicated in Table 7.

	Percent of Total Solids
Pollutant	**by Weight**
Volatile solids	5.1
BOD	0.23
COD	5.4
Grease	0.64
Total P	0.06
TKN	0.016
Nitrate	0.008
Asbestos	3.6 10^5 fibers/g
Lead	1.2
Chromium	0.008
Copper	0.012
Nickel	0.019
Zinc	0.15
Emission rates of total solids	0.671 g/axle-km

Tablo 7. Traffic Emissions[g]

Among heavy metals, Fe, Cu, Zn and Mn are trace elements which should be taken as low amounts in plant, animal and human nutrition. Their absence in the body may also cause important health problems. However, the reason why these elements are mentioned in scope of heavy metals is that their toxilogical effects they create when they are taken in the body in high amounts are heavy just like other heavy metals. Environmental concerns of today necessitate determination of toxilogical levels of heavy metals in drinking waters,

[f] USPA, (1993)

[g] Novotny & Olem, (1994)

purification sludge, soil, foods and especially foods consumed fresh. Heavy metal standards determined in drinking waters by different institutions like WHO and EPA are indicated in Table 8.

	WHO	EC DIRECTIVE	UK REGUL'NS	US EPA
	1993 Guideline	1980	1989	1992
	Value	MAC	Max.	MCL
	P=Provisional	GL=guide level		P=Proposed
		mg/l		
Antimony (Sb)	0.005 P	0.01	as EC	0.006
Arsenic (As)	0.001 P	0.05	as EC	0.05
Barium (Ba)	0.7	(GL 0.1)	1.0 (Av.)	2
Beryllium (Be)	NAD	No value set	as EC	0.001
Boron (B)	0.3	(GL 1)	2.0 (Av.)	
Cadmium (Cd)	0.003	0.005	as EC	0.005
Chromium (Cr)	0.05 P	0.05	as EC	0.1
Copper (Cu)	2 P	no MAC	3.0	1.3
Cyanide (CN)	0.07	0.05	as EC	0.2
Fluoride (F)	1.5	1.5@8-12 ^{0}C 0.7@25-30 ^{0}C		4
Lead (Pb)	0.01	0.05 in running water		0.015
Manganese (Mn)	0.5 P	0.05	as EC	no MCL
Mercury (Hg)	0.001	0.001	as EC	0.002
Molybdenum (Mo)	0.07	not listed	as EC	
Nickel (Ni)	0.02	0.05	as EC	0.1

Tablo 8. Drinking Water Standards[h]

3.3 Fertilizers

Another input applied in order to obtain more harvest from a unit of area is fertilizers. Increase in world population forces the limits of agricultural areas in one hand, leads to excess use of fertilizers on the other hand. Fertilizers upgrade the quality of agricultural products as well as the increase in productivity. In addition, negative effects of application of fertilizers on the environment have emerged in the countries where consumption of fertilizers is high. Even though harmful effects of excess application of fertilizers on human health and environment in industrialized countries are clearly seen, consumption of fertilizers in these countries increases more and more.

3.3.1 Environmental Effercts of Excessive Aplication of Fertilizers

Environmental pollution due to fertilizers not only depends on soil characteristics but also climatic and geographical conditions. Washing and erosion events are among fundamental

[h] Twort, et al., (1994)

factors in pollution of water resources. Fundamental negative effects created by excessive application of fertilizers are as summarized below: Causing eutrophication, its contribution in greenhouse effect, its harmful effects on soil microorganisms, its harmful effects on plants, Its effects on people's health, its negative effects on soil pH.
Macro and micro elements required in plant nutrition are provided through chemical fertilizers applied in the soil in different chemical forms for long years. Mostly macro-food elements are applied to the soil. These are nitrogen, phosphorus, potassium, calcium, magnesium and sulphur. The macro-elements causing most environmental pollution among them are nitrogen and phosphorus. Two nutrients called as eutrophication in water resources and removing beneficial using possibilities of water resources are nitrogen and phosphorus. Eutrophication is the event off algae and moss bloom and accumulation of toxic compounds in aquatic atmospheres as a result of nitrogen and phosphorus enrichment. It generally occurs due to human activities like land use, sewage and reach of industrial wastewaters to water atmosphere. Nitrogen and phosphorus loads coming with fertilizers from agricultural areas have an important share among nitrogen and phosphorus loads reaching to water resources from different areas. Nitrogen and phosphorus loads coming from those areas are indicated in Table 9.

Source		kg/ha/yr	
	Suspended Solids	**Nitrogen**	**Phosphorus**
Untreated dry weather wastewater flow	995	939	62
Wet weather diffuse urban loads	1241	223	26
Average agricultural loads		44-66	4-9

Tablo 9. Unit Loads of Pollutants from Diffuse Sources[i].

An important amount of NH_4^+ nitrogen come out as a result of mineralization of compounds applied to the soil with chemical fertilizers and organic nitrogen compounds are uptaken by plants. Remaining amount is adsorbed by clay minerals or used by soil microorganisms. For that reason, environmental risk in terms of NH_4^+ nitrogen is less compare to NO_3^- nitrogen. However, NO_3^- is not stable in the soil and cannot be adsorbed by clay minerals since it has negative valence. Therefore, NO_3^- nitrogen which cannot be adsorbed but pushed by soil colloids drains to groundwaters easily. For that reason, NO_3^- concentration in underground waters is essential. If nitrate concentrations in drinking waters exceed the value of 500 g NO_3^-/m^3, it leads to digestive and urinary system infections in adults. According to Winton et al., (1971) high concentration of nitrate cause the disease called "methemoglobinemia" and deaths in little babies. While the limit value of NO_3^- nitrogen in drinking waters permitted by USA Public Health Service was determined as 45 mg NO_3^-/l, it was determined as 5-10 mg NO_3^-/l by World Health Organization (WHO), and as 50 mg NO_3^-/l by the European Union.
Nitrogen fertilizers used in product cultivation made in anaerobic atmospheres like rough rice farming and having the characteristic of dissolving rapidly cause denitrification event. (Duxbury & McConnaughey, 1986). Denitrification event is an important environmental problem which leads nitrogen to be sent to the atmosphere from the soil in the form of NO_2, N_2O, NO and N_2. It contributes increase in industrial origin NO_x concentrations.

---

[i] Novotny & Olem, (1994)

It may create an amount of heavy metal pollution due to phosphoric fertilizers applied to the soil and Cd amoun in their content. Concentration of cadmium in phosphoric fertilizers is higher than total Cd concentration in the soil. Main source of heavy metal pollution created by phosphoric fertilizer applications is phosphate rock. A dangerous amount of cadmium accumulation may occur in the soil and plant as a result of phosphoric fertilizer application continuously (Syers et al., 1986; Laegreid et al., 1999). The most important ones among heavy metals contaminated with fertilizers are Cd, Hg, Ni, Pb and Zn.

An important part of phosphate ions is uptaken by plants, and the remaining part is hardly adsorbed by soil colloids or forms insoluble compounds with Ca, Fe, Al or Mn depending on pH of the soil. Therefore, phosphate ions are fixed in the soil and do not create water pollution risk unlike nitrate ions. Main reason why phosphate ions applied with fertilizers cause eutrophication in surface water resources is the erosion. Rainfall especially just after the application of fertilizers accelerates this process. Fertilizers should be applied by dividing in a few sections with doses appropriate with scientific rules and some form of fertilizers which are dissolved in the soil slowly should be selected in order to prevent barrages and lakes from being loaded with excessive amounts of nitrogen and phosphate and not cause environmental pollution. According to Zabunoğlu & Karacal, (1992) nitrogen fertilizers made slow effective by covering with sulfur should be used or according to Pauly et al., (2002) slowly dissolving phosphoric fertilizers which prevent phosphor from converting into an unbeneficial form by dissolving the whole phosphate and fixing in the soil should be ensured to be used.

The number of animals has increased in order to meet nutrition requirements of increased human population, so animal wastes have also increased and these wastes caused a significant degree of pollution in soil, water and air. According to estimations, 10 000 bovine animals produce approximately 300 ton/day manures. Gases deteriorating soil air like CO_2, NO_2, N_2O, and N_2 are formed during the process of decomposition and breaking up of manure. Excessively used manure cause salt and then Na accumulation in the soil, add destructive organisms in the soil like a large number of bacteria, fungi and virus, and accumulation of organic and inorganic compounds which are toxic for plants.

3.4 Agricultural Residues

Waste amounts also increase in parallel to increase in world population. Organic wastes have an important position among pollutions created by wastes. A large amount of organic wastes is composed of residues appeared as a result of agricultural production. Dispersed structure of these residues, and causing transport and workmanship costs are important problems. Agricultural residues can be examined in 3 groups: Residues remained as a result of vegetable production. Residues remained as a result of animal production. Residues come out as a result of processing agricultural products.

Field crops and animal breeding represent almost 90% of the sector. The amount and types of products constituting the basis of agricultural sector (wheat, sunflower, tobacco, cotton, corn, plant of greenhouse etc.) cause a large amount of agricultural residues. These wastes are processed in an uncontrolled way; they are either burned up in outdoor or left for corruption in solid waste storing areas. These wastes cause serious environmental pollutions in any case. Modern biomass resources are listed as energy forestry products and tree industry wastes, energy agriculture products, plant and animal wastes of agricultural sector, urban wastes, and agricultural industry wastes. Said biomass materials are processed

through low and high biomass fuel techniques and turned into solid, liquid and gas fuels (Exploitation of Agricultural Residues in Turkey, 2005).
Obtaining biogas through corroding agricultural wastes in controlled atmospheres, alternative and beneficial using forms of agricultural wastes, and cultivation of potential energy plants should be popularized and encouraged. Converting these wastes into wood or woody forms will also decrease burning of them in outdoor spaces.

4. Conclusions

Natural resources such as soil, water and air play an important role in preserving the existence as well as the development of our planet and its people. Currently, pollution of the agricultural environment is one of the serious environmental concerns in our planet. Proposed strategies for the protection of water, soil and air are as follows: At first, environmental education for all people is necessary. Growing ornamental plants, such as flowers, grasses, and woody plants in heavily polluted lands. The application of biodegradable pesticides in agriculture should be encouraged. In sustainable, or ecological agriculture, rather than only chemical pesticides is advocated, and fertilization is recommended with an emphasis on organic matter cycling. To facilitate positive advances in remediation, development of appropriate methods and efficient pollutant removal technologies is necessary. Control of total amount of pollutants discharged and treatment of all the waste laws, scientific management of pollutants and its perfect the legal system, are of primary importance. Broader objectives for environmental policy based upon the concept of sustainable development, and focus upon resource conservation as well as pollution control. We must understand only exist one world.

5. References

Charlesworth, S., Everett, M., McCarthy, R., Ordonez, A. & de Miguel, E., 2003. A Comparative Study of Heavy Metal Concentration and Distribution in Deposited Street Dusts in a Large and a Small Urban Area: Birmingham and Coventry, West Midlands, *UK. Environmental International. 29: 5. P. 563-573.*

Cheng, G.S. 1985. Ecological Effect of pesticide on microbe in the soil-The influence of fungucides on VA Mycorrhiza. *China Environ. Sci., 5(1), pp.21-25,*

Colume, A., Cardenas, S., Gallego, M. & Valcarcel, M. 2001. Semiautomatic Multiresidue Gas Chromatographic Method for the Screening of Vegetables for 25 Organochlorine and Pyrethroid Pesticides. *Analytica Chimica Acta. 436: 153-162.*

Duxbury, J.M., & McConnaughey, P.K., 1986. Effect of fertilizer source on denitrification and nitrous oxide emissions in a maize-field. *Soil Science Society. Am. J. 50:644-648.*

Ekmekyapar, F. & Kaykıoglu G., 2007. Application of industrial treatment plant sludge and heavy metal accumulation in lettuce plant (Lactuva sativa). *Asian Journal of Chemistry. 19, 5, 4093-4101.*

Exploitation of Agricultural Residues in Turkey, 2005. Training Course, Funded by the European Commission under the LIFE Programme. Notes of Course. pp. 124-125.

Haktanır, K., 1983. Çevre Kirliliği. A. U. Ders Notu. 107: pp. 54-55, Ankara.

Huang, P.M. & Iskandar, I.K., 2000. Soils and Groundwater Pollution and Remediation. Asia, Africa, and Ocenia. *Lewis Publishers. Boca Raton London, Newyork, Washington, D.C. pp. 196-231.*

Jiang, X.L. & Cai, D.J., 1990. Volatilization of pesticides from water and soil surfaces. *China Environ. Sci., 10(3), pp. 171-176.*

Laegreid, M., Bockman, O.C. & Kaarstad, O. 1999. Agriculture, Fertilizers and the Environment. *CABI publishing in association with Norsk Hydro ASA. Oslo, Norway, p. 144-157.*

Novotny, V. & Olem, H., 1994. Water Quality. Prevention, identification, and management of diffuse pollution. Van Nostrand Reinhold, pp. 456, New York.

Padron-Sanz, C., Halko, R., Sosa-Ferrera, Z. & Santana-Rodriguezb, J.J. 2005. Combination of microwawe assisted micellear extraction and liquid chromatography for the determination of organophosphorus pesticides in soil samples. *Journal of Chromatographya A. 1078: 13-21.*

Pauly, D.G., Malhi, S.S. & Nyborg, M., 2002. Controllede release P fertilizer concept evaluation using growt and P uptake of barley from three soils in greenhouse. *Canadian Journal of Soil Science, 82: 201-210.*

Scanlon, P.F., 1991. Effects of highway pollutants upon terrestrial ecosystems, In: Hamilton RS, Harrison RM, editors, Highway Pollution, Elsevier, Oxford, *Studies in Enviromental Science, 44:281-338.*

Sezgin, N., Özcan, H. K., Demir, G., Nemlioğlu, S. & Bayat, C., 2004. Determination of heavy metal concentrations in street dusts in Istanbul E-5 highway. *Environment International. 29:7. P. 979-985.*

Shan, Z.J., Zhu, Z.L. Hua, X.H. Jiang, X.M. & Cai, D.J., 1994. Mobility of three pesticides in soil. *Rural Eco-Environ., 10(4), pp. 30-33.*

Syers, J.K., Mackay, A.D., Brown, M.W. & Curie, L.D., 1986. Chemical and Phsical Characteristics of Phosphate Rock Materials of Varying Reactivity. *J. Food Agric. 37, pp.1057-1064.*

Taylor, A.W. & Spencer, W.F., 1990. Volatilization and vapour transport processes, in pesticides in the soil environment : Processes, impacts and modelling, Cheng, H.H., Ed., *Soil Science Society of America, Madison*, WI, pp. 213-269.

Twort, A.C., Law, F.M., Crowley, F.W. & Ratnayaka, D.D., 1994. Water Suuply, International Student Edition. Fourt Edition. P. 206.

Uslu, O. & Türkman, A. 1987. Water Pollution and Control. Prime Ministry General Directorate of Environment Publications Educational Series, 260-261, Ankara.

U.S. Environmental Protection Agency, 1985. National Primary Drinking Water Regulations. Synthetic organic chemicals, inorganic chemicals, and microorganisms; proposed rule, Fed. Regist. 50:46936.

U.S. Environmental Protection Agency, 1993. Standards for the use or disposal of sewage sludge; final rules (40 CFR Parts 257, 403 and 503). *Fed. Reg.* 58, pp. 9248-9415.

Wang, J.h. Zhang, L.Z. & Dai, G.M., 1989. Adsorbtion of some pesticides in soil. *Environ. Chem., 8(5), pp. 21-27.*

Winton, E., Tardift, R., Gand, McCabe, L. J., 1971. Nitrate in drinking water. *J. Am. Water Norks Assoc, 63:95.*

Zabunoğlu, S. & Karaçal, I., 1992. Gübreler ve Gübreleme. Ders Kitabı III. Baskı, A.Ü. Ziraat Fakültesi Yayınları, No:1279, Ankara.

Zhu, Z.L., 1994. The safe and effective use of pesticide, prevention of contamination of groundwater. *Pest. Sci. Admin., 13(3), pp. 34-37.*

4

Emission Sources and Their Contributions to Ambient Air Concentrations of Pollutants

Dragana Đorđević
IChTM – Centre of Chemistry, Serbia

1. Introduction

Understanding the chemical and physical processes in the atmosphere and emission sources of various technologies is the key to the development of cost-effective and health-protective air pollution control strategies. A number of species have been designated "hazardous air pollutants" or toxic air contaminants. Most are directly emitted into the air, but some also have significant secondary sources, i.e., are formed by chemical reactions in air. Furthermore, the ultimate health impacts are determined not only by the emission and formation of such compounds in air but also by their atmospheric fates. In short, some pollutants react in air to form less toxic species, whereas other form more toxic compounds (Barbara et al., 2000).

Today, no region of the global atmosphere is unaffected by anthropogenic pollution. Urban regions are affected by inorganic, light organic and heavy organic gases (Jacobson, 2002) and aerosols. Import heavy organics in urban air, such as toluene and xylenes, break down chemically over hours to a few days; thus, most of the free troposphere is not affected directly by these gases, although it is affected by their breakdown products.

Heating and cooling of the surface affect the stability of the atmosphere. The Earth's surface is a much more efficient absorber and emitter of radiation than the atmosphere above. During daytime, heating of the surface increases air temperatures close to the surface, resulting in an unstable atmosphere where air moves freely up and down. At sunset the land surface begins to cool, setting up stable conditions near the surface. Diurnal variation in atmospheric stability over land surfaces has important implication for urban air pollution; ventilation of cities tends to be suppressed at night and facilitated in the daytime (Jacob, 1999).

The mathematical models based on fundamentals of atmospheric chemistry and physics are essential tools in tracking emissions from many sources, their atmospheric transport and transformation, and finally their contribution to concentrations at a given location (receptor). The receptor models are enabling to attack the source contribution identification problem in reverse order, proceeding from particulate concentrations at the receptor site backward to responsible emission sources. The corresponding tools, named receptor models, attempt to relate measured concentration at a given site to their sources without reconstructing the dispersion patterns of the material (Seinfeld and Pandis, 1998).

2. Emission Sources in the Industrial Town Pančevo

Pančevo (20^0 40′ N, 44^0 53′ E) an industrial town of about 80,000 inhabitants is located about 20 km northeast of Belgrade. A major industrial complex includes petrochemical plant a fertilizer plant and a major oil refinery. Industries for refining oil and for the production of petroleum derivatives and fertilizer manufacture are situated in the South Industrial Zone of Pančevo (Fig. 1).

The dominant wind flows from the south-eastern segment, i.e. from the direction of the Southern Industrial Zone, towards the town of Pančevo. Industrial furnaces and oil refinery torches, which are emission sources of SO_2 and NO_x and other products of burning, are also situated within the South Industrial Zone.

Fig. 1. Map of Pančevo

The Vatrogasni dom receptor is located in the urban part of Pančevo - in the traffic area, at about 5 km from its industrial zone (Fig. 1 red point a). The "Vojlovica" receptor (Fig. 1, red point b) is at about 300 m from the nearest traffic artery, which minimizes the influence of traffic.

The factories for refining oil are emission sources of oil-type NMVOCs, including benzene, toluene, xylenes, as well as sulphur compounds, methyl-mercaptan and TRS (total reduced sulphur). NMVOCs are emitted from sources of technological processes for the production of petroleum derivatives. However, non-hermetic equipment and installations at the oil refinery and the petrochemical industry, from which NMVOCs can freely evaporate, contribute more significantly to the emissions. These are mainly non-hermetic oil- and oil derivative-tanks, dispatch facilities for petroleum products at the oil refinery, dispatch facilities for pyrolytic gasoline at the petrochemical industry and open-air waste water treatment plants at the oil refinery and petrochemical industry, where NMVOCs freely evaporate from the surface of the waste water. There is also a significant contribution of NMVOCs from traffic.

The fertilizer plant produces NH_3 using natural gas, which is first desuphurized with ZnO and subsequently, after mixing with water vapour, reformed and methanized to give a mixture with following composition: CO and CO_2, CH_4, H_2, N_2 and Ar. This gas mixture is compressed for the synthesis of NH_3 from the H_2 and N_2. One part of the recycled gas is driven to the reforming section, as a fuel, to keep the partial pressures of N_2 and H_2 constant. The production of NH_3 and its use in the manufacture of synthetic fertilizers is an only one emission source of NH_3 to the ambient air of Pančevo, while the combustion of the recycled gas is one of emission source of NO_x, besides traffic and other industrial furnaces.

2.1 The Contents of Specific Pollutants in the Ambient Air of Pančevo

In this study the results of continuous monitoring obtained from the municipal monitoring system of the industrial town Pančevo were used. The air pollutants in the urban area of the industrial city Pančevo was continuously monitored for one month (August, 2004). The hourly concentration of ammonia, Total Reduced Sulphur (TRS), NO_2, SO_2, PM_{10}, and hydrocarbons recorded minute-by-minute fluctuations were used.

For example, in August 2004, the average fuel consumption of the energy plant of the petrochemical industry, which produces steam for the industrial processes, was: 135 538 Nm^3 day^{-1} of natural gas, 70 t day^{-1} of fuel oil and 42 t day^{-1} of pyrolytic oil. The content of H_2S in the natural gas was between 0.86 and 5.40 mg l^{-1}, the content of sulphur in the fuel oil was 2.3 % and in the pyrolytic oil up to 4.5%. Assuming that combustion at the energy plant was complete, and that oxidation of the reduced forms of sulphur was complete, the maximum possible emission of SO_2 from high altitude emitters at this industrial firebox alone was approximately up to 350 kg h^{-1}, in August 2004. The emission sources of high altitudes originating from the combustion of liquid and gaseous fuels in industrial furnaces are expected to be the dominant contributors to the SO_2 and NO_2 in the ambient air but, near ground concentrations of these pollutants were low.

Routine measurements for several years showed that the NO_2 emission from the fertilizer plant was approximately 290 kg h^{-1}. However, NH_3 was also emitted, but the amount was not precisely quantified, either from technological balance or by measurements.

		Hourly			
		Min	**Max**	**Mean**	**1σ**
Vatrogasni dom	**SO_2**	10.8	143.1	19.3	10.2
	Benzene	0.1	160.5	18.0	25.3
	Toluene	0.1	180.0	18.8	22.9
	Me-me	0.2	20.0	2.3	3.1
	NH_3	0.3	401.5	10.8	26.2
	NO_2	1.5	159.3	17.0	14.3
	NMVOCs	0.6	497.8	75.4	110.7
Vojlovica	**SO_2**	0.7	63.2	5.5	4.5
	Benzene	0.4	150.2	12.5	22.5
	Toluene	0.7	184.7	15.9	27.8
	Xylenes	0.1	114.7	5.2	10.9
	TRS	0.7	12.3	2.7	1.8
	PM10	0.1	232.9	50.7	35.0

Table 1. The concentrations of the measured pollutants at the receptors Vatrogasni dom and Vojlovica in August 2004

Sulphur dioxide and nitrogen oxides as a combustion products, are emitting from industrial furnaces but it also originate from traffic. The high altitude industrial stacks in the Industrial Zone emit combustion products continuously, but traffic too.

2.2 Daily Variation of the Ambient Air Concentration

Based on the analysis of the mean values and standard deviations it is noticeable that no statistically significant differences existed in the concentrations of PM_{10} at Vojlovica receptor; 48.6 ± 35.8 μgm^{-3} and 49.6 ± 34.2 μgm^{-3} during the night and daytime periods as well as SO_2 during the day; 19.4 ± 10.5 μgm^{-3} and night; 19.4 ± 9.1 μgm^{-3} (Fig. 1 and Tab. 1).

Statistically significant differences between the concentrations during day and night times periods are noticeable in the cases of benzene; 11.5 ± 14.5 μgm^{-3} and 22.8 ± 29.2 μgm^{-3} at Vatrogasni dom receptor; and 7.4 ± 14.8 μgm^{-3} and 17.0 ± 25.4 μgm^{-3} at Vojlovica receptor; respectively.

Statistically significant differences between the concentrations during day and night time periods are noticeable in the cases of toluene too; 11.6 ± 14.5 μgm^{-3} and 22.8 ± 29.2 μgm^{-3} at Vatrogasni dom receptor, and 8.5 ± 15.8 μgm^{-3} and 22.8 ± 32.6 μgm^{-3} at Vojlovica, xylenes; 2.5 ± 4.6 μgm^{-3} and 7.8 ± 13.4 μgm^{-3} as well as total reduced sulphur (TRS); 2.1 ± 1.3 μgm^{-3} and 3.3 ± 1.9 μgm^{-3} at Vojlovica receptor respectively.

The mean concentrations of NH_3 were higher during the daytime (12.3 ± 31.3 μgm^{-3}) than those during the night period (7.5 ± 11.6 μgm^{-3}) at Vatrogasni dom receptor, while the mean of daily concentrations of NO_2 during the daytime (12.9 ± 14.8 μgm^{-3}) were statistically significantly lower than those during the night (20.1 ± 11.5 μgm^{-3}) at Vatrogasni dom receptor, unlike the daily variations of the concentrations measured in the ambient air of the Kaohsiung Petroleum Refinery in Taiwan (Chiu et al., 2005a; Chiu et al., 2005b). The results showed that the concentrations of pollutants originating from low altitude emission sources like organic pollutants were higher twice and more at night. However, SO_2 and NO_2 are also emitted from a near-ground source – traffic.

Statistically significant differences (mean ± 1σ) between the concentrations during day and night time periods are noticeable in the cases of benzene (7.4 ± 14.8 μg m^{-3} and 17.0 ± 25.4 μg m^{-3}), toluene (8.5 ± 15.8 μg m^{-3} and 22.8 ± 32.6 μg m^{-3}), xylenes (2.5 ± 4.6 μg m^{-3} and 7.8 ± 13.4 μg m^{-3}) and TRS (2.1 ± 1.3 μg m^{-3} and 3.3 ± 1.9 μg m^{-3}). Thus, the concentrations of these pollutants were higher during the night than during the daytime period, which is in accordance with the results of measurements in the ambient air of the Kaohsiung Petroleum Refinery in Taiwan (Chiu et al., 2005a; Chiu et al., 2005b). These pollutants were emitted from low altitude sources located at the oil refinery and petrochemical industry. Low altitude sources combined with surface temperature inversion of the atmosphere represent extremely favourable conditions for pollution of the near ground atmosphere. An investigation of the frequency of surface inversion in the Belgrade region showed that average incidence of night time inversions was the highest (as many as 26 cases) in August (Vukmirović et al., 2003).

When the boundary layer of surface inversion moves towards the ground, the combustion products inside a plume of smoke emitted from high altitude industrial stacks spreads above the boundary layer and therefore near ground concentrations of these pollutants is lower. This phenomenon is more pronounced at night and as a result, the concentrations in the near-surface atmosphere are lower at night. The upper edge of a boundary layer is a natural barrier to the vertical dispersion of pollutants by diffusion.

The primary particles (Seinfeld and Pandis, 1998) are emitted from industrial furnaces and traffic, but they also can be formed as the products of atmospheric reactions between NH_3, emitted from the fertilizer factory, and acid oxides (SO_2 and NO_2), emitted from industrial furnaces and traffic, whereby in observed ambient conditions are enabled $(NH_4)_2SO_4$ and $(NH_4)NO_3$ aerosol production in the neutralisation processes (Olszyna et al., 2005). In atmospheric reactions, through SOA mechanisms, atmospheric aerosols are formed from organic air pollutants (Hamilton et al., 2005; Jenkin and Clemitshaw 2000; Knippinga et al., 2004 etc.).

2.3 Statistical Cluster Analysis with a View to Source Identification

In order to identify the main contribution sources of some pollutants in the receptor region (sampling site), the most commonly implemented subset of clustering method, which is generally referred to as an agglomerative hierarchical method, was employed (Facchinelli et al., 2001; Đorđević et al., 2004; Lee et al., 2004). This method is appropriate to evidence correlations between variables. The parameter dendrogram based on Pearson's correlation coefficients is summarized in the dendrogram shown in Fig. 2. The smaller the value on the axis is, the more significant is the association.

Hierarchical cluster analysis was used to obtain dendrograms (Fig. 2) established using average linkage between groups according to the Pearson correlation of combinations of pollutants.

By cluster analysis of output data of measurements, the clusters of benzene (cluster 2) and NMVOCs (cluster 7) variables from the »Vatrogasni dom« are related to each other with the highest coefficient of correlation ($r_{Benzene\text{ - }THMNC}$ = 0.899) in the first stage of combinations in the day time, representing the strongest association (Fig 2a, Tab. 2). The second strong association is between toluene (10) and me-me (11) measeured in the night time with the correlation coefficien $r_{Toluene\text{-}me\text{-}me}$ = 0.887. The next strong association is between benzene (9) and NMVOCs (14) measured in the night time, with $r_{Benzene\text{-}NMVOC}$ = 0.870. In the fourth stage of agglomeration appears the association of toluene (3) and me-me (4) measured in day time with $r_{Toluene -me-me} = 0.864$ followed by association between benzene (9) and toluene (10) measured in the night time related to $r_{Benzene-Toluene} = 0.851$. The association between benzene (2) and toluene (3) ,measured in the day time, is related to $r_{Benzene-Toluene} = 0.831$. NH_3 (5) and NO_2 (6) ,that are agglomerated in the seventh stage, are bound with $r_{Benzene-Toluene} = 0.803$.

Vatrogasni dom (a)				Vojlovica (b)			
Stage	Cluster combined		Correlation coefficient	Stage	Cluster combined		Correlation coefficient
	Cluster 1	Cluster 2			Cluster 1	Cluster 2	
1	2	7	0.899	1	8	9	0.975
2	10	11	0.887	2	2	3	0.965
3	9	14	0.870	3	2	4	0.929
4	3	4	0.864	4	8	10	0.854
5	9	10	0.851	5	2	5	0.592
6	2	3	0.831	6	8	11	0.513
7	5	6	0.803	7	2	6	0.414
8	2	5	0.523	8	8	12	0.300
9	8	13	0.523	9	2	8	0.179
10	9	12	0.410	10	1	7	0.161
11	1	2	0.396	11	1	2	0.103
12	8	9	0.359				
13	1	8	0.265				

Table 2. Agglomeration Schedule

It is important to highlight that the clusters of benzene (2) and the NH_3 (5) measured in the day time, and the clusters of benzene (9) and NH_3 (12) measured in the night time are bound in both cases. Even the coefficient of correlation between the clusters of benzene and the NH_3 measured in the night time is lower ($r_{Benzene-NH3} = 0.410$). These parameters are linked in considerable correlation, this fact indicating the sources of benzene in the industrial zone beacuse the NH_3 originatig from just one sorce located in the »Azotara« Fertilizer plant . On the other hand the smallest values on the axis (Fig. 2) are agglomearated for benzene, toluene, NMVOCs and me-me measured in the day time, and for the same variables in the night time on the »Vatrogasni dom« receptor, indicating most considerable associations. Ammonia and nitrogen dioxide form a minor association , that becomes considerable in the day time and shows the influence of the »Azotara» Fertilizer plant .

As far as the results obtained on Vojlovica receptor are concerned, the associations are similar (Fig. 2b, Tab. 2). The most considerable and strongest associations are between benzene, toluene, xylenes, regardless of whether in the day time or in the night time.

The results of the cluster analysis show the existence of common strong emission sources having the effect of evaporation, because benzene, toluene and xylenes are within the scope of petrogenic hydrocarbons from the Oil Refinery and Petrochemical plants, but also in the scope of NMVOCs from the traffic. The prevailing winds in the region come from the south east direction. These winds from the South Industrial Zone carry pollutants to the town Pančevo (Fig. 1).

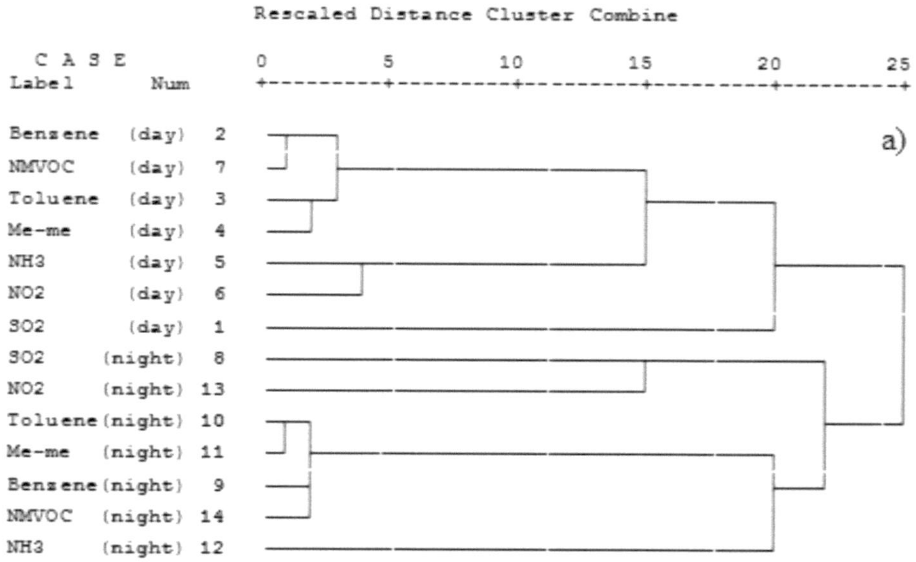

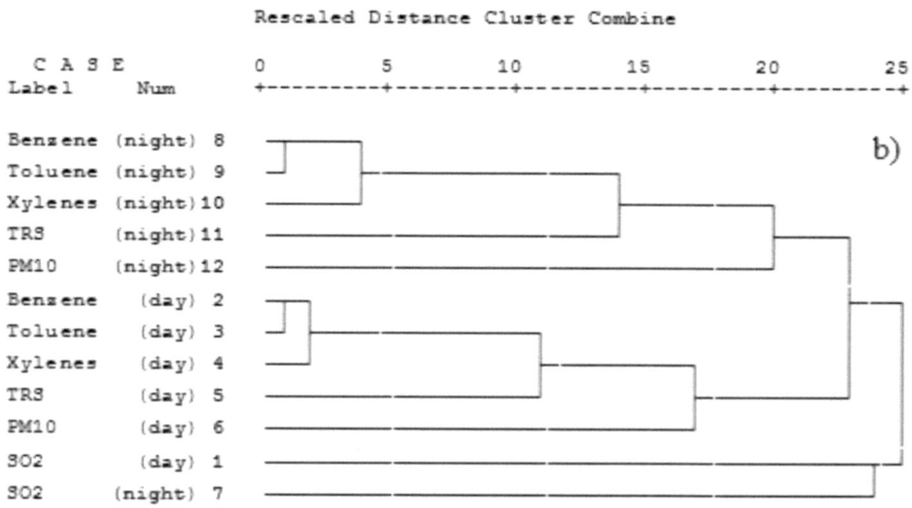

Fig. 2. Dendrograms established of variable combinations of pollutants at the receptors: a)Vatrogasni dom, and b) Vojlovica

The production processes of the oil refinery and petrochemical industry involve the use of volatile hydrocarbons as basic raw material as well as end products. Inevitably, processes with volatile hydrocarbons result in their evaporation into the atmosphere, especially when the production and storage equipment is non-hermetic. There are devices and parts of old technologies in both the oil refinery and petrochemical industry which enable the undisturbed evaporation of significant amounts of hydrocarbons into the atmosphere. An example of such devices are decanter pipes which are used for the shipping of petroleum products at the loading facilities for *van-* and *truck*-decanters in the oil refinery and for the shipping of pyrolytic gasoline at the loading facilities for *truck*-decanters at the petrochemical factories. When filling a tank with gasoline and other products, there is undisturbed evaporation of saturated vapour from the interior of the tank into the atmosphere.

Oily waste water treatment plants at the oil refinery and petrochemical industry are a second significant emission source of volatile organic compounds into the atmosphere.

For example, Fig. 3 shows the influence of the once-for-all pouring of pyrolytic oil on the occasion of shipping 500 m^3 from the petrochemical plant that was taking place on August 23 in the afternoon's and the night's shifts, as well as on the occasion of shipping 100 m^3 of pyrrolytic oil that was taking place on August 24 in the morning's shift. On the "Vatrogasni dom" measurement site, situated in the direction of the dominant southeast wind in relation to the Petrochemistry, considerable concentration growth of benzene, together with the concentration growth of toluene and methylmercaptans, was registered.

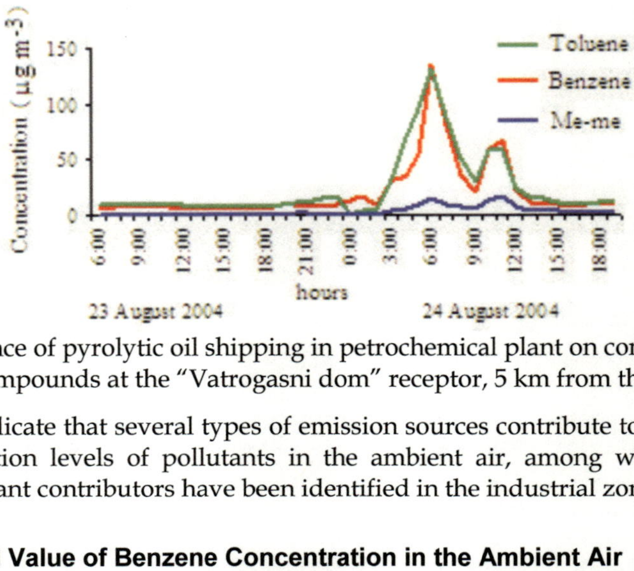

Fig. 3. The influence of pyrolytic oil shipping in petrochemical plant on concentrations of organic compounds at the "Vatrogasni dom" receptor, 5 km from the source

All these facts indicate that several types of emission sources contribute to different degrees to the concentration levels of pollutants in the ambient air, among which low altitude sources as dominant contributors have been identified in the industrial zone.

2.4 Valid Limited Value of Benzene Concentration in the Ambient Air

According to Directive 2000/69/EC, ANNEX I, relating to limit values for benzene and carbon monoxide in ambient air the limited value for benzene standardized at a temperature of 293 K and a pressure of 101,3 kPa, for the protection of human health is 5 μg/m^3 on the averaging period of calendar year with margin of tolerance of 5 μg/m^3 (100%) on 13 December 2000, reducing on 1 January 2006 and every 12 months thereafter by 1 μg/m^3 to reach 0 % by 1 January 2010. Date by which limit value of 5 μg/m^3 on the averaging period of calendar year, is to be met is 1 January 2010, except within zones and agglomerations within which a time-limited extension has been agreed in accordance with Article 3(2).

According to Article 3(2) of Directive 2000/69/EC, when the limit value laid down in Annex I is difficult to achieve because of site-specific dispersion characteristics or relevant climatic conditions, such as low wind speed and/or conditions conducive to evaporation, and if the application of the measures were to result in severe socio-economic problems, a Member State may ask the Commission for a timelimited extension. The Commission, acting in accordance with the procedure laid down in Article 12(2) of Directive 96/62/EC, may, at the request of a Member State and without prejudice to Article 8(3) of this Directive, grant one extension for a period of up to five years if the Member State concerned:

- designates the zones and/or agglomerations concerned;
- provides the necessary justification for such an extension;
- demonstrates that all reasonable measures have been taken

to lower the concentrations of the pollutants concerned and to minimise the area over which the limit value is exceeded, and — outlines the future developments with regard to the measures which it will take according to Article 8(3) of Directive 96/62/EC.
The limit value for benzene to be granted during that timelimited extension shall, however, not exceed 10 µg/m3.

2.5 Emission Control and Decreasing Benzene Concentration in the Ambient Air Pančevo

High concentrations of evaporable carbon hydrogen and benzene having been found out in the ambient air of Pančevo, and taking into consideration the limitations set by the Directive 2000/69/EC, there have been made priorities of overhaul of key sources of emissions within the industrial zone. In support of the model Chemical Mass Balance (Seinfeld and Pandis, 1998) were identified installations and engines in both the Oil Refinery and the Petrochemistry, whose contribution to the concentration of carbon hydrogen in the ambient air is considerable. These are first of all tanks for storing evaporable products originating from oil, then pouring stations for the shipping oil derivatives, and also installations for the refinment of waste waters in the above mentioned factories. These installations make part of ground floor sources of emissions, altitudes up to 10 m. Firstly was made the plan of activity in factories of the industrial zone (Đorđević et al., 2004), which excludes pouring of evaporable derivatives during unfavourable meteorological conditions (feeble southeast wind which blows the clouds of pollution over the city), and sets new location for pouring of pyrolytic oil, at about 9 km from the previous pouring location, toward Danube (Fig. 1). The system of loading pyrolytic oil in shipping wagons was moved out from the Petrochemistry factory in 2004. This was followed by numerous actions, such as hermetic closure of tanks in both the Oil Refinery and the Petrochemistry, then substitution of one part of obsolete systems for pouring of oil derivatives on the occasion of their loading for the shiping from the Oir Refinery, then hermetic closure of installations for the refinement of waste waters in the Petrochemistry, and filally harmonization of activities of pouring evaporable carbon hydrogen with meteorological conditions. When meteorological conditions are unfavorable, i.e. when in concrete case a feeble southeast wind blows in the direction Idustrial zone - the town, and the near ground floor inversions of atmosphere are considerable (that usualy take place in the night time). In these cases the activities of pouring have been stopped, i.e they have been planned in harmony with weather forecast. These measures resulted in considerable fall of concentration of benzene on the Vatrogasni dom receptor (Fig. 1a) as of 2004.
With the exception of the period of time 1994 - 1996 when, due to heavy economic crisis, the production was reduced to the minimum or completely suspended, the average annual concentration of benzene in the period of time 1997 - 2003 surpassed border line values set in the Directive 2000/69/EC (Fig. 4). New temporary measures of controlling key sources after their identification, undertaken with the aim to reduce emission of evaporable organic compounds, began to be applied as of 2004.

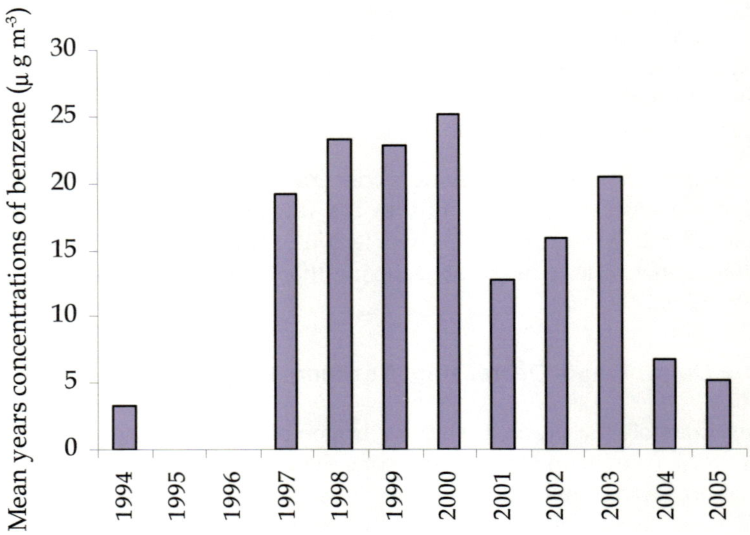

Fig. 4. Trend of annual benzene concentration in the ambient air at Vatrogasni dom receptor

Some investigations in the European cities indicate the presence of benzene originating from traffic in ambient urban air. Measurements have shown that mean annual value of the concentration of benzene in urban areas is the lowest in Copenhagen (around 3 μg m^{-3}); while in Athens the mean annual value is 20 μg m^{-3}. In these cities, the benzene derives mainly from traffic and the differences in the concentrations are the consequence of different meteorological conditions (Cocheo et al., 2000).

3. Conclusion

Harmful substances in the ambient air can be reduced only on condition that their emiters are kept under control. It is indispensable, for that purpose, to know technological processes, the nature of emiters, the link between the emission flux on the emiter and the concentration of pollutants in one part of receptors, as well as meteorological base of the area. Main emiters of pollutants have been identified, on application of statistical methods, upon analysis of technological processes in factories of the South industrial zone in Pančevo (Refinery, Petrochemistry and Fertilizer factory), by measurement of pollutants' concentrations in the ambient air and meteorological base of the area. The measures of control of emission - such as extinguishing key sources (pouring station of pyrolytic oil in the Petrochemistry), planning of emiters' activities in harmony with meteorological conditions, and reconstruction of one part of sources in both the Refinery and Petrtochemistry - had as the result the reduction of benzene concentrations in the Vatrogasni dom receptor's part for more than twice. The benzene concentrations on the measurement site which is under high influence of industrial sources reached in the years 2004 and 2005 the values acceptable with regard to the limit value of the Directive 2000/69/EC. It is important to highlight that the identification of emission sourcs on the base of measurement of pollutants' concentrations in the ambient air will be more exact and more precise if a grater number of various kinds of pollutants are measured, and if among them there are some specific emiters' tracers. It is also importnant to highligh that the control of emissions need not alwasy be followed by high investments in order to make the effect on the ambient air more visible.

4. Acknowledgements

I am very grateful to the Ministry of Science and Environment of the Republic of Serbia for financial support. I would also like to thank the Municipality of Pančevo for the measurement data used in this paper, as well as the Oil refinery, and the factories Petrochemical and Fertilizer "Azotara" from Pančevo for the technological data.

5. References

Chiu K.H.; Sree U.; Tseng S.H.; Wu C.H.; Lo J.G., (2005a). Differential optical absorption spectrometar measurement of NO_2, SO_2, O_3, HCHO and aromatic volatile organics in ambient air of Kaohsiung Petroleum Refinery in Taiwan, *Atmospheric Environment* 39, 941-955, ISSN 1352-2310.

Chiu K.H.; Wu B.Z.; Chang C.C.; Sree U.; Lo J.G., (2005b). Distribution of volatile organic compounds over semiconductor industrial park in Taiwan, *Environmental Science and Technology* 39, 937 - 983.

Cocheo,V.; Sacco, P.; Boaretto, C.; De Saeger, E.; Perez Ballesta; P., Skov, H.; Goelen, E.; Gonzalez, N.; Baeza Caracena, A., (2000). Urban Benzene and population exposure, brief communications, Nature 404, 141,

Directive 2000/69/EC of the European parliament and of the council of 16 November 2000 relating to limit values for benzene and carbon monoxide in ambient air

Đorđević D.; Radmanović D.; Mihajlidi-Zelić A.; Ilić M.; Pfendt P.; Vukmirović Z.; Polić P.; (2004). Principal associations of trace elements in airborne particulate matter at the South Adriatic Coast, *Environmental Chemistry Letters* 2 (3), 147-150.

Đorđević. D.; Šolević. T.; Arsić P., Petrović S., (2004). Final Report: *Investigation of causes and the level of ambient air pollution by harmful substances in the urban area of Pančevo*,http://www.ekoserb.sr.gov.yu/dokumenti/izvestaji/Konacni%20izvestaj_Pancevo.pdf

Facchinelli A.; Sacchi E.; Mallen L., (2001). Multivariate statistical and GIS-based approach to identify heavy metal sources in soils, *Environmental Pollution* 114, 313-324, ISSN 0269-7491.

Finlayson - Pitts Barbara J.; Pitts, Jr., James N. (2000). Chapter 16: Application of Atmospheric Chemistry, Air Pollution Control Strategies and Risk Assessment for Tropospheric Ozone and Associated Photochemical Oxidants, Acids, Particles, and Hazardous Air Pollutants in: *Chemistry of the Upper and Lower Atmosphere, Theory, Experiments and Applications*, Academic Press, ISBN 1-12-257060-x, California, USA

Hamilton J. F., Webb P. J., Lewis A. C., Reviejo M. M., (2005). Quantifying small molecules in secondary organic aerosol formed during the photo-oxidation of toluene with hydroxyl radicals, *Atmospheric Environment* 39, 7263–7275, ISSN 1352-2310.

Jacob J. Daniel (1999). Chapter 4: Atmospheric transport in: *Introduction to Atmospheric chemistry*, Princeton University Press, ISBN 0-691-00185-5, Chichester, West Sussex, United Kingdom

Jacobson, Mark, Z. (2002). Chapter 4: Urban air Pollution in: *Atmospheric Pollution, History, Science, and Regulation*, Cambridge University Press, ISBN 0 521 01044 6, United Kingdom

Jenkin M. E.; Clemitshaw K. C., (2000). Ozone and other secondary photochemical pollutants: chemical processes governing their formation in the planetary boundary layer, *Atmospheric Environment* 34, 2499-2527, ISSN 1352-2310.

Knippinga E. M.; Griffinb R. J.; Bowmanc F. M.; Pund B.; Seigneurd C.; Dabdube D.; Seinfeldf J. H., (2004). Comment on "Instantaneous secondary organic aerosol yields and their comparison with overall aerosol yields for aromatic and biogenic hydrocarbons" by Weimin Jiang, *Atmospheric Environment* 38, 2759–2761, ISSN 1352-2310.

Lee W.S.; Chang-Chien G.P.; Wang L.C.; Lee W.J.; Tsai P.J.; Wu K.Y.; Lin C., (2004). Source identification of PCDD/F for various atmospheric environments in a highly industrialiyed city, *Environmental Science and Technology* 38, 4937-4944.

Olszyna K.J.; Bairai S. T.; Tanner R.L., (2005). Effect of ambient NH_3 levels on $PM_{2.5}$ composition in the Great Smoky Mountains National Park, *Atmospheric Environment* 39, 4593-4606, ISSN 1352-2310.

Seinfeld, J. H.; Pandis, S. N., (1998). Chapter 24: Statistical Models in: *Atmospheric Chemistry and Physics From Air Pollution to Climate Change*, John Wiley & Sons, Inc., ISBN 0-471-17816-0, New York, USA

Vukmirović, Z.; Đorđević, D.; Vuković, T., (2003). Chapter: Structure of the Lower Troposphere above the Territory of Belgrade (Vuković T.), In the study: "*The Solutions of Ecological Problems Caused by the Thermal Power Plant Nikola Tesla A and B*", Book 1, Belgrade (p.p. 372 – 382).

5

Qualitative Environmental Health Impact Assessment in Veles, Republic of Macedonia

Vladimir Kendrovski and Dragan Gjorgjev
Republic Institute for Public Health, Medical Faculty
Republic of Macedonia

1. Introduction

The environment in which people live, work and play is an important determinant of health and well being, but the extent of its importance in developed economies is difficult to quantify. The non-communicable diseases present the biggest burden to public health analyzed by direct cost to the society as well as to the governance from aspect of disability adjusted life years (DALY) indicator. Due to fact that more and more citizens are elderly and because of that are more exposed to non-communicable diseases and disability, the needs for data, which will reflect the life quality, including the influence of environmental risk more precisely is essential. The summary measure of the population health and the methodology for the burden of diseases estimation nowadays is extraordinary indicators for the public health policy development as well as for actions needs for its reduction. The total burden of most frequent diseases in the Republic of Macedonia: circulatory, malignant and respiratory are estimated to 60,7% to DALY from all cause mortality, which is different than the percentage from the year of life lost, i.e. 52,4, respectively (Kendrovski& Gjorgjev, 2004).

The most common diseases in the Republic of Macedonia - heart and circulatory diseases, cancer, respiratory diseases, injuries and non defined symptoms - have many causes which are often interconnected; including genetics, the condition people are in (via diet, exercise etc.), and the environmental circumstances to which they are exposed. Identifying cause-and-effect relationships is therefore very difficult, especially if the impact of the environment on health is delayed, or is the product of many, perhaps small, environmental factors acting together. The country has a few environmental hot spots, characterized by high levels of pollution (air, water and soil), due to emissions from industrial facilities. According the Second National Environmental Action Plan there is a direct correlation between the air pollution and human health based on a number of research works performed in the period 1997-2002. The positive correlation was found between the monthly average concentrations of the black smoke and SO_2 and the increased respiratory morbidity for the children at the age 0-6 and 7-14. The air pollution problem is more acute in the winter period due to the effects of temperature inversion and climate circumstances in the country; this is supported by the 32% of all sold drugs being under the respiratory diagnosis, as reported by the Pharmacy Information System. Air pollution affects approximately 60% of the population, in particular those living in the cities of Skopje, Veles, Bitola and Tetovo. In regard to the industry sector, obsolete equipment and non-existent modern technologies result that this sector represents a major air polluter. The main pressure on environment (in particular air quality) originates from the metallurgy sector (until 2003 the lead and zinc smelter MHK Zletovo in Veles and ferro-alloy SILMAK in Jegunovce, Maksteel, MAKSTIL, FENI Industries-Kavadarci),and the chemical industry- refinery OKTA, OHIS chemical complex and TITAN cement factory. Until closure of MHK Zletovo in 2003, the lead and zinc smelting plant located in Veles operates a single absorption sulphuric acid plant with no additional treatment of the exhaust gas leading to an emission quantity of about 2100 t/y of SO_2. Additional 1800 t/y of SO_2 and 2.5 tpa of lead have been emitted through the ventilation system and the fugitive sources. In Veles a significant correlation has been found between the emissions of lead, zinc and cadmium as well as SO_2 in the air and the health of inhabitants. The higher concentration of the lead in the blood was registered and it has been connected with occurrence of cancer, respiratory diseases, miscarriages and birth defects. The direct correlation has been found by the medical experts between the particulate matters with small dimensions ($PM_{2.5}$) in the form of dust originated by the REK Bitola Power Plant and health problems with respiratory system at adults as well as bronchia at the children age. Although there are no any study in Macedonia presenting the direct correlation between the lead from mobile sources and human health, the medical experts uses the series of publications worldwide that confirms the harmful effect of lead (NEAP 2, 2005).

There is a serious lack of data and information on exposures, effects and biological models that connect them. Therefore considerable uncertainty surrounds many issues of concern, such as air pollution, noise, water contamination, waste, climate change, chemicals (including endocrine disruptors and antibiotics), ionising and non-ionising radiation.

Risk and hazard are two distinct, but interrelated, concepts. A hazard represents a chemical, physical, or biological substance that has the potential to produce harm to health if it is present in the environment and comes into contact with people. The hazardous properties of an environmental agent are defined according to the nature and severity of its harmful consequences. Fortunately, many hazards can be either contained or avoided, so not every potential environmental hazard poses an actual health risk. A risk, in turn, is defined as the likelihood of adverse health effects arising from exposure to a hazard in a human population, which is conceptually expressed as the product of two factors: the probability of exposure and the severity of the consequences.

Environmental health risk assessment is an essential element in environmental management and an important condition in precise priority-setting to the necessary actions for its sanitation. At present there are not sufficient scientific data available for a large number of health-related environmental hazards representing risk on human health. In addition, even with the best possible information available on the nature and level of pollutants in the environment and about population exposure to different pollutants, environmental health risk assessment may not be complete because of difficulties in analyzing the complexity of possible interactions in the case of multiple exposures. Even more complex is the assessment and comparison of costs and benefits of health risk elimination. This is partly because environmental health risk assessment is still limited in its effectiveness by the inadequacy of the information available, especially on exposure. In addition, even with the best possible information, an environmental health risk assessment may not be complete because of difficulties in analysing the complexity of possible interactions in the case of multiple exposures (NEPC,1999)

2. Health Risk Assessment Methodologies

In the context of environmental health, the risk management process can be organized into several distinct activities. The three core activities that constitute the essential decision-making steps in the risk management process are each involved in examining different aspects of the risk problem:

2.1 Risk estimation

The use of science-based risk information and analytical methods to characterize the nature and extent of environmental health risks in the human population;

2.2 Risk evaluation

Consideration of the economic, social, political, and legal factors that influence a decision to adopt a particular course of action to reduce health risks - in some risk frameworks, the quantitative economic analysis of the benefits and costs of risk reduction is combined with results of the risk estimation process, so that a, risk assessment may subsume part or all of risk evaluation;

2.3 Risk control

The selection of options and the commencing of actions intended to reduce risk to an acceptable or tolerable level; this activity is often referred to as risk management, but the term risk control is more specific and better reflects the objectives of the activities it denotes. Risk assessment is the process of estimating the potential impact of a chemical, physical, microbiological or psychosocial hazard on a specified human population or ecological system under a specific set of conditions and for a certain timeframe. Risk assessment is intended to provide complete information to risk managers, specifically policymakers and regulators, so that the best possible decisions are made. There are uncertainties related to risk assessment and it is important to make the best possible use of available information.

3. Methods

Retrospective epidemiological method was used in order the following four distinct and essential components of the risk assessment paradigm to be addressed:

(1) Hazard identification - identification of the inherent capability of a substance to cause adverse effects by sides;

(2) Assessment of dose-response relationships involves characterization of the relationship between the dose of an agent administered or received and the incidence of an adverse effect;

(3) Exposure assessment is the qualitative and/or quantitative assessment of the chemical nature, form and concentration of a chemical to which an identified population is exposed from all sources (air, water, soil and diet);

(4) Risk characterization is the synthesis of critically evaluated information and data from exposure assessment, hazard identification and dose-response considerations into a summary that identifies clearly the strengths and weaknesses of the database, the criteria applied to evaluation and the validation of all aspects of methodology, and the conclusions reached from the review of scientific information (EPA,1992).

The logical consequence of the process of assessment of potential risk is the application of the information to the development of practical measures (risk management) for the protection of human health. All available studies, articles and reports related to defined hotspots in the Republic of Macedonia from 2000 to 2007 are included in assessment. The migration pathways and sensitive receptors are as following:

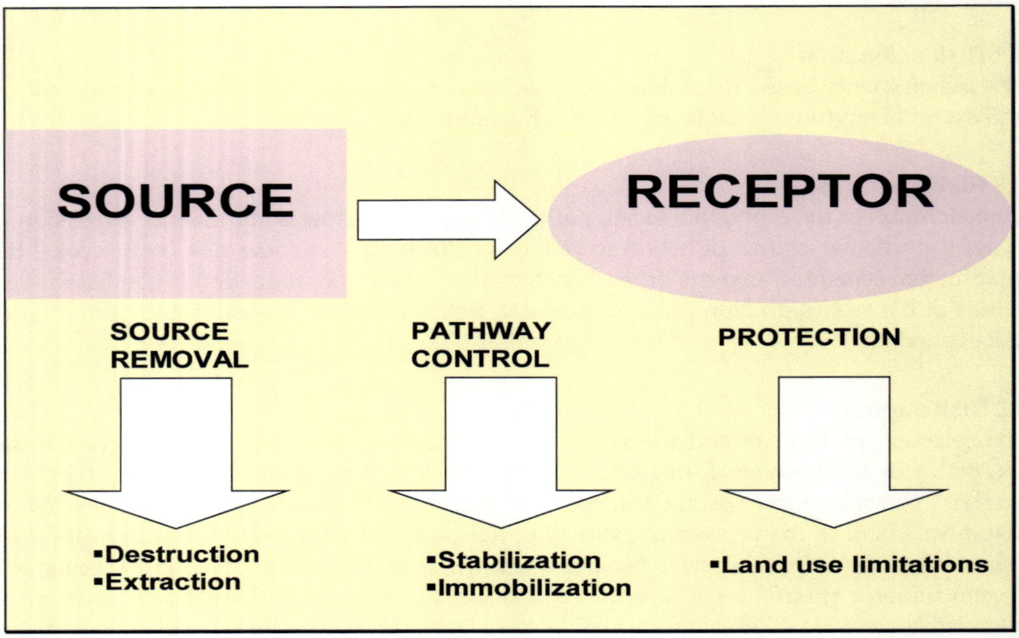

Fig. 1. Contaminated land management

The Figure 1 showed the principles of contaminated land management as general approach for the identification of sources of contamination and associated hazards.

3.1 Advantage of Various Methods

Many organisations are now actively involved in Environmental Risk Assessment, developing methodologies and techniques to improve this environmental management tool. Such organisations include OECD, WHO and ECETOC. One of the major difficulties concerning the use of risk assessment is the availability of data and the data that is available is often loaded with uncertainty (WHO,1994). The risk assessment may include an evaluation of what the risks mean in practice to those effected. This will depend heavily on how the risk is perceived. Risk perception involves people's beliefs, attitudes, judgements and feelings, as well as the wider social or cultural values that people adopt towards hazards and their benefits. The way in which people perceive risk is vital in the process of assessing and managing risk. Risk perception will be a major determinant in whether a risk is deemed to be "acceptable" and whether the risk management measures imposed are seen to resolve the problem. The procedures, methods and techniques for regulatory risk assessment of chemicals in the EU is described in both legislation and supporting Technical Guidance Documents. Implementation is supported by the European Chemicals Bureau, part of the Joint Research Centre, in Ispra.

Most methodologies for human health risk assessment of chemicals are based on the NAS model. A number of methodologies exist due to differences in the toxic mechanisms exerted by different classes of chemical and the toxicological end-point being assessed. The end-point being assessed could be death, or a specific pathological condition relating to exposure to a chemical. When attempting to assess the risks from an immuno-suppressant toxin, specific end-points may be difficult to determine, as may be the role of other agents and stressors on the body. This will lead to risk assessment methodology for immuno-suppressants being different from assessments for irritants for instance. All human health risk assessments of chemicals include hazard identification, dose-response assessment, exposure assessment and risk estimation/characterisation. If the assessment is site-specific, then a release assessment would be required in the absence of good data of environmental levels or to account for non-routine, accidental releases.

3.2 Risk and Constrains of Various Methods

Risk assessments may assess individual or population risks. Individual risks may be for the average (i.e. typical) individual or the highly exposed or particularly susceptible individual and the risks may be estimated for various duration of exposure (e.g. per year or per lifetime) or for different locations. Individual risk can only be assessed for a hypothetical individual with assume characteristics. Assessing the risk for any real individual will be frustrated by the fact that risk predictions for an individual can never be validated by experience. Any real individual will either experience the negative outcome or will not. Neither of these results can validate any risk prediction other than a probability of one or zero.Population risk may relate to the number of adverse health effects (eg. fatalities, cancers, or illnesses) in a population over a specified period of time or the rate of adverse effects for a given location or sub-population .

The UNEP/ILO/WHO International Programme on Chemical Safety (IPCS), in collaboration with the US Environmental Protection Agency (US EPA), the European Commission (EC), the Organization for Economic Cooperation and Development Cooperation, and other international and national organizations developed a working partnership to foster the integration of assessment approaches to evaluate human health and ecological risks. The overall goal of this project was to promote international understanding and acceptance of the integrated risk assessment process. Three specific objectives were identified to meet this goal: 1) enhance understanding of the benefits of integration, 2) identify and understand obstacles to integration, and 3) engage key scientific organizations to promote discussion of an integrated approach to risk assessment

A generic framework and associated documentation were developed to communicate how an integrated risk assessment could be conducted. Recognizing the similarities in risk assessment frameworks currently in use internationally, the integrated risk assessment framework is based on US EPAs framework for ecological risk assessment and its associated terminology (EPA, 1998). Ecological risk assessment frameworks have greater general applicability than do human health frameworks (or those environmental frameworks derived directly from human health frameworks) in that they 1) were developed to deal with a range of environmental stressors beyond toxic chemicals, 2) must describe the nature and role of the environment in the risk assessment process, and 3) must explicitly identify the endpoint to be assessed. Further, a well-developed body of concepts and terminology exist in the literature treating ecological risk assessment that supports integration. The integrated framework consists of three primary assessment phases. During the first of these, ***Problem Formulation***, the overall goals, objectives, scope, and activities of the assessment are delineated. The *Analysis* step consists of data collection and modelling exercises to characterize exposure in time and space, and to define the effects on humans and ecological systems resulting from exposure. The methods appropriate for the Analysis step may be stressor-specific, but also depend upon the nature of the systems identified to be at risk. Exposure and effect information are synthesized as estimates of risk in the ***Risk Characterization*** step. Ideally, these estimates are quantitative with respect to the level of risk expected under different exposure scenarios, although only qualitative estimates of risk may be possible in some circumstances. The integrated risk assessment framework treats the relationships among risk assessment, risk management, stakeholder input, and data collection activities in a general parallel and concurrent manner. Essentially, risk characterisation is a summary of the data compiled in the risk assessment process including the uncertainties associated with each stage and the presentation of a risk estimate.

3.3 Parameters and Indicators for Choosing a Methodology

Risks can be managed in many ways. They can be eliminated, transferred, retained or reduced. Risk reduction activities reduce the risk to an "acceptable" level, derived after taking into account a selection of factors such as government policy, industry norms, and economic, social and cultural factors. It is important to note that although risk assessment is used extensively in environmental policy and regulation it is not without controversy. This is also true for risk management. There are various criteria for assessing risk assessment including: The logical soundness of the method is eg. its justification based on theoretical arguments or scientific knowledge, and the validity of the underlying methodological assumptions.

Completeness - (e.g. whether it can address all aspects of the problem and the degree to which it excludes issues because they are hard to accommodate).

Accuracy - (e.g. the precision reflected in the confidence level associated with the results; biases resulting from undue weight given to specific interests or considerations; and the sensitivity of results to untested or untestable assumptions).

Acceptability - (e.g. compatibility with existing processes; whether it is viewed as rational and fair; the level of understanding for all parties affected by it; and the confidence and familiarity of those who will use it).

Practicality - (e.g. the level of expertise, time and input data required).

Effectiveness - (e.g. usefulness of results; range of applicability across different risks and problem areas; the generalisability of the conclusion to other problem areas; and effectiveness and efficiency of linkage with other types of methods).

The level of risk can be described either qualitatively (ie by putting risks into categories such as 'high', 'medium' or 'low' as we used in our study as matrix showed in Figure 2) or quantitatively (with a numerical estimate). Current risk assessment methods do not enable accurate quantitative estimates of risk for low levels of exposure to environmental hazards. Numerical estimates of risk will rarely be feasible because of variability in the agent and population and limitations in toxicological and exposure data which will be reflected in the uncertainty assessment, but a degree of quantification may be possible for some components such as data collection and exposure assessment.

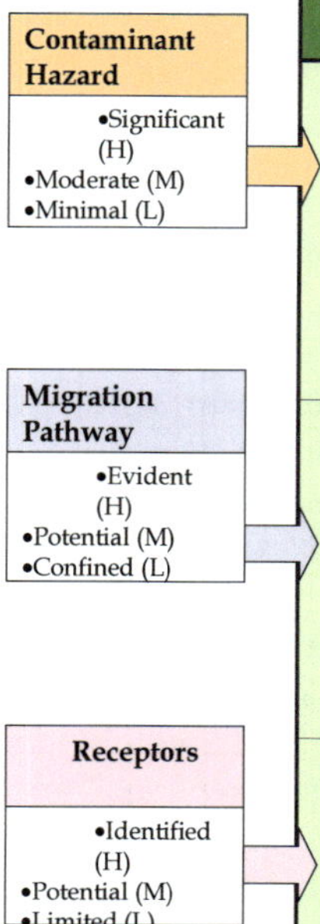

Contaminant Hazard Factor	Receptor Factor	Migration Pathway		
		Evident	Potential	Confined
Significant	Identified	HHH	HHM	HHL
	Potential	HHM	HMM	HML
	Limited	HHL	HML	HLL
Moderate	Identified	HHM	HMM	HML
	Potential	HMM	MMM	MML
	Limited	HML	MML	MLL
Minimal	Identified	HHL	HML	HLL
	Potential	HML	MML	MLL
	Limited	HLL	MLL	LLL

Fig. 2. Qualitative environmental health risk matrix

Regarding human health, the assessment was focused on exposure routes, both direct and indirect. For each of the two sites we established what exposure routes exist and what routes are significant. We also established the number and type of people that may be affected to a significant extent, depending upon location, age and profession. Finally, using an expert judgement method the conclusions are based of principles used during a qualitative risk assessment for each of 2 hotspot sides.

4. Results-Side Specifics Risk Assessment

The Municipality of Veles is located in the central part of the Republic of Macedonia and covers approximately 465 square kilometres, supports a population of over 55.000 inhabitants, making it the 8th largest municipality in the country. Established in 1973, MHK Zletovo was a lead and zinc smelter employing 1,100 workers. Each year it was uses lead and zinc concentrates to produce 30,000 tons of lead, 60,000 tons of zinc and 250 tons of cadmium, as well as smaller quantities of silver-gold and copper dross, and bismuth alloy. The process was producing 100,000 tons per year of sulphuric acid as a by-product. The same company also was owned and operates a nearby fertilizer plant. The smelter emits into the atmosphere large quantities of sulphur dioxide, and dust bearing lead, zinc and cadmium. MHK Zletovo-Veles during its activities has operated as the country's smelter and refinery for the production of lead, zinc and associated metals. About 45% of the feedstock was came from domestic lead and zinc mines (Sasa- Makedonska Kamenica,

Zletovo-Probistip, and Toranica-Kriva Planca); the balance was imported concentrate. The zinc refinery had a production capacity of 14.000 Mg/yr and the lead refinery a capacity of 40.000 Mg/yr. (UNEP, 2000)

4.1 Hazard Identification

The identification of the inherent capability of a substance to cause adverse effects. The purpose of hazard identification is to evaluate the weight of evidence for adverse effects in humans based on assessment of all available data on toxicity and mode of action:
Lead - is a very toxic element, causing a variety of effects at low dose levels. Brain damage, kidney damage, and gastrointestinal distress are seen from acute (short-term) exposure to high levels of lead in humans. Chronic (long-term) exposure to lead in humans results in effects on the blood, central nervous system (CNS), blood pressure, kidneys, and Vitamin D metabolism. Children are particularly sensitive to the chronic effects of lead, with slowed cognitive development, reduced growth and other effects reported. Reproductive effects, such as decreased sperm count in men and spontaneous abortions in women, have been associated with high lead exposure. The developing fetus is at particular risk from maternal lead exposure, with low birth weight and slowed postnatal neurobehavioral development noted. Human studies are inconclusive regarding lead exposure and cancer.
Cadmium- The main sources of cadmium in the air are the burning of fossil fuels such as coal or oil and the incineration of municipal waste. The acute (short-term) effects of cadmium in humans through inhalation exposure consist mainly of effects on the lung, such as pulmonary irritation. Chronic (long-term) inhalation or oral exposure to cadmium leads to a build-up of cadmium in the kidneys that can cause kidney disease. Cadmium has been shown to be a developmental toxicant in animals, resulting in fetal malformations and other effects, but no conclusive evidence exists in humans. An association between cadmium exposure and an increased risk of lung cancer has been reported from human studies, but these studies are inconclusive due to confounding factors. Animal studies have demonstrated an increase in lung cancer from long-term inhalation exposure to cadmium. EPA has classified cadmium as a Group B1, probable human carcinogen. Zinc is is essential for the function of more than 300 enzymes, including alkaline phosphatase, alcohol dehydrogenase, Cu, Zn-superoxide dismutase, carboxypeptidase, delta-aminolevulinic acid dehydratase, carbonic anhydrase, ribonucleic acid polymerase, and reverse transcriptase. Zinc is also involved in DNA and RNA synthesis and cell proliferation. Zinc coordinates with cysteine and histidine residues of certain peptides and produces a tertiary structure which has an affinity for unique segments of DNA in promoter gene regions, including zinc finger protein domains, the most common zinc motif, and the zinc thiolate cluster. Other physiological roles of zinc include enhancement of the affinity of growth hormone for its binding receptors, modulation of synaptic transmissions by interacting with specific sites on ionotrophic neurotransmitter receptor proteins, and induction of metallothionein (ACGIH, 1999; U.S. Department of Health and Human Services, 1993; WHO, 2004).

4.2 Assessment of Dose-Response Relationships

It involves characterization of the relationship between the dose of an agent administered or received and the incidence of an adverse effect:

For lead the acute effects are as follows:

- Death from lead poisoning may occur in children who have blood lead levels greater than 125 µg/dL and brain and kidney damage have been reported at blood lead levels of approximately 100 µg/dL in adults and 80 µg/dL in children.
- Gastrointestinal symptoms, such as colic, have also been noted in acute exposures at blood lead levels of approximately 60 µg/dL in adults and children.
- Short-term (acute) animal tests in rats have shown lead to have moderate to high acute toxicity.)

The chronic Effects (Non-carcinogen) are as follows:

- Chronic exposure to lead in humans can affect the blood. Anaemia has been reported in adults at blood lead levels of 50 to 80 µg/dL, and in children at blood lead levels of 40 to 70 µg/dL.
- Lead also affects the nervous system. Neurological symptoms have been reported in workers with blood lead levels of 40 to 60 µg/dL, and slowed nerve conduction in peripheral nerves in adults occurs at blood lead levels of 30 to 40 µg/dL
- Lead also affects the nervous system. Neurological symptoms have been reported in workers with blood lead levels of 40 to 60 µg/dL, and slowed nerve conduction in peripheral nerves in adults occurs at blood lead levels of 30 to 40 µg/dL.
- Children are particularly sensitive to the neurotoxin effects of lead. There is evidence that blood lead levels of 10 to 30 µg/dL, or lower, may affect the hearing threshold and growth in children.
- Other effects from chronic lead exposure in humans include effects on blood pressure and kidney function, and interference with vitamin D metabolism.
- Animal studies have reported effects similar to those found in humans, with effects on the blood, kidneys, and nervous, immune, and cardiovascular systems noted.

The Reproductive/Developmental Effects are as follows:

- Studies on male lead workers have reported severe depression of sperm count and decreased function of the prostate and/or seminal vesicles at blood lead levels of 40 to 50 µg/dL. These effects may be seen from acute as well as chronic exposures.
- Occupational exposure to high levels of lead has been associated with a high likelihood of spontaneous abortion in pregnant women. However, the lowest blood lead levels at which this occurs has not been established. These effects may be seen from acute as well as chronic exposures.
- Exposure to lead during pregnancy produces toxic effects on the human fetus, including increased risk of preterm delivery, low birth weight, and impaired mental development. These effects have been noted at maternal blood lead levels of 10 to 15 µg/dL, and possibly lower. Decreased IQ scores have been noted in children at blood lead levels of approximately 10 to 50 µg/dL.
- Human studies are inconclusive regarding the association between lead exposure and other birth defects, while animal studies have shown a relationship between high lead exposure and birth defects.

Human studies are inconclusive regarding lead exposure and an increased cancer risk. Four major human studies of workers exposed to lead have been carried out; two studies did not find an association between lead exposure and cancer, one study found an increased incidence of respiratory tract and kidney cancers, and the fourth study found excesses for lung and stomach cancers. However, all of these studies are limited in usefulness because the route(s) of exposure and levels of lead to which the workers were exposed were not reported. EPA considers lead to be a Group B2, probable human carcinogen. Human exposure to lead occurs through a combination of inhalation and oral exposure, with inhalation generally contributing a greater proportion of the dose for occupationally exposed groups, and the oral route generally contributing a greater proportion of the dose for the general population. The effects of lead are the same regardless of the route of exposure (inhalation or oral) and are correlated with internal exposure, as blood lead levels. For this reason, this fact sheet will not discuss the exposure in terms of route but will present it in terms of blood lead levels.

For cadmium the acute effects are as follows:

- Acute inhalation exposure to high levels of cadmium in humans may result in effects on the lung, such as bronchial and pulmonary irritation. A single acute exposure to high levels of cadmium can result in long-lasting impairment of lung function.
- Cadmium is considered to have high acute toxicity, based on short-term animal tests in rats.

The chronic Effects (Non-carcinogen) are as follows:

- Chronic inhalation and oral exposure of humans to cadmium results in a build-up of cadmium in the kidneys that can cause kidney disease, including proteinuria, a decrease in glomerular filtration rate, and an increased frequency of kidney stone formation.
- Other effects noted in occupational settings from chronic exposure of humans to cadmium in air are effects on the lung, including bronchiolitis and emphysema. Chronic inhalation or oral exposure of animals to cadmium results in effects on the kidney, liver, lung, bone, immune system, blood, and nervous system. The Reference Dose (RfD) for cadmium in drinking water is 0.0005 milligrams per kilogram per day (mg/kg/d) and the RfD for dietary exposure to cadmium is 0.001 mg/kg/d; both are based on significant proteinuria in humans. The RfD is an estimate (with uncertainty spanning perhaps an order of magnitude) of a daily oral exposure to the human population (including sensitive subgroups) that is likely to be without appreciable risk of deleterious noncancer effects during a lifetime. It is not a direct estimator of risk, but rather a reference point to gauge the potential effects. At exposures increasingly greater than the RfD, the potential for adverse health effects increases. Lifetime exposure above the RfD does not imply that an adverse health effect would necessarily occur.

The Reproductive/Developmental Effects are as follows:

- Limited evidence exists for an association between inhalation exposure and a reduction in sperm number and viability in humans.
- Human developmental studies on cadmium are limited, although there is some evidence to suggest that maternal cadmium exposure may result in decreased birth weights.
- Animal studies provide evidence that cadmium has developmental effects, such as low fetal weight, skeletal malformations, interference with fetal metabolism, and impaired neurological development, via inhalation and oral exposure.
- Limited animal data are available, although some reproductive effects, such as decreased reproduction and testicular damage, have been noted following oral exposures.

Several occupational studies have reported an excess risk of lung cancer in humans from exposure to inhaled cadmium. However, the evidence is limited rather than conclusive due to confounding factors. EPA considers cadmium to be a probable human carcinogen (cancer-causing agent) and has classified it as a Group B1 carcinogen.
Regarding the zinc there are no reports on the possible carcinogenicity of it and compounds per se in humans. Case studies have been used to evaluate the effects of zinc administered for therapeutic reasons. There are reports which compare zinc levels in normal and cancerous tissue. Studies of occupational exposure to zinc compounds have also been conducted, but have limited value because they do not correlate exposure with cancer risk. Either zinc deficiency or excessively high levels of zinc may enhance susceptibility to carcinogenesis, whereas supplementation with low to moderate levels of zinc may offer protection . For example, zinc deficiency enhanced carcinomas of the esophagus induced by methylbenzylnitrosamine but retarded the development of cancer of the oral cavity induced by 4-nitroquinoline-N-oxide Thus, zinc's modifying effect on carcinogenesis may depend both on the dose of zinc and the identity of the carcinogen being affected. The mutagenicity of zinc, particularly in Salmonella typhimurium, appears to depend greatly on the chemical form. (CalEPA, 1997,1999; IARC, !979)

4.3 Exposure Assessment

Is the qualitative and/or quantitative assessment of the chemical nature, form and concentration of a chemical to which an identified population is exposed from all sources (air, water, soil and diet); Hazardous waste generated in industry and mining create serious potential for soil and groundwater contamination and still is an important issue. Monitoring and regulation of industrial waste is inadequate. Evaluations are that at least this waste has contaminated 6.000 hectares in the country. There is no hazard assessment made on the risks these mine tailings impose on the soils, groundwater and surface water downstream. Neither the impact on public health through wind and water erosion is studied. Their potential danger is eminent and large. Hazard assessment of these hotspots is recommended, both to protect the water resources and the people. Hazard assessment of the hot spots is planned under the National Waste Management Plan. (NWMP, 2005) The UNEP did some hazard assessments for mine tailings in regard of their risk to soil, groundwater and surface waters downstream. (UNEP, 2001)
Regarding human health, the assessment is focus on exposure routes, both direct and indirect as well as establishes the number and type of people that may be affected to a significant extend.. Although the smelter was undoubtedly polluting the groundwater beneath it, the major source of soil and groundwater contamination is the disposal of more than 850,000 tons of solid waste containing heavy metals. This waste is deposited at a dump approximately one kilometer from the smelter. There has been no detailed investigation or monitoring of soil or groundwater contamination in the area, including private wells located downstream. Nevertheless, it is very likely that groundwater and nearby areas are being contaminated with heavy metals as a result of percolate from the dump. Due to the direction of groundwater flow, the private wells are probably being affected. Wastewater containing sulfuric acid and other pollutants is also a source of serious concern. During the activity, the MHK has a treatment plant that was designed to treat 135 m^3 of effluent per hour. The plant, however, was generates 1,500 m^3/hour of wastewater. The NEAP reported that cadmium, lead and zinc levels were 10-15 times higher in vegetables grown in Veles relative to control regions. As much as 4 to 10 times

the acceptable levels for lead and cadmium were found in spinach and lettuce due to soil contamination. According to the World Health Organization, blood lead levels in children of 100 to 150 μg/l have been consistently reported as having a negative effect on measures of cognitive functioning, such as the psychometric intelligence quotient (WHO, 1993. The obtained results from Veles Study have shown slightly increased blood lead levels in randomly tested children (mean value 16,51 μg/dl), reflected correspondingly in the level of intelligence and graphomotor ability (randomly, n=31). In addition, the autonomous nervous system studied by peripheral biofeedback appeared to function normally, reacting adequately in stress situations. However, the EEG results have shown that only in two children the theta-beta ratio has normal values, while in others the obtained ratios correspond to increased attention deficit. Moreover, the two children from 31 with the highest theta-beta ratio showed pathological ADHD findings and also the highest blood lead levels (> 20 μg/dl). The suspected correlation between increased blood lead levels due to industrial pollution and changes in EEG, toward increased attention deficit in tested children has been confirmed, implying the need for corresponding health care and environmental response measures. Compared to other applied psychometric instruments, neurofeedback appeared to be the most sensitive and discriminative modality. Analyses of monthly morbidity reports produced by the Public Health Institutes show that both preschoolers (under 6 years of age) and schoolchildren (aged between 7 and 14 years) living in polluted cities, such as Skopje and Veles, have a higher (up to 2 – 3 times) level of morbidity from respiratory diseases (J00-J99) (excluding influenza and pneumonia (J10-J18)) than children living in relatively less polluted villages (RIHP,2007). The data for 2006 has showed the distribution of heavy metals in Veles as follows:

Hazard	Measuring point	Average yearly concentration (mg/m^3)	Minimum – maximum (mg/m^3)	(WHO,1999) Recommended Value (mg/m^3)
Pb	Nova Naselba	0.117 (361)	0.000-0.640	0.5
	Biro za Vrabotuvawe	0.125 (24)	0.000-0.200	
Cd	Nova Naselba	0.039 (361)	0.000-0.190	0.05
	Biro za Vrabotuvawe	0.006 (24)	0.000-0.026	
Zn	Nova Naselba	0.038 (361)	0.000-3.850	/
	Biro za Vrabotuvawe	0.251 (24)	0.000-0.501	

Source: RIHP, 2007

Table 1. Monitoring of air pollution in Veles during the 2006 for Pb, Cd and Zn

In accordance with results obtained from the Institute for Public Health, Veles- the average yearly concentration for Pb and Cd didn't exceed WHO recommended value. The average concentration of aero- sediment for heavy metals in Veles were for Pb-41,6(mg/m^3); Cd-0,92(mg/m^3) and Zn-121,1(mg/m^3) with higher values for Pb in measuring points: "Zdravstven Dom" and "Nova naselba" compared with "nas. Tunel" or the lowest value was detected in control area, i.e. village Ivankovci. The highest value for Cd were detected in measuring points: "Biro za vrabotuvanje" and "Nova naselba", and 10 times less in the control area (v.Ivankovci).

Veles/Month	Preschool children		Schools children	
	Urban	Rural	Urban	Rural
I	411.04	183.06	66.32	61.87
II	419.35	128.34	83.97	48.44
III	387.43	436.48	105.48	47.48
IV	435.32	142.02	71.07	35.01
V	420.69	203.26	63.6	23.фев
VI	564.68	226.06	45.04	25.42
VII	298.30	126.38	34.4	23.50
VIII	351.18	96.42	42.78	16.79
IX	464.92	115.96	77.86	18.23
X	517.46	118.57	66.09	29.26
XI	390.09	418.89	92.35	34.05
XII	555.70	137.16	110.68	43.17
Average	**434.68**	**183.71**	**71.64**	**33.85**

Table 2. The monthly rate (°/₀₀) of registered patient with respiratory diseases (J00-J99) without (J10-J18) among preschool and school children in Veles for 2006

The average monthly rate registered patient with respiratory diseases (J00-J99) without (J10-J18) among preschool and school children in Veles in 2006, shows that in urban area the rate is 2,5 higher than in rural area, especially in winter due to air pollution

2004 (n=55)	Pb μgr/dl	Cd μgr/dl	Zn mg/l
Average	16.4	4.34	7.92
Maximum	46.70	13.30	10.20
Minimum	3.32	0.00	5.90
St. dev.	9	2.54	1.10
MPL	10 μgr/dl	9.5 μgr/dl	7 mg/l

Table 3. The monthly rate (°/₀₀) of registered patient with respiratory diseases (J00-J99) without (J10-J18) among preschool and school children in Veles for 2006

In accordance with investigation has been carried out in 2004 by the Republic Institute for Public Health the average value for Pb in blood samples among former employees (N=55) in MHK Zletovo was 64% higher than MPL as well as average Zn value for almost 15%.The Cd in blood samples is 4,34μgr/dl or twice less than MPL. During the same Study, 51 blood samples from mothers and 61 ones from newborns in Obstetric Clinic in Veles has been investigated for heavy metals, respectively. Even the average Pb value was less than MPL, it was reported that several mothers had high Pb contain in blood samples which posses high health risk for their newborns. Also several newborns with high Pb contain in blood samples was detected.

	1995	1997	1999	2001	2003	2005
Macedonia	843.5	830.6	829.1	830.8	849.5	839.5
Veles	917.2	887.2	869.3	905.7	929.9	911.2
Maximum	1569.6	1582.8	1495.8	1472.3	1680.2	1540.1

Fig. 3. General Mortality in the Republic of Macedonia, Veles Region and regions with lowest and higher mortality for the period 1995-2005

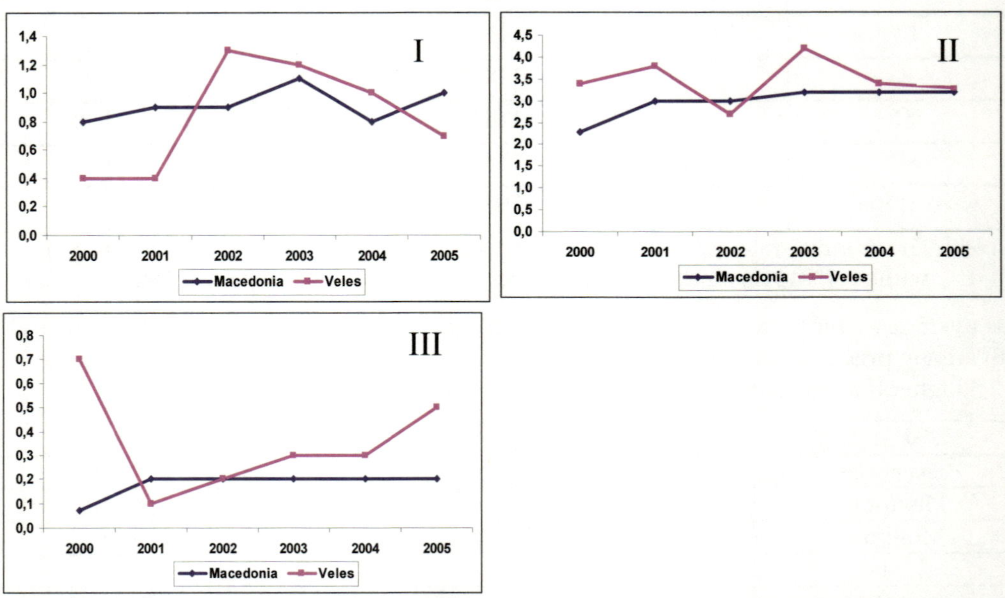

Fig. 4. Distribution of Mortality from malignant neoplasm of liver (I), lung and bronchial tubes (II), and kidneys (III) in the Republic of Macedonia and Veles Region for the period 2000-2005 (rate/10.000)

During the period 1995-2005 in Veles region has been registered higher general mortality rate per 100.000 than Macedonian average. Also, during the period 1995-2005 in Veles region has been registered higher mortality rate due to cancers per 100.000 than Macedonian average, especially for 2003 and 2005 (RIHP, 2006). The distribution of Mortality from malignant neoplasm of liver in the Republic of Macedonia and Veles Region for the period 2000-2005 (rate/10.000) has increasing especially in Veles in 2002 compared to 2000 and 2001. The distribution of Mortality from malignant neoplasms of lung and bronchial tubes in the Republic of Macedonia and Veles Region for the period 2000-2005 has showed higher values for Veles region compared to average Macedonia value. The distribution of Mortality from malignant neoplasms of kidneys in the Republic of Macedonia and Veles Region for the period 2000-2005 has showed higher mortality rate per 10.000 compared to Macedonia (RIHP, 2006).

4.4 The Fertilizer Plant

The fertilizer plant was uses the sulfuric acid from the MHK Zletovo smelter and phosphate from Morocco to produce phosphoric acid and, in turn , mono-ammonium phosphate (MAP) and NPK fertilizer. The plant has closed in 2003 with the last reported producing of about 60,000 tons of fertilizer per year. Careless handling of raw materials and products was causing severe dust problems in the vicinity of the plant. Some areas are covered by deposits of up to several centimeters. According to studies carried out with support from the EU's Phare Programme, the plant's wastewater loadings of phosphorus and nitrogen were equivalent to those that would be generated by population centers of 4.6 million and 0.4 million people, respectively. Some diffuse soil and groundwater contamination can be expected on and around the plant facilities due to the use of impure raw materials containing heavy metals. (UN, 2002)
For each ton of phosphate produced, 5.5 tons of gypsum waste is generated. A mixture of 20 % gypsum and 80 % acidic process water (pH 2 - 3) was pumped to a special landfill that currently holds five million tons of gypsum waste. This waste was formerly deposited in the bottom of the valley, and sludge used to float directly into the Vardar River. It is now being deposited in the upper part of the valley, thereby reducing the direct risk to the Vardar. However, wastewater from the sludge dewatering process continues to drain into the river, and sludge can still be flushed into the river in flood conditions.
Our analyses from the gypsum special landfill have been carried out in the Republic Institute for Health Protection in June, 2007 showed higher natural radioactivity due to phosphate material and technological process.

5. Summary of Assessment Results

Large quantities of industrial waste are generated in the mining, metallurgical, fertilizer, and chemical industries, as well as in the coal-fired power plants. Most of the larger industries have their own industrial waste sites.
During the 3 week of June 2007 we investigated environmental health conditions in 2defined hotspots.
The Table shows distribution of hazards, possible health effects and potentional number of excised people by hazardous sides.

Hazardous site	Hazards	Possible health effects	Potential number of excised people
Former Smelter Company Veles + Fertilizer company	Lead, nickel and cadmium contamination during past years of smelter company activities	Lead: toxic and carcinogenic effects in pregnant women and children; CNS damages	47.000 inhabitants

Table 4. The Distribution of hazards, possible health effects and potential number of excised people by hazardous sides

The defined health impacting hazards for Veles-Lead Smelter were as follows:
Toxic solid waste, airborne particulate matter and SO_2; Particulate matter: lead/zinc and iron oxides, As, Sb, Cd, Cu, metallic sulphates. Air emission for processes with few controls may be of the order of 30 kg Pb or Zn/t lead of zinc produced. Emissions of arsine, chlorine, and hydrogen chloride vapours and acid mists are associated with electro refining. Water effluents: Pb, Zn, As etc including dissolved and suspended solids, metals and oil and grease. Discard slag up to 0,7& lead/zinc up to 3 tons of solid waste per ton of lead/zinc produced.

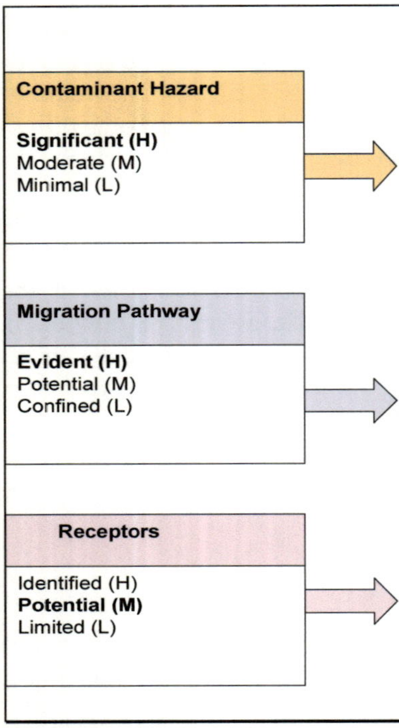

Contaminant Hazard Factor	Receptor Factor	Migration Pathway		
		Evident	Potential	Confined
Significant	Identified	HHH	HHM	HHL
Significant	**Potential**	**HHM**	HMM	HML
Significant	Limited	HHL	HML	HLL
Moderate	Identified	HHM	HMM	HML
Moderate	Potential	HMM	MMM	MML
Moderate	Limited	HML	MML	MLL
Minimal	Identified	HHL	HML	HLL
Minimal	Potential	HML	MML	MLL
Minimal	Limited	HLL	MLL	LLL

Fig. 4. Qualitative risk assessment- Smelter Company- Veles

The key release or exposure vectors were as follows: to air (dust and smelter emission) and to water/groundwater Regarding human health, the qualitative assessment is focus on exposure routes, both direct and indirect as well as establishes the number and type of people that may be affected to a significant extend:
According to recent studies, emissions were having a significant impact on the health of the population. Increasing morbidity, especially from respiratory diseases, and frequent occurrence of lung cancers and anemia, have been noted in the local population and among factory workers. Children are especially affected. Workers have been observed to have experienced blood in urine, suggesting possible kidney disease. They have also had elevated concentrations of lead and cadmium in blood relative to control populations .Analyses of monthly morbidity reports produced by the Public Health Institutes show that both preschoolers (under 6 years of age) and schoolchildren (aged between 7 and 14 years) living in polluted cities, such as Veles, have a higher (up to 2 – 3 times) level of morbidity from respiratory diseases (excluding influenza and pneumonia) than children living in relatively

less polluted villages. The difference is particularly high in winter, when heating and climatic factors (including temperature inversion) contribute to an increase in air pollutants (especially SO2 and black smoke).The health effects of particulate air pollution depend on particle size, composition and concentration, and can fluctuate with daily changes in PM10 or PM2.5 levels. This is the particulate fraction of the greatest concern for health, as it penetrates the respiratory system. Particulate matter may have acute health effects, such as increased mortality, increased hospital admissions because of the exacerbation of respiratory disease, fluctuations in bronchodilator use, cough and peak flow reductions. According to the World Health Organization blood lead levels in children of 100 to 150 µg/l have been consistently reported as having a negative effect on measures of cognitive functioning, such as the psychometric intelligence quotient. The Institute of Occupational Health, Skopje in workers occupationally exposed to cadmium, has diagnosed kidney diseases. There is no official recording of the use of occupational carcinogens. There is insufficient information on dose-effect relations in some segments of the chemical industry. Obtained recent data for morbidity and mortality showed that even the factory has stop with operation in 2003, existing health risk among vulnerable population and former employees is still high. The eventual remediation of a smelter dump will have first of all direct environmental benefit and indirect one for citizens of Veles.

Environmental health risk assessment is an essential element in environmental management and an important condition in precise priority-setting to the necessary actions for its sanitation. Risk assessment is intended 'to provide complete information to risk managers, specifically policymakers and regulators, so that the best possible decisions are made'. There are uncertainties related to risk assessment and it is important to make the best possible use of available information. Environmental Health Risk Assessment provides a tool for appraising health risks in the broader process of Health Impact Assessment.

6. References

American Conference of Governmental Industrial Hygienists (1999). TLVs and BEIs. Threshold Limit Values for Chemical Substances and Physical Agents, Biological Exposure Indices. Cincinnati, OH

California Environmental Protection Agency (1997). Technical Support Document for the Determination of Noncancer Chronic Reference Exposure Levels. Draft for Public Comment. Office of Environmental Health Hazard Assessment, Berkeley, CA.

California Environmental Protection Agency (1999). Air Toxics Hot Spots Program Risk Assessment Guidelines: Part II. Technical Support Document for Describing Available Cancer Potency Factors. Office of Environmental Health Hazard Assessment, Berkeley, CA.

Environmental Protection Agency (1992) Guideline for exposure assessment, Washington D.C.

IARC (1979). IARC monographs on the evaluation of the carcinogenic risk of chemicals to humans - Some halogenated hydrocarbons.

Kendrovski V., Gjorgjev D (2005). The burden of diseases in the Republic of Macedonia. I-st International Congress of Occupational Medicine, Ohrid

National Health Environmental Action Plan (1999) Government of the Republic of Macedonia, Skopje,

National Waste Management Plan 2006-2012 (2005). Government of the Republic of Macedonia, Skopje

National Environmental Protectoral Council (1999). Guideline of human risk assessment methodology. Canberra, Australia

National Institute for Occupational Safety and Health (2005). Pocket Guide to Chemical hazards. U.S. Department of Health and Human Services, Public Health Service, Centers for Disease Control and Prevention. Cincinnati, OH,149

Republic Institute for Health Protection (2006). Medical map for 2003,2004 and 2005 in the Republic of Macedonia, Skopje

Republic Institute for Health Protection (2007). Yearbook of preventive programs in the Republic of Macedonia for 2006, Skopje

The Second National Environmental Action Plan (2005). Government of the Republic of Macedonia, Skopje

UN (2002). Environmental Performed Review for FYR of Macedonia, Geneva

UNEP (2000).Post-Conflict Environmental Assessment—FYR of Macedonia, Geneva

UNEP (2001). Feasibility Study for urgent Risk Reduction Measures at hot spots In FYR of Macedonia, Geneva

U.S. Department of Health and Human Services (1993). Registry of Toxic Effects of Chemical Substances (RTECS, online database). National Toxicology Information Program, National Library of Medicine, Bethesda, MD.

WHO (1993). Biomarkers and risk assessment: Concept and principles. Environmental Health Criteria 155, Geneva

WHO (1994). Assessing human risk of chemical: Derivation of guideline values for health based exposure limits., Environmental Health Criteria 170, Geneva

WHO (1994) International Programme on Chemical Safety, The WHO recommended classification of pesticides by hazard and guidelines to classification 1994-1995, UNEP/ILO/WHO 1994.

WHO (1999). Guidelines for Air Quality, Regional Office for Europe, Geneva

6

The Role of Adaptive Environmental Management in Sustainable Development Case Study Assessing the Economical Benefits of Sustainable Construction in Greece

Odysseus G. Manoliadis
Democritus University of Thrace
Greece

1. Introduction

Recent public awareness of resources and its competing and conflicting use has prompted the sustainable development as a desired goal in the decision making process. The need for new approaches to sustainable management such as adaptive environmental management arises from the relatively new demands posed by the concept of sustainable development. The importance of information, integration and participation are necessary since science has changed to involve people more closely in research and development. Particular attention is paid to the challenges being posed for science as it seeks to more explicitly deal with the human dimension of sustainable management. The potential for adaptive management as an approach to more closely link research with management and policy is examined. Effective collaborative management requires different stakeholders associated with environmental problems to develop and enact solutions co-operatively, as opposed to acting as advocates purely in their own interest. However, because environmental issues are generally characterised by conflicting social perceptions, it is often difficult to ensure adequate stakeholder participation in developing and managing information to support collaborative decision making and subsequent change 'on-the-ground'. Everything has been said about development, but almost everything remains to be said and therefore to be explored or rediscovered, because incontestably, almost everything remains to be done. (Cosmao. 1984) Sustainable management issues continually change over time in response to co-evolving social, economic and ecological systems. Under these conditions adaptive management, offers an opportunity for more proactive and collaborative approaches to resolving environmental problems.

This research represents an inquiry into how an adaptive management that supports the concept of sustainable development can be initiated and implemented in complex, regional or large-scale contexts. An action research inquiry process is used to find improved ways of managing collaborative or multi-stakeholder approaches to environmental management, and to develop an integrated information framework to underpin subsequent decision making. The research involves one main case study economic assessment of sustainable construction.

The main case study around which this research began with a research and development initiative in 2007 to address the problem of the economical benefits of sustainable construction in Greece (Manoliadis and Vatalis 2007). Consistent with an action research process, an initial framework for supporting a collaborative approach –a version of Delphi method approach - addressing this problem is outlined at the beginning of this study. What is significant about this case study is not that it has resulted in a regional adaptive management approach to assessing the economic benefit of sustainable construction but rather that those involved (researchers, managers and local government staff representatives) have learnt more about the issues involved, and continue to seek ways to resolve them within the framework of adaptive management.

One of the key characteristics of this approach is decision making and common action. The process that the researcher uses to guide those involved can be described as a spiral of action research cycles consisting of phases or the so called rounds. In this research these rounds can be seen to have taken place at a number of levels each involving different groups. The main group should be seen as the core research team involved in the implementation of the framework within the case studies described here. What follows is best regarded as an illustration of actions 'plan--act--reflect' using systems approach. This is, as Checkland (1981a,b;1985a,b,) points out, 'the best that can be hoped for, given the impossibility of capturing the actual richness of an intervention in human affairs'.

In the rest of this Chapter first background information about sustainability sustainable development and sustainable construction is presented. Then the research methodology based on adaptive management and Delphi method is analysed. The case study in terms of assessing the economic benefits of sustainable construction is following. Finally the results and conclusions of this research are presented together with proposals for further research.

2. Background

2.1 Sustainability

Sustainability is defined as the ability to carry forward or support or maintain for a prolonged period approaching perpetuity. Sustainability means meeting the needs of today without compromising the ability of future generations to meet their needs (Brundtland report, 1987). In business terms this is something that has traditionally been aspired to in terms of economic factors as businesses strive to balance inputs and outputs to maintain a profitable enterprise.

2.2 Sustainable Development

Sustainable Development has become a widely used term that goes beyond such economic security to include issues of environmental impact and sustainable use together with social effects. The triple bottom line created by this approach is what is now commonly meant when the term is used and this is the definition adopted. However, pressures are growing rapidly to adopt such an agenda and indeed to assess performance and report against such issues. Key business decisions are being increasingly taken with environmental and social concerns alongside economic ones as crucial parts of total value.

2.3 Sustainable Construction

The term sustainable construction was originally proposed to describe the responsibility of the construction industry for attaining sustainability. Kibert's definition of sustainable construction as "...creating a healthy built environment using sustainable-efficient, ecologically based principles". Hill and Bowen, 1997 divided Kibert's principles in four pillars: social, economic, biological and technical. Social sustainability highlights improvements in the quality of human life, and human living environment, which include culture, health, education, and intergenerational equity. Economic sustainability includes the use of full-cost accounting methods and real-cost pricing to set prices and tariffs for goods and services and achieve more efficient use of sustainable. Biological sustainability includes the motion that sustainable construction needs to protect the natural environment rather than pollute, encourages the use renewable sustainable and reduce the use of water, energy, materials and land in each stage of a project. Technical sustainability requires high performance, durability, quality and mixed use of a building (Hill and Bowen, 1997).

3. Methodology

The research methodology is based on Delphi method. Delphi concept was developed by the American defense industry. A project named Delphi was a study undertaken by the Rand Corporation for the US Air Force in the early 1950s (Helmer, 1966; Helmer 1970 Robinson, 1991). The Delphi method is a technique for obtaining forecasts from a panel of independent experts over two or more rounds. Experts are asked to predict quantities. After each round, an administrator provides an anonymous summary of the experts' forecasts and their reasons for them. When experts' forecasts have changed little between rounds, the process is stopped and the final round forecasts are combined by averaging. Delphi is based on well-researched principles and provides forecasts that are more accurate than those from unstructured groups (Rowe and Wright 1999).

The procedure was designed to obtain the most reliable consensus of opinion of a group of experts by a series of intensive questionnaires interspersed with controlled opinion feedback, and with the results of each round being fed into the next round (Linstone and Turoff 1975). This method involves the selection of procedures for suitable experts, development of appropriate questions to be put to them and analysis of their answers (Outhred 2001; Cabanis 2001). The desired outcome is that, by using such an iterative forecasting procedure, on reaching the final round the experts will have achieved unanimity on the issues put before them.

The main features of the Delphi method are: anonymity; iteration with controlled feedback; and statistical response (Dickey and Watts 1978; Adnan and Morledge 2003). Panel members remain unknown to one another and respond to a series of questionnaires. The iterative nature of the procedure allows them to modify their assessments and project them beyond their own subjective opinions. It can represent the best forecast available from consensus of experts (Corotis, et. al. 1981). The process is continued until a consensus is reached on the various issues under consideration, or until it becomes evident that no further consensus can be achieved. Generally, the number of rounds varies between two and seven while the number of participants ranges from three to fifteen (Rowe and Wright 1999; Adnan and Morledge 2003).

The method involves posing a number of rounds (Figure 1) of appropriate questions on the specific subject area to a group of suitable pre-selected experts. Each one of the experts responds expressing his/her opinion and the collected results (answers) of this first stage are analyzed.

These results are then collated and fed back to the respondents in an anonymous or non-attributable way, along with the second round questions. Each one of the participants is then able to take into account the opinions of the other participants when responding to the second round of questions. This process is then repeated for each subsequent round and, it is hoped, that by the final round a consensus of opinion on the issues put before them will have been reached. The process is typically carried out by remote correspondence, such as mailed questionnaires, rather than involving face-to-face group discussions. This enables all participants to respond individually and reduces the impact of group dynamics on the resulting consensus.

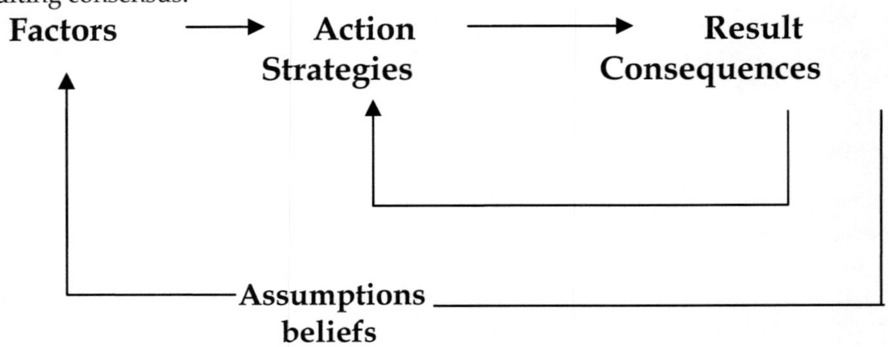

Fig. 1. The process of the proposed methodology

The method is based on the assumptions and beliefs of the selected experts, and does not rely on previous historical data being available. Moreover, the method is typically intended to provide a judgement or opinion on the specific subject area, rather then producing a quantifiable measure or result (Figure 1). Because of this, the method can easily work well in new areas which are frequently subject to unpredictable forces, which are not easily quantifiable in most of the cases. Another key factor is the anonymity of the participants, which permits them to express their opinions freely.

The major difficulties of Delphi, however, lie in maintaining the high level of response and in reaching and implementing a consensus. It is important that panel members treat the work seriously, devote the time necessary to provide thoughtful and reasoned responses to the questions and that they contribute to each round of Delphi. (Robinson, 1991).

3.1 Preparation of the Delphi Study of the Greek Construction Industry

The key issues in preparing the study are:

- The experts definition and selection
- the number of rounds and
- the questionnaire structure (i.e. number of questions) in each study round.

An expert may be defined as someone with special skills or knowledge evident through his or her leadership in professional organizations, or someone holding office in a professional organization, a presenter at national conventions or who has published in recognized journals (Cabanis 2002).

Once an expert panel has been identified, an additional problem is to maintain their input throughout the rounds of the study. The study is organized into a greater number of rounds in order to distil greater consensus of the participants and gain better forecasting accuracy.

Moreover, the number of questions in each round is closely related to the time required from the participants to complete each round.

4. Case Study

4.1 The Problem: Is Sustainable Construction More Economical?

As sustainability issues enter in building construction the corresponding sustainability criteria are introduced in the building design. The organizations are empowered to accelerate the sustainable development; regulations both on national and local basis (Sha et al., 2000). Questions of economic assessment of the sustainable solutions have been arisen. For instance sustainably-designed buildings have such great economic benefits, in terms of the buildings life cycle then why more stakeholders aren't aware of them. The reason for this is likely due to the widely held belief that sustainable design costs more. This belief persists in spite of a growing body of research that point to numerous economic benefits from sustainably-designed buildings only in countries in which the construction is in decline. Many of the current examples in the private and public sector are often perceived as willing to take the initiative toward in strategic levels sustainable solutions without risk economic savings (Ofori, 1998).

Building occupiers, managers and others involved in making decisions on the maintenance and procurement of buildings are apt to adopt current best practice in their decision making through an improved understanding of the benefits that can be achieved through extra value to the value of inputs (Pearce, 2006). There is a perception amongst these groups that more sustainable solutions in maintenance and procurement inevitably result in increased capital costs and /or reduced market / staff appeal by adopting innovation in sustainable construction (Bossink, 2002; Manoliadis et al., 2006). They are interested firstly for the economic benefits of implementing sustainability solutions and secondly for environmental and social benefits. While experts on the topic claim that overall the economic assessment is to the sustainable solution's benefit no research has been made to substantiate the claim. The potential economic benefits may be difficult to see in the construction phase but there are returns of investment at the operation maintenance and the restoration phase. Each project has unique economics and a different performance and it's difficult to generalise that sustainable solutions through the life cycle will be economically viable project specific issues include site location, local conditions, national infrastructure taxes and fees, permit requirements.

It is actually a whole systems approach to design and implement that may influence the economics of the project in question. Since design is the first step to question if sustainable solutions will be applied without a overall study of the life cycle convincing stakeholders (owners operators public) is very hard attempt. The UK strategy on sustainable construction focuses 10 action points in order to achieve more sustainable construction. They are: the reuse of built assets, design for minimum waste, the aim of lean construction, minimizing energy in construction, minimizing energy in use, do not pollute, preserving and enhance biodiversity, conservation of water sustainables, the respect of people and their local environment and target setting (DETR, 1999).

The present trend in sustainable construction research focuses on the environmental and technical issues, via design and developing new technical, materials and components to achieve more efficient and better quality building. However, the achievement of marginal improvements in sustainable performance per unit of production will be made insignificant when compared to a higher future production rate (Uher, 1999). Therefore, it is essential to seek the economic principles of sustainable construction in order to create a profitable market for sustainable construction. Some concepts in the economic terms can be concluded in the below: Value for money, maximum output with minimum input, integration of short term return and long term benefits, stakeholder partnership between the demand and supply sides of the industry, business pattern changes from a linear process to a cyclic process (Khalfan et al, 2001).

4.2 Sustainable Construction In Greece

In Greece, the complex problems shared by cities are evidence of the impacts of urban sprawl: increasing traffic congestion and commuting times, air pollution, inefficient energy consumption, loss of open space and habitat, non-optimal allocation of economic sustainables and the loss of a sense of community. These combined pressures, along with the challenges faced specifically by stakeholders of the built environment, have led to a growing awareness of the need for change. In response to these pressures the concept of sustainability is beginning to permeate the Greece construction industry as a possible strategy to better meet the needs of clients and owners while ensuring business success in an increasingly competitive and constrained operational environment. While a variety of initiatives have been implemented to initiate the change toward increased sustainability, some critics have begun to realize that these initiatives are insufficient to bring about the change that is needed (Manoliadis and Vatalis 2007).

According to an extensive survey at the national level, the total annual energy consumption of buildings varies from 60–500 kWh/m2, depending on their utilization and type. More specifically, the annual energy consumption for heating varies between 30– 170 kWh/m2; for cooling between 15–90 kWh/m2; for lighting between 10–50 kWh/m2; and for the operation of various appliances and equipment between 10–80 kWh/m2 (Bikas and Milonas, 1998).

Having this in mind a questionnaire survey was conducted to assess the economical benefits of sustainable construction in Greece. In doing so a questionnaire survey to Greek as well as other European Nationality respondents was conducted participating the EUREKA workshop.

Twenty two questionnaires were returned during the period of April 2007 in two rounds acoording to the proposed Delphi method.

4.3 Experts Selection

The following criteria were devised in order to identify eligible participants for the present Delphi study:

1. Practitioners who have extensive working experience in the construction industry in Greece.
2. Experts should be involved in the management of construction projects in Greece.
3. Experts should have a detailed knowledge of the whole construction process. A list of the panel members and their type of occupation are shown in Table 1.

Profession of the experts	Number of experts
Academics	4
Architects	5
Construction engineers	7
Contractors-developers	4

Economists	2
Total	22

Table 1. Delphi experts

4.4 First and Second Round Questions

As mentioned above the Delphi method adopted in this study consisted of the following two rounds.
In the questionnaire of first round there is an introductory text regarding the purpose of this research, such as:

> *"...Thank you for being willing to complete the following questionnaire disseminated through the internet, which is being carried out for the research conducted in terms of the paper "Assessment of the economical benefits of sustainable construction in Greece. .... in this matter the questionnaire sent to you below is an effort of scientific communication process trough Delphi tecnique which tries to highlight your opinion on the subject of the economic benefits of Sustainable Construction. The survey is based on the Delphi principles is fully anonymous and will be used only in scientific purposes. Please fill the information below (tick the appropriate boxes) and send it as a reply to our internet message"*

Q1a. Are you aware of the economic benefits of sustainable construction during the project life cycle?

Q1b. If yes, did you manage to adopt sustainable development issues in construction practice?

Q2. What is your involvement in the projects of your organization?

Q3. According to your experience and in the basis of international literature what are the main economic benefits of sustainable construction? Please fill bellow.

In Table 2 the results of first round of the Delphi questionnaire sent via email to the panel experts (Table 2)

	YES	NO
Q1a	22	2
Q1b	13	10

Table 2. Awareness and adoption of economic benefits in sustainable construction

In Table 3 the Participant's involvement in the projects are presented

Director	Manager	Staff	Consultant
4	8	6	4

Table 3. Participant's involvement in the projects

In Table 4 the common and not common economic benefits on Sustainable construction are tabulated.

1	Reduce energy usage
2	Protection natural / social environnent
3	Providing healthy comfortable living environment
4	Success for developers and occupiers
5	Building innovation management/technologies/policy

Table 4. List of the main economic benefits on Sustainable construction

The second round questionnaire consisting of 4 questions was sent out in March 2007 via email to the experts and was collected by the end of May 2007. At this stage, most of the experts had reconsidered and made adjustments to their score. The questionnaire refers to the drivers for change that have emerged as a result of the international and national construction industry's response to sustainability. The final drivers of change selected for consideration are presented in Table 2. Twenty two (22) questionnaires were returned, representing 66% of the issued questionnaires and these were used in the analysis presented in the following section. The questionnaire refers to the assessment of the economic benefits of sustainable construction that have emerged as a result of the international and national construction industry's response to sustainable construction. The challenges selected for consideration are those presented in Table 1. Persons employed in target population firms, organizations/authorities were asked to answer the following questions:

Question 4. In what phase of Life Cycle Assessment (LCA) do you believe that sustainable construction adopted in your field is more economical than standard practice? During the design phase; During the construction phase; During operation – maintenance phase; During the demolition phase

Question 5. How the sustainable construction adopted in your field is compared over the entire project life cycle to the standard practices? More economical; Equally economical; Less economical.

Question 6. How are the economic benefits of sustainable construction prioritized?
(Please use the numbers 1 to 4 in order to declare preference)
Reduce energy usage; Protection of natural and social environment; Providing healthy and comfortable living environment; Success for developers and occupiers; Building innovation technologies.

Question 7. Did you realize economic advantages in the future of sustainable construction? If so how in what is the most important factor? Efficiency/Renewable Energy issues; Reuse/Recycle Water conservation; Conservation Materials and Techniques; Waste management.

In Table 5 the phase of LCA economic benefits were significant in sustainable construction are presented.

	Votes	Rank
Design	10	1
Operation Maintenance	7	2
Construction	3	3
Demolition	2	4

Table 5. Phase of LCA adoption of sustainable construction

In Table 6 the question *How the sustainable construction adopted in your field is compared over the entire project life cycle to the standard practices?* in terms of three scales More economical; Equally economical; Less economical is presented.

	More economical	**Rank**	**Equally economical**	**Rank**	**Less economical**	**Rank**
votes	**19**	**1**	**2**	**2**	**1**	**3**

In Table 7 the prioritization in terms of the importance (scale 1 to 10) of the economic benefits in sustainable construction is presented.

	Economic benefits *(priority selection)*	1st	2nd	3st	4th	5th	final ranking
1	Reduce energy usage, REU	10	2	0	1	0	1
2	Protection natural/social environment, NSE	2	3	6	0	1	3
3	Providing healthy comfortable living environment, HLE	1	5	4	1	2	4
4	Success for developers and occupiers, SDO	0	2	0	6	7	5
5	Building innovation management/technologies/policy BIM	5	6	3	2	2	2

Table 7. Importance degree of the economic benefits in sustainable construction (prioritized)

This is best illustrated in Figure 2 where these factors are presented in terms of their rank (1st 2nd ..).

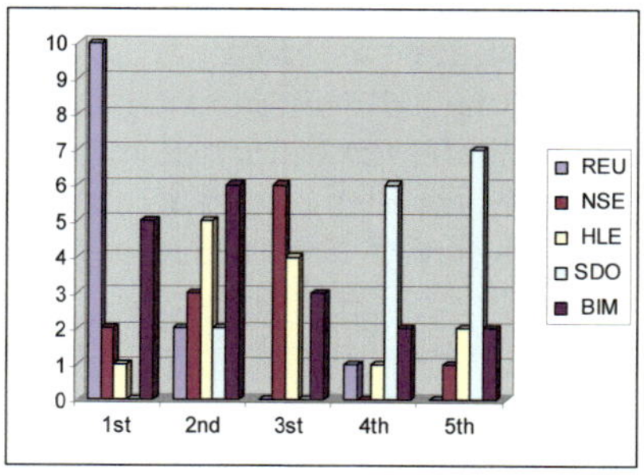

Fig. 2. Prioritization of the economic benefits of sustainable construction

The second round shows the results of the experts perception on which phase (design, construction, maintenance and demolition) was more economical than the standard practice. Table 5 presents the perception of the experts on how the sustainable construction adopted in their field is compared over the entire project life cycle to the standard practices. Table 8 presents the prioritization of the economical benefits in terms reduction of energy usage Protection of natural and social environment, Providing healthy and comfortable living environment and success for developers and occupiers. As can be seen reduction of energy ranks first (voted by 10 participants) followed by protection and natural and social environment (voted by 2 participants) while only one voter has selected providing healthy
The majority of the participants 93% are aware of the benefits of sustainable construction during the project life cycle and most of them (64%) have managed to adopt sustainable development issues in construction practice.

The novelty of this approach than the classical Delphi method is the possibility of feedback between stages where the factors or actions may be modified to include changes in the research. For instance question 7 can be reconsidered to the participants as follows

Q7. From the following factors Efficiency/Renewable Energy issues; Reuse/Recycle Water conservation; Conservation Materials and Techniques; Waste management what is the most important factor?

The same information can be used to state the factors affecting the futures of sustainable construction

	votes	Rank
Efficiency/renewable Energy Issues	10	1
Conservation Material and Techniques	5	2
Waste Managemet	4	3
Reuse/Recycle Water Conservation	3	4

Table 8. Factors affecting the futures of sustainable construction

5. Conclusion

Action research is seen as a process which both helps the development of this social capital, and provides lessons into how it can be expanded. Moreover, building capacity for the use of participatory learning processes should be part of the method, that capacity cannot be assumed to be there. The role of eadaptive environmental management for participation and measuring process success is highlighted Finally the research points to the need to draw out lessons from across action research case studies, and suggests some challenges for action research to help in large-scale collaborative learning.

A version of Delphi study embracing a development paradigm that places people at the centre and seeks to empower stakeholders to influence and share control over development initiatives and the decisions which affect them. Even as macroeconomic policies and trends continue to shape sustainable development opportunities, the move away from centralised planning by governments is requiring R&D initiatives to work towards empowering communities to deal with their own needs. In this regard, the challenge for researchers is to work with communities and undertake inquiries that begin with the search for solutions to social (community) problems -- placing an emphasis on problem context and identification.

The outlined version of Delphi, however, it is suggested that this should be implemented in an environment characterized environment by high social capital.

Because one of the main issues related to establishing such a collaborative approach within the wider social and institutional contexts of catchments and regions is one of implementation, an action research approach provides an appropriate methodology. This is directly applicable to the study of how individuals and groups design and implement action in relation to one another.

What is significant about case study is not that it has resulted in a regional adaptive management approach to sustainable construction but those involved (researchers, managers and local government staff representatives) have learnt more about the processes and issues involved in working together and sharing information, and continue to seek ways to implement adaptive management. In doing so. a questionnaire survey based on the Delphi method was conducted in order to assess the benefits of sustainable construction. The majority of the participants 93% are aware of the benefits of sustainable construction during the project life cycle and most of them (64%) have managed to adopt sustainable development issues in construction practice. Also the majority of the participants believe that sustainable construction is overall more economical than standard practice specially is more economical at the operation phase of LCA (Life Cycle Assessment) and important to be adopted at the design phase. By using feedback in the process the factor *Reduce energy usage*, finally is ranking first between the economic benefits of sustainable construction followed by *Protection of natural and social environment*, and *Providing healthy and comfortable living environment*. The competitive advantages in the future of sustainable construction are realized by the majority of the participants and the most important factor for this is assigned to the *Efficiency/Renewable Energy issues* followed by *Conservation Materials and Techniques*.

Future areas of activity with the potential to leverage improved information flows within environmental research and management are suggested as improving participation and the use of local knowledge in the research process improving the dissemination and use of this knowledge in the wider community through improved networking and collaboration; and supporting these approaches through participatory monitoring and evaluation. Moreover, there is an increasing realisation that new sources of 'expert knowledge' and databases are needed to identify persistent sustainable management practices more clearly. In many cases, the knowledge that is required about the past and present state and about the relationships between social and environmental systems is held within local communities and implementation and policy agencies. Accordingly, it follows that the task of organising information to understand better the links between sustainable management is needed, social realities dynamics should be a collaborative venture between research scientists and the different stakeholders involved.

6. References

Adnan, H. and Morledge, R. (2003) "Application of Delphi Method on Critical Success Factors in Joint Venture Projects in Malaysian Construction Industry" Proceedings CITCII Conference, held in Hong Kong 10-12 December 2003. Bawden, R.J. (1987) Learning systems and technological change. Proceedings of the

Bossink B.A. (2002) Dutch public–private strategy for innovation in sustainable construction Construction Management and Economics, Volume 20, Number 7, 1:633-642.

Brundtland report, 1987

Beemans, P. (1996) Culture, spirituality and economic development: Foreword in Culture, spirituality, and economic development: Opening a dialogue.(Ryan, W.F.) Ottawa, Canada: International Development Research Centre. (Available from <http://www.idrc.ca/books/focus/782/782fore.html> Accessed 4 October 2000).

Checkland, P.B. (1981a) Systems thinking, systems practice. Chichester, UK: Wiley.

Checkland, P.B. (1981b) Towards a systems-based methodology for real-world problem solving. Pp. 288-314 in Systems behaviour 3rd edition (Ed.: Open Systems Group)London: Harper and Row.

Checkland, P. (1985a) From optimizing to learning: A development of systems thinking for the 1990s. Journal of the Operational Research Society 36:757-767.

Checkland, P. (1985b) Achieving 'desirable and feasible' change: An application of soft systems methodology. Journal of the Operational Research Society 36: 821-831.

Cosmao, V. (1984) Un monde en development! Guide de reflexion. Paris: Editions Ouvrieres.

Bikas, D. and Milonas, S. (1998) Sustainable construction in Greece. Web: http://bativille.cstb.fr/CIB_Reports_pdf/gr.pdf

Cabanis, K (2002) Computer-Related Technology Use By Counselors in the New Millennium- A Delphi Study, Journal of Technology in Counseling, Vol. 2-2, pp. 23-34 Web: http://jtc.colstate.edu/Vol2_2/cabanis/cabanis.htm

Cabanis, K. (2001). Counseling and computer technology in the new Millennium—An Internet Delphi study [Online dissertation]. Retrieved May 4, 2001, from the Virginia Tech Digital Library and Archives on the World Wide Web: http://scholar.lib.vt.edu/theses/available/etd-03072001-175713/

Corotis, R. Fox, R. Harris (1981) Delphi Methods: theory and design load application. Journal of the Structural Division, ASCE, 107 (6), 1095-105

Dahlberg, K.A. (1979) Beyond the green revolution. New York: Plenum Press.

Dahlberg, K.A. (1991) Sustainable agriculture - fad or harbinger? BioScience 41: 337-340.

DETR, (1999), Sustainable Development Education Panel: First Annual Report 1998, HMSO, London.

Dickey, J & Watts, T (1978) Analytic Techniques in Urban and Regional Planning, McGraw-Hill, New York.

Helmer, O. (1966). Social technology. New York:

Helmer, O. (1970). Long-range forecasting: Role and methods. Paper presented at the Conference on Forecasting the Future, Harrogate, England.

Hill, R., and Bowen, P., (1997) Sustainable construction: Principles and a framework for attainment, Construction Management and Economics 15, 223 -239

Khalfan, M. M. A., Bouchlaghem, N. M., Anumba, C. J., & Carrillo, P. M. (2003),

Knowledge Management for Sustainable Construction: The C-SanD Project, 2003 Construction Research Congress, held in Honolulu, Hawaii on 19 – 21 March 2003,

Linstone, H and Turoff, M (1975) The Delphi Method: Techniques and Application, Addison, Wesley, Reading, MA, pp. 3-12

Manoliadis, O. and Tsolas, I. (2004) "Sustainability initiatives undertaken by EU in procurement policies and practices awarding construction products", 18th IPMA World Congress on Project Management 19-20 June 2004, IPMA -FOVOSZ, Budapest.

Manoliadis O. and Vatalis K. (2007). Assessment of the Economic Benefits of Sustainable Construction Ecosummit Conference Sanghai 12-18 May 2007.

Ofori, G. (1998). "Sustainable construction: principles and a framework for attainment - comment," Construction Management & Economics, Taylor and Francis 16(2):141-145

Outhred, G.P (2001) The Delphi Method: A Demonstration of its Use for Specific Research Types, Proceeding of the RICS Foundation, Construction & Building

Robinson,.B.L (1991) "Delphi Technology for economic impact assessment". Journal of Transportation Engineering, 117(3)

Rowe, G. and Wright, G. (1999) "The Delphi Technique as a Forecasting Tool: Issues and Analysis", International Journal of Forecasting, Volume 15, Pages 353-37

Sha, K., X. Deng, C. Cui. 2000. "Sustainable construction in China: status duo and trends", Building Research & Information, 28(1): 59-66

Uher T.E., (1999) Absolute Indicators of Sustainable Construction RICS Research Foundation available at http://www.rics.org/NR/rdonlyres/5A12CAFF-13FD-4D31-9C62-D5DC1958A745/0/ absoluteindicators of_sustainableconstruction_ 19990901.pdf

Vanegas, J. A. and Pearce, A., R. Drivers for Change (2006): An Organizational Perspective on Sustainable Construction Web: www.maven.gtri.gatech.edu/sfi/sustainables/pdf/ RCP/RCP001. PDF

7

Indoor Air Pollution in the Romanian Homes

Anca Maria Moldoveanu
University of Medicine and Pharmacy"Carol Davila" Bucharest
Romania

1. Introduction

Environmental pollution is one of the most important issues in the world today. Environmental pollution includes outdoor pollution and indoor pollution. For many decades the scientists have been studied outdoor pollution. This area of interest includes the pollution of ambient air, the pollution of water, soil, housing, the effects of ionizing and non-ionizing radiation, wastes.

The indoor environment has several aspects that are of importance. One aspect is linked to the chemical pollution of the indoor air. Other aspects can be linked to the biological contamination of air and surfaces, or to radiation pollution of indoor air linked especially to the presence of radon and radon daughter.

Health can be negatively affected by all types of environmental pollution. The outdoor and the indoor environments are linked. The two elements cannot be separated. Man spends around 85% -90% of a day indoor. There are category of population like infants and young children, elderly people, sick people and convalescent which spend 100% of a day time indoor.

Vulnerable groups are represented by children, pregnant women, elderly persons over 65 years of age, persons suffering of asthma or other respiratory diseases, persons suffering from cardiovascular diseases. Immunodeficiency may render people more vulnerable especially for biologic pollutants (SCHER, 2007). In the indoor environment are various sources of pollutants like: heating and cooking appliances, open fires, building and insulation materials, furniture, fabrics and furnishings, glue, cleaning products and other consumer products (P.T.C. Harrison, 2002).

A lot of diseases can originate in the indoor environment. Environmental immunologic diseases are both allergic and non-allergic (Ledford K Dennis, 2002). Diseases categorized under the generic name of "building related illnesses" were identified, as well as another group of health problems that can be linked to the buildings. This category includes the so called "sick-building syndrome" or "tight-building syndrome".

The group of diseases in the "building related illnesses" category includes diseases like humidifier fever, hypersensitivity pneumonitis, some infectious diseases like Legionnaire's disease, Pontiac fever, tuberculosis, viral illnesses like flu, diseases of the childhood like measles, chicken pox, etc (Burge H.A. , 1995, Berglund at all,1992). The contamination of indoor air can induce allergic problems like rhinitis and asthma that can be related to indoor allergens such as dust mite, cockroaches, animals and mold. Some molds can release disease-causing toxins. These toxins can damage a variety of organs and tissues in the body, including the liver, central nervous system, digestive tract, and immune system. In the case of certain diseases, like humidifier fever, it is not certain whether the disease is an allergic reaction or a toxic response.

Indoors there are several sources of biological contaminants like:

- air handling system condensate,
- cooling towers,
- water damaged materials,
- high humidity indoor areas,
- damp organic material,
- porous wet surfaces,
- humidifier,
- hot water systems,
- outdoor excavation,
- plants,
- animal excreta,
- animal and insects,
- food and food products.

Symptoms of exposure to biologic contaminants include sneezing, watery eyes, coughing and shortness of breath, dizziness, fever and digestive problems.

Irritants such as certain volatile hydrocarbons may aggravate existing asthma via non-immunologic mechanisms.

Hypersensitivity pneumonitis is a non-allergic, immunologic pulmonary disease resulting from the inhalation of organic dusts from contaminated humidifiers, moldy organic debris, animal proteins and certain chemicals. This disorder is often misdiagnosed as an acute infectious respiratory condition because of the fever, chest infiltrates and cough. Interstitial fibrosis with permanent scarring of the lung may result if repeated acute or chronic episodes occur.

The etiology of the symptoms of "sick building syndrome" is usually due to some environmental features of the building, but the exact cause is difficult to ascertain. Potential causes include poorly functioning ventilation systems with inadequate fresh air, the presence of inorganic chemicals such as ozone or nitrogen dioxide and organic chemicals such as mycotoxin or formaldehyde. The symptoms often include mucosal irritation, skin irritation, neuropsychiatric complaints of fatigue and limited ability to concentrate, odor and taste complaints.

Another important pollutant that can be identified indoors is radon and radon daughters from soil, construction materials used for building the house, water, and the use of natural gas. This element can induce lung cancer in a long term exposure taking into account the lag time until the appearance of the disease.

That makes the indoor environment very important in connection to human health.

2. Purpose of the Study

The purpose of this study is to make an assessment of the indoor condition in the homes in Bucharest and the possible impact on health of the inhabitants.

For that purpose 34 apartments in city of Bucharest were studied. All the investigated homes were in blocks of flats. All flats were situated on different floors, starting from the ground floor until the 11-th floor. The geographical orientations of the flats were in all direc-

tions N, S, E, W. There were flats with a single room as well as flats with two, three and four rooms.
In all the flats the chemical pollution was investigated.
The biological pollution of indoor air was investigated in 10 apartments.
Indoor chemical pollution
The levels of several chemical pollutants were investigated: nitrogen dioxide, total aldehydes, formaldehyde, total suspended particles, carbon monoxide and benz (a)pyrene.
All the pollutants were measured in the kitchen and in the living room, in the winter and in the summer time.
Indoor biological pollution
The bacteriological and fungal pollution was also investigated in the flats, in the summer and in the winter time, in three compartments of each flat respectively: the living room, the bedroom and the bathroom.
Health assessment
The impact of this type of pollution on health of the inhabitants of those flats was investigated as well.
Methods used in the study
The chemical evaluation focused on several chemical pollutants.

- Nitrogen dioxide was measured by a colorimetric method.
- Total aldehydes and formaldehyde were measured through colorimetric methods.
- Total suspended particle were identified with a reflectometric method.
- Carbon monoxide was identified using a colorimetric method.
- Benz (a) pyrene was identified using a chromatographic method.

The chemical evaluation was made twice: once in the summer time and once in cold season.
For this study a biological investigation of the indoor air was performed. The biological evaluation included a bacteriological evaluation and an assessment of fungal contamination of the air.
The sampling method used was a passive method with exposure of a Petri dish with blood-agar for 10 minutes in the investigated rooms of each home.
This study investigates: the total number of bacteria, expressed in CFU/m^3 (CFU/m^3 - colony-forming units/cubic meter), the presence of Streptococcus alpha and beta hemolytic, Staphylococcus aureus and non-aureus with and without hemolytic proprieties and coliform bacteria as an indicator of enteric contamination of air.
For the assessment of fungi contamination of indoor air, the sampling method was a passive method with exposure of a Petri dish with Chapek with antibiotics support, for 10 minutes in each home, in the kitchen, in the bathroom and in the living room.
This study investigates the total number of fungi, expressed in CFU/m^3 (CFU/m^3 - colony-forming units/cubic meter) and the presence of different types of fungi.For an accurate assessment of the biological contamination of indoor air, the measurements were made twice in a year, in the winter and in the summer.

3. Results

The apartments can be divided into two categories: the first category was formed of apartments in which you can pass from one room in another and the second type was formed of

apartments with independent rooms separated by a hall. 38.2% apartments belonged to the first category and 61.8% apartments belonged to the second category.
The geographical orientation of the flats can divide the apartments into the following categories: 11.8% of the apartments were orientated EV, 14.7% of the apartments were oriented N, 23.5% NE, 5.9% NS, 14.7%NV, 8.8% S, 2.9%SE, 11.8%V and 5.9% SV.
70.6% of the investigated flats were situated on the inferior floors and 29.4% on the superior floors of the blocks of flats.
Taking into account the number of rooms in each flat, the situation was: 5.9% of flats had one room, 23.5% of flats had two rooms, 50% of flats had three rooms and 20.6% were flats with four rooms.
The building material used was: concrete - 58.9%, bricks - 29.4%, preformed walls - 11.7%.
The heating system was: a centralized system in 94.1%, gas heating in 5.9% of investigated flats.
The fuel used in the kitchen was gas in all investigated apartments.
In some of the apartments, the owners made some improvements like the use of wall paper for the living rooms and bedrooms, some dye work, faience and sandstones for the kitchen and bath wall and floors.
In 30% of apartments, the inhabitants smoked under 10 cigarettes per day and 70% of the inhabitants smoked over 10 cigarettes per day or more.
The inhabitants of the investigated apartments were divided by age as follows: 8.1% of the inhabitants were with age between 0 – 10 years old, 12.9% of the inhabitants were with age between 10 – 20 years old, 12.9% of the inhabitants were with age between 20 – 30 years old, 19.4% of the inhabitants were with age between 40 – 50 years old, 21% of the inhabitants were with age between 50 – 60 years old, 17.7% of the inhabitants were with age between 60 – 70 years old and 8% of the inhabitants were with age over 70 years old.
The inhabitants of the investigated apartments were divided by sex as follows: 51.6% were female and 48.4% were male.
The distribution according to the two types of apartments was equal.
The distribution of the inhabitants of the studied apartments according to the number of hours spent in the kitchen, where the main source of gas is located, showed the following situation: 22.6% of the inhabitants spend one hour per day in the kitchen, 11.3% of the inhabitants spend two hours per day in the kitchen, 12.9% of the inhabitants spend three hours per day in the kitchen, 33.9% of the inhabitants spend four hours per day in the kitchen, 12.9% of the inhabitants spend five hours per day in the kitchen and 6.40% of the inhabitants spend six hours per day in the kitchen.
According to the number of cigarettes smoked per day, 67.7% of the inhabitants were non-smokers, 12 % of the inhabitants smoked under 10 cigarettes per day and 20.3% of the inhabitants smoked more than 10 or more cigarettes per day, even more than a pack per day.
The levels of the chemical pollutants identified in indoor air were:

Levels of nitrogen dioxide (μg/m^3)	Kitchen -summer	Living room - summer
Minimum values	81	55
Maximum values	2770	1631
Average values	478	33

Table 1. Levels of nitrogen dioxide in indoor air in summer

Levels of nitrogen dioxide (μg/m³)	Kitchen-winter	Living room- winter
Minimum values	164	80
Maximum values	6903	1698
Average values	1629	586

Table 2. Levels of nitrogen dioxide in indoor air in winter

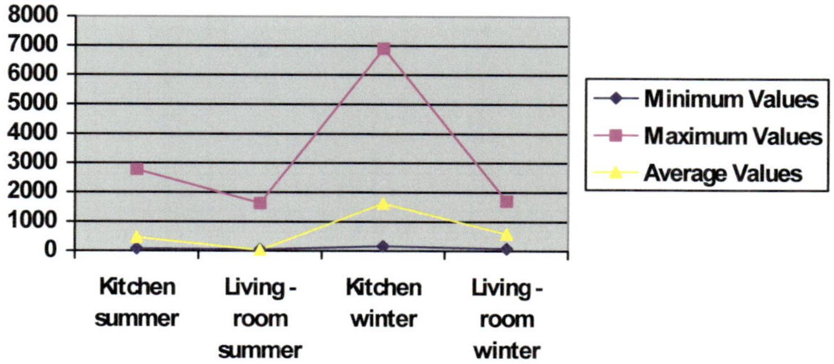

Fig. 1. Comparison between levels of nitrogen dioxide (μg/m³) measured indoor in summer and in winter

Levels of total aldehyde (μg/m³)	Kitchen -summer	Living room -summer
Minimum values	50	40
Maximum values	350	1280
Average values	108	148

Table 3. Levels of total aldehyde in indoor air in the summer

Levels of total aldehyde (μg/m³)	Kitchen-winter	Living room- winter
Minimum values	19	16
Maximum values	6769	240
Average values	459	136

Table 4. Levels of total aldehyde in indoor air in the winter

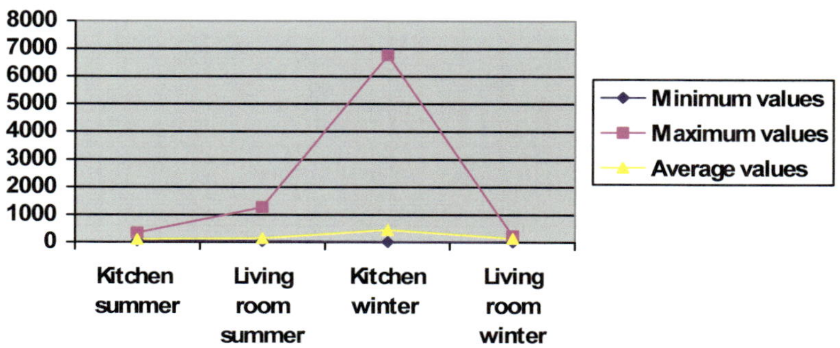

Fig. 2. Comparison between levels of total aldehydes (μg/m³) measured indoor in summer and in winter

Levels of formaldehyde (μg/m³)	Kitchen-summer	Living room -summer
Minimum values	3	8
Maximum values	91	122
Average values	35	34

Table 5. Levels of formaldehyde in indoor air in summer

Levels of formaldehyde (μg/m³)	Kitchen-winter	Living room- winter
Minimum values	2	4
Maximum values	860	188
Average values	142	65

Table 6. Levels of formaldehyde in indoor air in winter

Fig. 3. Comparison between levels of formaldehyde (μg/m³) measured indoor in summer and in winter

Levels of carbon monoxide (mg/m³)	Kitchen -summer	Living room -summer
Minimum values	3.38	2.480
Maximum values	58	49
Average values	9	6.76

Table 7. Levels of carbon monoxide in indoor air in summer

Levels of carbon monoxide (mg/m³)	Kitchen-winter	Living room - winter
Minimum values	4.28	3.38
Maximum values	10	9.45
Average values	6.65	5.60

Table 8. Levels of carbon monoxide in indoor air in winter

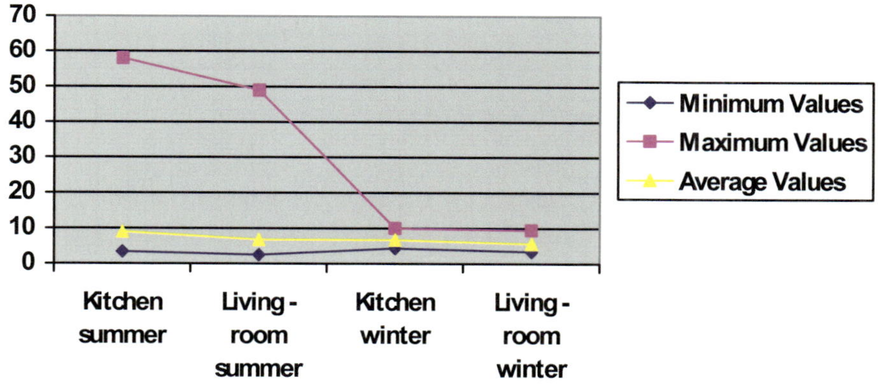

Fig. 4. Comparison between levels of carbon monoxide (mg/m^3) measured indoor in summer and in winter

Levels of total suspended particles (µg/m^3)	Kitchen-summer	Living room -summer
Minimum values	34	17
Maximum values	326	66
Average values	165	46

Table 9. Levels of total suspended particles in indoor air in the summer

Levels of total suspended particles (µg/m^3)	Kitchen-winter	Living room- winter
Minimum values	117	1
Maximum values	365	326
Average values	245	224

Table 10. Levels of total suspended particles in indoor air in the winter

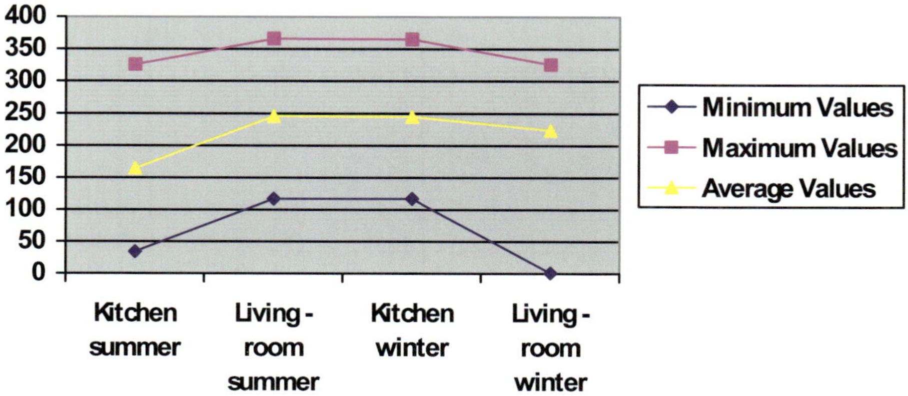

Fig. 5. Comparison between levels of total suspended particles (µg/m^3) measured indoor in summer and in winter

Levels of benzo(a)pyrene (ng/m³)	Kitchen -summer	Living room -summer
Minimum values	81	55
Maximum values	2770	1631
Average values	478	33

Table 11. Levels of benzo (a) pyrene in indoor air in the summer

Levels of benzo(a)pyrene (ng/m³)	Kitchen-winter	Living room-winter
Minimum values	164	0
Maximum values	6903	698
Average values	1629	86

Table 12. Levels of benzo (a) pyrene in indoor air in winter

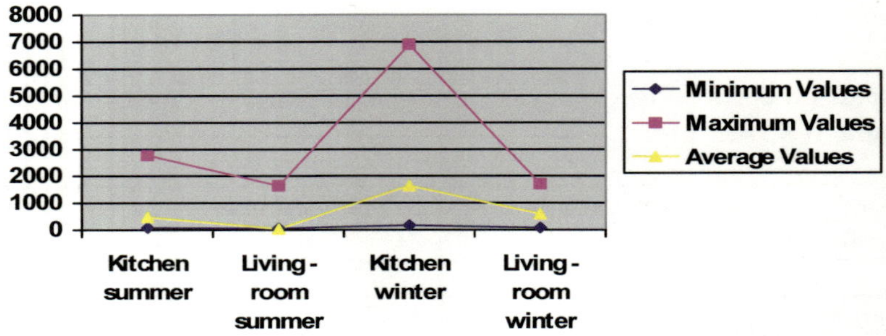

Fig. 6. Comparison between levels of benzo (a) pyrene (ng/m³) measured indoor in summer and in winter

Biological pollution of indoor air in the studied apartments from Bucharest city:
For the bacteriological evaluation the results were:

CFU/m³	In the kitchen	In the bathroom	In the living room
Minimum values	850	1500	120
Maximum values	57500	57500	32500
Average values	9367	13241	6090

Table 13. The total number of bacteria found in the summer time

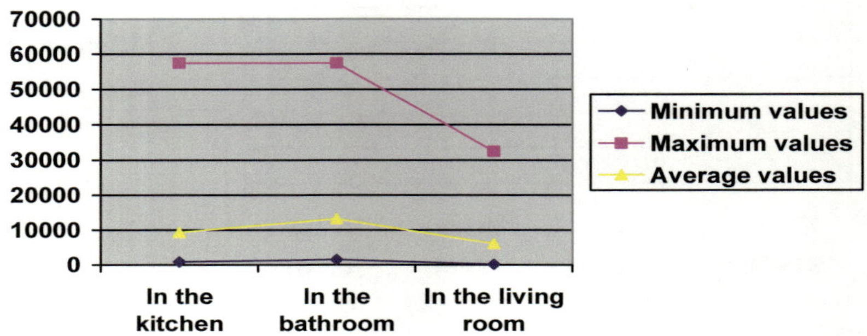

Fig. 7. The total number of bacteria found in the summer time

CFU/m³	In the kitchen	In the bathroom	In the living room
Minimum values	625	688	63
Maximum values	12600	50400	6300
Average values	3124	7903	2406

Table 14. The total number of bacteria found in the winter time

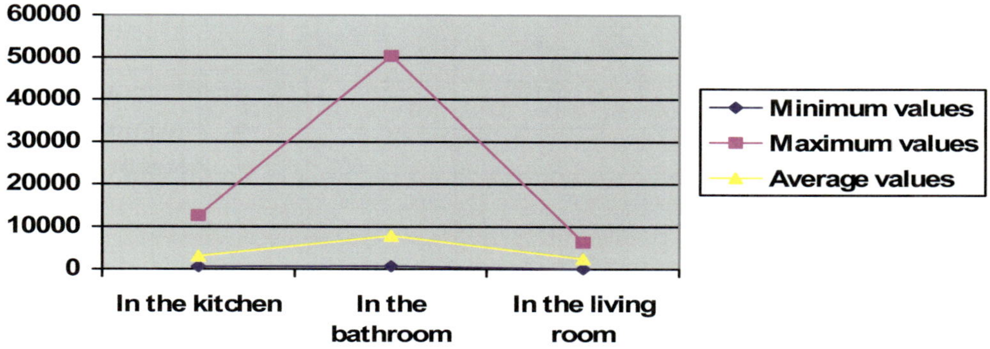

Fig. 8. The total number of bacteria found in the winter time

CFU/m³	In the kitchen	In the bathroom	In the living room
Minimum values	250	134	200
Maximum values	38750	55000	26125
Average values	6338	7200	4308

Table 15. The total number of Staphylococcus found in the summer time.

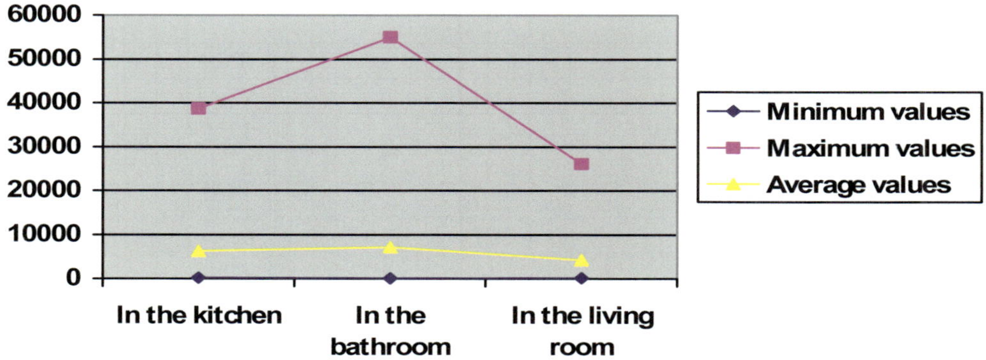

Fig. 9. The total number of Staphylococcus found in the summer time

CFU/m³	In the kitchen	In the bathroom	In the living room
Minimum values	188	313	0
Maximum values	10250	55000	36500
Average values	3219	6421	4488

Table 16. The total number of Staphylococcus found in the winter time

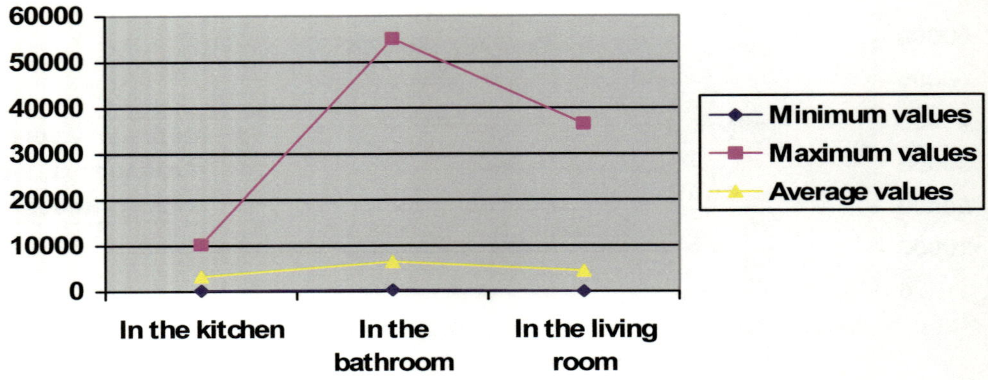

Fig. 10. The total number of Staphylococcus found in the winter time

The presence of Staphylococcus with hemolytic elements was identified in one home, in the bathroom, at a level of 313 CFU/m³.
The presence of Streptococcus beta hemolytic was found in one house at the level of 938 CFU/m³ in the kitchen, 700 CFU/m³ in the bathroom and 1188 CFU/m³ in the living room.
The presence of hemolytic Staphylococcus and Streptococcus beta hemolytic was confirmed in diferent houses.
The fungi contamination was assessed in 10 flats. The obtained results were:

Number of the house	The fungi level in the kitchen (CFU/m³)	The fungi level in the bathroom (CFU/m³)	The fungi level in the living room (CFU/m³)
1	1142	Layer	728
2	908	990	776
3	Layer	Layer	1271
4	0	0	0
5	180	110	52
6	0	250	0
7	1142	Layer	728
8	600	400	290
9	908	990	776
10	580	Layer	350

Table 17. The level of fungi measured in houses in the summer time

Number of the house	The fungi level in the kitchen (CFU/m^3)	The fungi level in the bathroom (CFU/m^3)	The fungi level in the living room (CFU/m^3)
1	728	Layer	350
2	378	28	12
3	Layer	1500	1550
4	0	0	0
5	160	80	36
6	29	941	36
7	3313	Layer	1820
8	0	0	47
9	1217	630	1710
10	633	3600	720

Table 18. The level of fungi measured in houses in the winter time

The species of fungi identified were:

- Penicilium,
- Aspergilium,
- Rhizopus,
- Mucor,
- Cladosporium,
- Fusarium.

Health evaluation

Health evaluation of those living in the investigated apartments showed the presence of respiratory symptoms like coughing, wheezing and respiratory diseases like bronchitis and pneumonia. The presence of irritations localized to the nose, throat and eyes and facial pains were linked to the indoor exposure.

The frequency of appearance of different symptoms to the inhabitants of the investigated apartments was:

- The cough was present in 12.1% of investigated inhabitants,
- Wheezing was present in 7.3% of investigated inhabitants,
- Shortness of breath was present to 4% of investigate inhabitants,
- Bronchitis and pneumonia were present in a percentage equal to 6.5%,
- Asthma was not identified among inhabitants of the investigated houses,
- Ocular irritations were present in 25% of investigated inhabitants,
- Nose and throat irritations were present in 22.6% of the investigated people,
- Irritations of the skin were not identified among the investigated inhabitants of the flats,
- Headaches were present in a percentage of 21.8% of investigated inhabitants,
- Facial pains were present in a percentage of 16.1% of investigated inhabitants,
- Infection of sinuses were present in a percentage of 6.5% of investigated inhabitants,
- Ischemic cardiopaty was present in a percentage of 11.3% of the inhabitants.
- Cancer was not identified among the inhabitants of the investigated houses.

Risk assessment

- Ocular irritation correlated with nitrogen dioxide exposure had an OR = 5.90 (95%CI: 1.67 - 22.88) for summer time and an OR = 1.19 (95%CI: 0.40 - 3.55) for winter.
- Ocular irritation correlated with total aldehyde exposure had an OR = 1.61 (95%CI: 0.45 – 6.36) for winter.
- Ocular irritation correlated with fomaldehyde exposure had an OR =1.26 (95%CI: 0.43 – 3.66) for summer time and an OR = 1.13 (95%CI: 0.36 - 3.69) for winter.
- Nose and throat irritations in correlation with nitrogen dioxide exposure had an OR = 2.76 (95%CI: 0.99 – 7.91) in the summer time and an OR = 1.70 (95%CI: 0.61 – 4.87) in the winter.
- Nose and throat irritations in correlation with total aldehyde exposure had an OR = 2.14 (95%CI: 0.61 – 8.22) in the winter.
- Nose and throat irritations in correlation with formaldehyde exposure had an OR = 0.54 (95%CI: 0.19 – 1.49) in the summer time and an OR = 2.81 (95%CI: 0.81 – 10.71) in the winter.
- Facial pains in correlation with nitrogen dioxide exposure had an OR = 3.18 (95%CI: 1.02 – 10.28) for the summer and an OR = 3.70 (95%CI: 1.05 -14.30) for the winter.
- Facial pains in correlation with total aldehyde exposure had an OR = 0.92 (95%CI: 0.17 – 7.02) for the summer and an OR = 1.62 (95%CI: 0.45 -6.36) for the winter.
- Facial pains in correlation with formaldehyde exposure had an OR = 1.26 (95%CI: 0.43 – 3.66) for the summer and an OR = 2.12 (95%CI: 0.59 -8.23) for the winter.
- For ischemic cardiopaty the correlation with exposure to carbon monoxide an OR = 1.15 (95%CI: 0.30 – 4.41) for the summer and an OR = 2.50 (95%CI: 0.64 – 2.74) for the winter.

Results discussion

The chemical substances identified in the indoor air reached very high levels of pollution, with positive correlations to effects on human health due to exposure to nitrogen dioxide, total aldehyde, formaldehyde, carbon monoxide WHO (c). A positive correlation between exposure to benzo (a) pyrene, formaldehyde and cancer could not be established, possibly because of the lag time the cancer has in inducing effects on human health.

No standards exist for acceptable levels of indoor air contamination with microorganisms, since the infectivity of pathogens is extremely species dependent, although a number of guidelines exist for indoor fungi levels and a few exist for indoor bacterial levels. As limit to refer to, it is possible to use the limit proposed by ACGIH (American Conference of Government Industrial Hygienists) and AIHA (American Industrial Hygiene Association) that stipulates 1000 of viable CFU/m^3 of air as an upper limit for concentration in indoor environment (Rao et.al., 1995). CEC (Commission of the European Communities) (2004) and WHO (a, b, d, e) defines as a "very high" level the value equal to 2000 CFU/m^3. A value of 10,000 CFU/m^3 of nondescript airborne microbes could therefore be considered a hazardous level for indoor environment.

In this respect, the values that were obtained in this study are much higher, both in the summer and in the winter time, for the maximum values. For Staphylococcus the obtained levels were over the limit presented above, also in the case of the average value.

For fungi in non-industrial spaces the limit that can be used to compare the results to is

equal to 100 CFU/m^3 (CEC).
The results obtained in the study were much higher than 100 CFU/m^3 and in the bathrooms, where there are areas with high humidity, a layer of fungi was frequently found.
The health of inhabitants of investigated houses was affected, especially due to the presence of respiratory symptoms and diseases and the presence of irritations.
It can be concluded that a high biological contamination in indoor spaces can be a threat to human health.
Further research must to be done to develop guidelines and to propose measures to reduce the possibility of having a high level of indoor pollution and to protect in this way the human health. Regarding the development of the guidelines, World Health Organization (e) as well as other organization like US-EPA (Environmental Protection Agency of The United States of America), IARC (International Agency for Research on Cancer) and other scientific organizations propose developments of guidelines for formaldehyde, benzene, naphthalene, nitrogen dioxide, carbon monoxide, radon, particulate matter ($PM_{2..5}$ and PM_{10}), polycyclic aromatic hydrocarbons (PAH), especially Benzo-a-pyrene (BaP) and halogenated compounds like tetrachloroethylene. For this group of substances a number of relevant systematic reviews and risk assessment for pollutants present in indoor environments is identified. There is another group of substances for which there is uncertain or insufficient evidence for development of guidelines. This group includes toluene, styrene, xylenes, acetaldehyde, hexane, nitric oxide (NO), ozone (O_3), phthalates, biocides, pesticides, asbestos, glycol ethers, carbon dioxide, limonene, pinene.
For biologic agents development of quantitative guidelines requires research for all agents (WHO, e). Guidance can be provided for pathogenic agents like viruses and bacteria, bacteria (non pathogenic), fungi, toxin and other microbial products, microbial compounds, algae and amoebae, mites, pet allergen (cats and dogs), pest allergens (cockroaches and other insects, rats), pollen, fungal allergens (e.g. Alternaria, Cladosporium), ventilation (humidity control, pollutant removal), ventilation (components - as sources of contamination), dampness, moisture control, condensation, chemical degradation and microbial growth, hygiene and lack of cleaning, use and mis-use of cleaning and disinfection products and use of air fresheners.

4. References

Berglund B., Brunekreef B., Knoppel H., Lindvall T., Maroni M., Molhave L., Skov P. (1992): "Effects of indoor air pollution on human health". In Indoor Air, 2: 2-25;

Burge H.A. (1995): "Biological contamination of buildings in temperate climates". Health buildings' 95, vol. I: p. 239-250;

European Commission, JRC, (2004), The INDEX Project: Critical Appraisal of Setting and Implementation of Indoor Exposure Limits in the EU,

Harrison P.T.C. (2002), Indoor air quality guidelines, Occupational and Environmental Medicine, vol.; 59, pp.73 -74,

Ledford K Dennis, (2002) *Building Related Illness Your health and how it relates to your indoor* environment, Pure Air Control Services, November, 20, 2002, vol. 2, Issue 50,

Rao C., Burge H., Chang J. (1995): "Review of concentration standards and guidelines for fungi in indoor air". In "Healthy Buildings'95", Vol.1: pp. 239-250;

SHER (Scientific Committee on Health and Environmental Risks) Preliminary report on risk assessment on indoor air quality, 31 January 2007,

US EPA. An introduction to indoor air quality, Formaldehyde,

http;//www.epa.gov/iaq/formaldehyde.html
US EPA. An introduction to indoor air quality, Respirable particles. http;//www.epa.gov/iaq/rpart.html
WHO a – Air quality guidelines for Europe, (1987), Regional Office for Europe, European series, No. 23, Copenhagen, Denmark,
WHO b - "Indoor air quality: Biological contaminants" (1988), European Series nr. 31,
WHO c – Carbon Monoxide, 1999, 2nd edition, Environmental Health Criteria 213,
WHO d – Air quality guidelines for Europe, (2000), second edition, Regional Office for Europe, European series, No. 91, Copenhagen, Denmark,
WHO e – Development of WHO Guidelines for Indoor Air Quality (2006),Regional office for Europe, Report on a working group meeting, Bonn, Germany, 23 -24 October.

8

Soil Pollution and Remediation Problems in Turkey

E. Burcu Özkaraova Güngör
Ondokuz Mayıs University
Turkey

1. Introduction

Turkey has been undergoing significant economic transformations since the 1980s. The adopted structural reform programme was based on outward-looking trade policies. The growth of industry and services, reflecting from time to time high rates among the OECD countries, resulted in a decline in the share of agriculture in the Gross National Product. This, however, does not mean that the importance of agriculture is continuing to decline, because investigations in rural infrastructure and implementation of agricultural policy, supporting improved productivity, are carried out by the government.

The rapid growth in industrial investments, unfortunately, increased the pressure on environment, which was recognized by the Turkish government. Thus, the Prime Ministry Undersecretariat for Environment was founded in 1978, later on replaced by the Ministry of Environment in 1991. This change led to a diversification of the Ministry's responsibilities and an expansion of its staff and empowered the administration power to implement and enforce policies for the protection and conservation of the environment (Okumuş, 2002). Besides developing environmental policies and strategies, other duties of the Ministry are to coordinate environmental activities on local, national and international levels, issuing environmental licenses and data collection. Currently, provincial branches of the Ministry are dealing with environmental tasks in all 81 provinces located in seven regions, namely the Mediterranean, Aegean, Marmara, Black Sea, Central Anatolia, Eastern Anatolia and South Eastern Anatolia. Besides many others, the main environmental tasks of the provincial branches are to take cautions for the prevention and minimization of pollution, to inspect all kind of activities that might threaten the environment and cause to possible pollution in soil and water resources, and to control the facilities approved by the local authorities. Responsibilities of provincial and local authorities, dealing with environmental matters like waste management and pollution prevention have expanded. Especially, the increased power of municipalities obligates them to improve the infrastructure of waste management systems, like sewage and other waste services (collection, treatment and disposal) according to the standards set by the Environmental Law and regulations.

The Environmental Law, which stems from the 1982 Constitution, came into force in 1983. According to the Constitution, each citizen has a right to live in a healthy and balanced environment and that the citizens and the Government are responsible in the protection and improvement of the environment. The Environmental Law, besides these basic principles, states that all measures should be taken to minimize pollution in every economic activity. For the implementation of the Environmental Law, several regulations, fully recognizing the principles of environmental management, were published in the official gazette since 1983. The first issued regulations are:

- Air Quality Control Regulation (1986)
- Noise Control Regulation (1986)
- Water Pollution Control Regulation (1988)
- Solid Waste Control Regulation (1991)
- Environmental Impact Assessment Regulation (1993)
- Hazardous Waste Control Regulation (1995)

After Turkey became a candidate state of European Union (EU) at the Helsinki European Council in 1999, the Accession Partnership, which was proposed by the European Commission in 2000, also identified the environmental priority areas. Besides many others, one of the priority areas was to ensure environmental protection, implementation of the acquis with particular attention to the framework legislation, horizontal legislation and legislation on nature protection, water quality and waste management. Thus, the adaptation of EU directives into the national legislation resulted in the revision of present legislations and enforcement of many others including those on soil protection and pollution control (Okumuş, 2002).

2. Environmental Policies and Strategies

Turkey, recognizing the pressures on the environment from rapid economic growth, took the first step for the development of an environmental policy in the Third Five-Year National Development Plan (1973-1977) with the inclusion of the 'Environmental Management' concept. For the implementation of environmental protection issues and creating solutions on local and national level, the Undersecretariat for Environment was founded. However, the enforcement of an environmental policy has been difficult for years, since priority was given to industrial investments. 'Sustainable Development' was adopted as the central concept for the period 1991-1996 (Sixth Plan), and 'Protection and Improvement of the Environment' was a major objective for the period 1996-2000 (Seventh Plan) (OECD, 2000). The major objectives, principles and policies of the Seventh Five Year National Development Plan were determined as (State Planning Organization, 1998):

- emphasizing pollution prevention rather than clean-up;
- using an appropriate combination of economic and regulatory instruments;
- developing regional and eco-basin strategies;
- strengthening the system of environmental management;
- ensuring that policies and solution are in accordance with EU norms and international standards;
- revising and enhancing the financing system for environmental protection, management and improvement;
- promoting environmental awareness through formal and non-formal channels,
- ensuring compatibility between economic development and environmental protection.

Nevertheless, the Seventh Five-Year National Development Plan noted that environmental concerns have not been adequately incorporated into all economic and social decisions or in legislative/organizational arrangements till then (State Planning Organization, 1998). Thus, to implement the objectives of the Seventh Development Plan, a National Environmental Action Plan (NEAP) was prepared by the State Planning Organization (DPT) in cooperation with the Ministry of Environment and financial support of the World Bank. Under the authority of the Prime Ministry, State Planning Organization takes strategic decisions on investment priorities based on investment requirements in all areas of economic activity. The State Planning Organization has the power to require that environmental considerations are incorporated into investment proposals, which are totally or partially financed from public funds (Okumuş, 2002).

The Preparation process of NEAP started in 1995 with tremendous efforts of nineteen working groups representing local and national governmental authorities, municipalities, the private sector, academia, non-governmental organizations, members of media and donors (the UNDP and World Bank). These groups produced reports and recommendations on air quality, natural, historical and cultural heritage; demography and public health; soil, water, solid waste, wastewater, marine resources, land use and coastal zone management; economic and financial issues; legal, institutional and regulatory frameworks; education and participation; and noise pollution (State Planning Organization, 1998). With the synthesis of NEAP from the working papers, the problem areas and action categories were set (Table 1).

Soil protection and pollution issues were included in the NEAP under 'Soil and Land Use' and 'Hazardous Wastes' headings. Soil protection, which is mainly considered as an agricultural concept, is mostly based on prevention of erosion, desertification, salination and misuse of agricultural land. For this purpose, the development of agricultural support policies and rural development policies; the preparation of action plans for the prevention of soil erosion, desertification etc. are accepted as mandatory actions (Annex 1). Land use decisions should be coupled with soil suitability classifications. Therefore, the preparation of more precise inventories of land and soil qualifications is one of the main targets. For the implementation of these action plans, the Soil Conservation and Land Improvement Law and the Regulation on Implementation of Soil Conservation and Land Improvement Law were enforced in 2005 after the enforcement of the Regulation for the Protection and Improvement of Agricultural Land in 2003. Each of these put an emphasis on the need of soil qualification inventories and maps to enable proper use of land.

Classification and mapping for Turkey's soil have been made by the General Directorate of Soil and Water, today known as the General Directorate of Rural Services since 1950s. The first prepared map was the Turkey General Soil Map (1:800 000 scale) followed by the Turkey Development Soil Map (1:25 000 scale). With the evaluation of data, like depth, slope, stoniness, erosion degree etc., two maps, the Soil Resource Inventory Map (1:100 000 scale) and the Watershed Soil Map (1:200 000 scale) and Report was developed. From 1987 onwards, maps were prepared from the results of the Turkey Development Soil Map at a scale of 1.100 000. With the consultation of the General Directorate of Rural Services and the surveys, a map named the Turkey Soil Zones Map (1:2 000 000 scale) was also made (Keskin, 2001). However, the need for the development of more precise smaller-scale maps continues.

¶ **Problem Areas**	**Coverage**
Population/Urbanization	Quality, Quantity, Density, Movements
Air Quality	Monitoring,
Energy Generation and Consumption	Production, Distribution, Consumption; Development of New/Renewable Sources, Conservancy
Water and Wastewater Management	Rivers; Seas; Lakes; Ponds, Groundwater, Wastewater.
Solid and Hazardous Wastes	Domestic; Industrial; *Hazardous*; Medical (Liquid, Solid, Gas)
Noise	Traffic; Industry
Soil and Land Use	*Erosion; Pollution; Misuse of Agricultural Lands; Desertification; Salination; Mining Industry, Agricultural Technology.*
Forests/Vegetative Cover	Forests; Steppes; Ranges and pastures
Marine Resources, Coastal Zones and Sensitive Environments	Wetland; Shore; Island; Aquatic products, Mountain; Highland, Sand Dunes
Historical, Cultural and Natural Heritage	Monuments; Ruins; Natural Formations; Endemic Plants Under the Threat of Extinction; Animals; Value Judgments
Action Categories	**Areas**
Policies	Economic, Social, Cultural and Sectoral
Institutional Reform	Ministries and Related Organizations; professional Organizations; Voluntary Organizations; Private and Public Sector Enterprises; Cooperatives; Management
Legislation	The Constitution, Laws, Regulations, By-laws; Circulars
Economic & Financial Measures	Taxes, Incentives, Subsidies, Funds; Fines, Voluntary Contributions
Education & Awareness	Non-formal and Formal Education; Arts; Religion
Participation	International, National and Local
Specific Tools	Planning (Urban Development, Environment, Sectoral and Action Plans); Monitoring and Supervision; Standardization; EIA, Data Collection
R & D	Research; Technique and Technology Development; Monitoring and Transfer

Table 1. Problem Areas and Action Categories Used as Basis for Sectoral Options. (State Planning Organization, 1998)

Causes and facts of soil pollution are considered in more detail in the Hazardous Waste Management Report of NEAP rather than in the Land Use Report, which only proposes survey about contaminated areas (Ongan, 1997; Zanbak and Bayazit, 1997). Zanbak and Bayazit Tugal state in their report on hazardous waste management, that the effect of improper disposal of hazardous wastes on soil quality is not known since measurements on possibly contaminated areas are not performed. Policies of NEAP on hazardous waste management, includes rehabilitation issues of old waste dumps, besides emphasizing minimization, recycling and recovery of hazardous wastes and hazardous waste treatment. NEAP also points to a preparation of programmes for the improvement of polluted areas and establishment of an inventory of polluted sites (Annex 2).

Soil and water resources commission report on 'Use and Management of Watershed,' which was also used for the Eighth Five-Year National Development Plan (2001-2005), reflects the causes and status of soil pollution and proposes development of programmes and policies for the improvement of soil resources (State Planning Organization, 2001). A more detailed strategy for soil pollution management was defined in the commission report of Ninth Five-Year National Development Plan (2007-2013). The strategy is 'to prepare a national programme, which enables the development of methods and techniques for the identification, investigation, classification and finally remediation of polluted areas in accordance with the Soil Protection Strategy of EU' (State Planning Organization, 2006a). Recommended tasks and precautions for the minimization of soil pollution are;

- determination of polluted water and soil areas including areas with contamination risk,
- risk classification of polluted areas,
- taking remedial actions for the clean-up of polluted areas,
- implementation of projects and programmes for soil and water pollution prevention,
- preparation of emergency plans for areas with high contamination risk.

For the prevention and control of soil pollution several regulations have been enforced after 2000, which were also a part of the Environmental Screening Process of EU.

3. Legal Framework

Turkey has many legislative elements that recognize the principles of pollution prevention and environmental management. The Environmental Law (1983) defines these principles and necessary activities, which are to protect and improve the environment, and to prevent and solve environmental problems. The Law prohibits the direct discharge of wastes into the receiving environment. It embodies the 'polluter pays' principle and set forth the concept of 'absolute liability' to enforce it. It sets monetary and imprisonment penalties in circumstances of violation of prohibitions and standards presented in the Law and regulations. According to the Law, procedures to be followed, plans to be created, standards to be met, and activities to be prohibited are specified in regulations. With the amendment of the Environmental Law in 2006, principles pertaining to the protection of soil and prevention of soil pollution are presented separately in a supplementary article. The introduction of higher fines and entry into force of the penal code's provisions, envisaging sanctions for environmental crimes are expected to improve inspection (Commission of the European Communities, 2006).

The publication of a regulation for the control of soil pollution took a very long time, since the importance of soil pollution was not early realized. However, the Solid Waste Control Regulation (1991), the Environmental Impact Assessment Regulation (1993), and especially the Hazardous Waste Control Regulation (1995) provided the necessary measures to inhibit soil pollution. Both the Solid Waste Control Regulation and Hazardous Waste Control Regulation provide a legal framework for the management of municipal solid wastes and hazardous wastes. The target of the Solid Waste Control

Regulation is to prohibit direct or indirect delivery, storage, transportation, removal and similar activities of any kind of wastes and residuals into the environment in a way damaging the environment; to prevent the pollutants which exhibit a permanent effect on air, water and soil, deteriorate the ecological balance. Thus, it only prevents soil pollution arising from the infiltration of leachate into the subsurface due to improper storage and disposal of municipal solid wastes.

The purpose of the Hazardous Waste Control Regulation is to set legal and technical bases for the control and minimization of hazardous waste production, technical and administrative standards for the construction and operation of disposal sites, waste recycle, recovery and treatment plants. Similarly, it aims the prevention of direct or indirect disposal of hazardous wastes into the receiving environment. The regulation puts an emphasis on the necessity of reuse and recycling of hazardous waste. Under circumstances were reuse and recycling can not be carried out, appropriate disposal method is determined according to the characteristics of wastes and convenient technology.

Both regulations, especially the Hazardous Waste Control Regulation, have no direct implications related to contaminated sites. The Hazardous Waste Control Regulation defines hazardous wastes with respect to the lists categorizing hazardous wastes in relation to their sources and chemical composition. Thus, any site, which is polluted by any of these categorized hazardous wastes, can be seen as a contaminated site. However, difficulties arise from the lack of information for most of the chemicals in these lists regarding specific maximum concentration levels (NATO, 2002). The consignor, the hazardous waste producer, is responsible from the remediation of polluted area. Expenditures to cover all types of environmental hazards are compensated by the consignor. The consignor is also obliged to annually fill out a waste declaration form, which includes knowledge about the type, amount and composition of produced hazardous waste, and the applied recovery and disposal method. On the other hand, the Hazardous Waste Control Regulation does not specify actions for the identification, remediation and control of polluted site. It only states that wastes are physically, chemically or biologically treated prior to the disposal in a controlled landfill, and for the minimization of their impact on the environment.

In December 2001, the Soil Pollution Control Regulation, which is currently the major regulation related to soil pollution, was enforced. The purpose of this regulation is to specify the basic principles for taking the necessary precautions to prevent soil pollution. The regulation covers the technical and administrative principles related to activities like discharge, disposal and leakage of hazardous substances and wastes causing soil pollution, and the application of compost and sludge, arising from the treatment of domestic and industrial wastewater with domestic character, on soil. According to the regulation, the identification, investigation and monitoring of polluted sites are carried out by the Provincial Branches of the Ministry, which also determine the necessary measures to be taken. All expenditures are compensated by the polluter. Principles and methods for the identification, investigation, monitoring and remediation of contaminated sites are set forth by the Ministry. For the determination of pollution, limiting values for some heavy metals, a kind of indicative values, are given in Annex 1-A a), and target values for some inorganic and organic soil pollutants, which should be met after clean-up of site, in Annex 1A-b).

The Soil Pollution Control Regulation has many deficiencies with respect to its practicality regarding the identification, investigation, monitoring and remediation of contaminated sites. The regulation has been amended in 2005, but still, it does not provide any principle and/or method to be followed with regard to these matters. The major revisions required should cover the following points:

- The scope of regulation should be restricted to prevention of soil pollution and remediation of contaminated sites; legislative framework for the application of sludge and compost on soil should be presented in a separate regulation.
- Principles and methods to be followed with regard to identification, investigation, monitoring and remediation of contaminated sites should be included.
- Principles of investigation for remediation and planning of remediation should be adopted.
- Indicative levels, soil remediation intervention values and target values for serious contamination of soil/sediment and groundwater should be specified.
- Inorganic and organic substances covered by the list in Annex 1A should be extended.
- Sampling, sample preservation, sample pre-treatment, analysis standard and analysis techniques for the substances in list for soil and groundwater medium should be defined in a new annex.
- Data required for determining the remediation urgency and the remediation deadline for listed substances should be presented in a new annex.
- A guideline covering the principles of selection of appropriate remediation technique should be issued.

Several other regulations, which are indirectly related to soil contamination, have been enacted in recent years. However, due to institutional problems, their implementation remains very low. One of the regulations is the Environmental Auditing Regulation (2002), whose objective is to regulate the principles and methods of environmental auditing of facilities, beginning with their construction, operation and production including waste management and disposal. The regulation presents a legal framework for the quality of work and practices carried out during environmental autiting, qualifications of auditors, responsibilities of facility owner, and tasks and authorities of auditing agencies. According to the regulation, the facilities are responsible to fill up a facility activity information form, which is periodically send to the Ministry, to keep related records and to maintain an internal environmental auditing programme on a yearly basis. The facility activity information form includes knowledge about the composition, amount of stored, consumed and produced raw material and chemicals, type of storage and disposal of wastes and their amount and composition, types and results of industrial accidents. Thus, with the implementation of the Environmental Auditing Regulation, the identification and especially registration of contaminated and suspected polluted sites will be easier.

Another regulation, which includes issues on monitoring and prevention on hazardous substances in water, sediment and soil is the Regulation on Conrol of Pollution Caused by Hazardous Materials in and Around the Aquatic Environment (2005). The objective of this regulation is to determine, prevent and progressively reduce the pollution in aquatic environment. Direct discharge of hazardous substances containing wastewater into the groundwater is forbitten. Other helpful regulations are the Regulation on Waste Batteries and Accumulators (2004), the Medical Waste Control Regulation (2005), etc.

4. Current Status of Contaminated Sites in Turkey

4.1 Sources of Soil Pollution

According to a working paper of Eighth Five-Year National Development Plan, main sources of pollution are illegal waste dumping sites, as well as municipal and industrial waste disposal sites (State Planning Organization, 2001). Especially industrial wastes, either buried illegal or disposed off in temporary storage sites close to the factory, cause to contamination of both soil and groundwater. Statistical data of the Turkish Statistical Institute (TUIK) presents that 47,3 % of disposed industrial waste was disposed into the sea, lake or river, 20 % was disposed off in municipal dumping sites, 10,9 % was controlled landfilled, 9,3 % was stored in the field of establishments, 2,4 % was incinerated and 10,1 % was disposed off by other methods in 2004. 60 % of total disposed industrial waste belonged to the basic metal industries, followed by the non-metallic mineral industry (14 %), manufacture of food and beverages industry (13 %) and manufacture of chemicals and chemical products industry (5 %) (Turkish Statistical Institute, 2006).

Considering hazardous wastes, the regulation encourages their reuse and recycling. However, the number of hazardous waste treatment and disposal plants is very low. Therefore, industrial hazardous wastes are generally legally or illegally disposed off, incinerated or sold, rather than recycled. According to the data of Turkish Statistical Institute, 67 % of the total amount of hazardous waste produced by the manufacturing industry was controlled landfilled, 21 % was incinerated, 7 % was stored within the establishment field, 2 % was disposed off in municipal dumping sites and 3 % was disposed off by other methods in 2004 (Turkish Statistical Institute, 2006).

The principal management option adopted for clinical waste management is incineration (Okumuş, 2002). Similarly, incinerators used for incineration of clinical wastes are mostly available for the waste management of metropolitan municipalities, due to their limited number (six such facilities operating in Turkey). Thus, clinical wastes, which are separately collected, are mostly disposed off with municipal refuse. 38,9 % of clinical wastes was disposed off in municipal dumping sites, 20,7 % was controlled landfilled, 20,5 % was incinerated, 12,4 % was buried, 7,5 % was burned in 2003 (State Planning Organization, 2006b)

The waste management in Turkey has improved especially in aspects of waste collection, transport and disposal. However, recycling rates are still low and recovery of wastes very limited, due to the limited numbers of waste treatment plants. The number of composting plants has increased from two in 1994 to five in 2004 (Turkish Statistical Institute, 2006). Additionally, there are three incineration plants and sixteen controlled landfills (State Planning Organization, 2006b; Turkish Statistical Institute, 2006). Since the awareness of sanitary landfilling of solid wastes increased in recent years, especially by metropolitan municipalities or municipalities in touristic regions, other sanitary landfills are under construction or are planned to be constructed. However, soil and groundwater pollution mainly occurs from uncontrolled dumping sites or other illegal burials. There are over 3200 municipalities, which dispose off their wastes in dumping sites with very limited or no leachate control. The amount of solid waste collected in municipalities in 2004 was 24,2 million tonnes, of which 65, 3 % was disposed off in municipal dumping sites, 28,9 % was disposed off in controlled landfills, 1,6 % was disposed off by burial, 1,4 % was treated in composting plants, 0,4 % was disposed off into lakes, and river and 0,3 % was disposed off by burning in open areas (Turkish Statistical Institute, 2006).

Besides industrial waste disposal sites, industrial facilities causes to pollution due to handling losses, leakages from raw material or product storage tanks and pipelines, and accidents. Among industries, the chemical industry and metal working industry are the most polluting industries followed by the oil industry (refineries). Other activities resulting in the contamination of soil in Turkey are mining and energy production activities. Generally mine tailing disposal sites have no lining or drainage systems. Considering power plants for energy production, disposal of their ashes is very problematic. For energy production, the amount of waste generated increased from 18,8 million tonnes in 2000 to 26, 4 million tonnes in 2004, despite the reduction of its share in waste production from 26 % to 24 %; the share of municipal waste and manufacturing industry is 40 % and 36 % respectively (Turkish Statistical Institute, 2006). Generally, in Turkey, both mine tailings and coal ashes are stored within the establishment under uncontrolled conditions.

4.2 Contaminated Sites in Turkey

The problem of polluted sites started to emerge especially in heavily industrialized regions of Turkey. However, an inventory of contaminated sites is not maintained (State Planning Organization, 2006b). Currently, identification of any contaminated site is not based on a certain systematic approach. Sites are mostly identified after some potential problems become obvious and public, as a result of the efforts of local authorities or concerned citizens (NATO, 2002). The number of contaminated sites is expected to be in the range of 1000-1500, of which 5-10 % are believed to be sites requiring remediation (Ünlü, 2006).

Soil pollution incidents, which became public, are mostly due to illegal disposal of industrial wastes, oil leakages resulting from accidental spill at oil storage tanks or pipelines, metal leaching from disposed metal ore processing residues and waste disposal sites. Remedial measures were carried out for very few of the contaminated sites. Therefore, some information about the contaminated sites and the remediation techniques used exists officially. However, there is no statistical data about formerly used remedial technologies and methods. Several publicly known soil pollution incidents are given below.

Tuzla Orhanli, Istanbul

The soil pollution incident in Tuzla Orhanli, one of the most publicly known incidents relies on the discovery of 640 toxic barrels in Tuzla Orhanli, a town near Istanbul, in March 2006. Samples taken from the site were analyzed by the Scientific and Technological Research Council of Turkey (TÜBITAK). The barrels, which also included hazardous phenolic compounds, were illegally buried two or three years ago by one or more chemical companies. Some barrels, which were found open on purpose, caused to the contamination of soil. Almost within a month, all 640 barrels and 2000 sacks full with contaminated soil were removed from site very carefully by a team from Izmit Waste Treatment, Incineration and Recyling Co. Inc. (IZAYDAS) and transported with special container trucks to the IZAYDAS hazardous waste treatment and disposal plant. IZAYDAS facilities are the only licensed hazardous waste disposal facility in Turkey. Remediation of contaminated site was mainly based on the incineration of both toxic barrels and highly contaminated soil.

Fig. 1. A view of the investigation at the contaminated site in Tuzla Orhanli (Zanbak, 2006)

Cayırova Gebze Incident

Cayirova Gebze incident was a subject of the media in the same period of the Tuzla incident. Five barrels, full with asbestos, were found together with some other wastes near a village close to Cayirova Gebze. The Environmental Protection Department of the Kocaeli Greater Municipality took the necessary precautions and the team from IZAYDAS, which were informed, transported the barrels for inspection.

BOTAS Pipeline Incident Near Ataturk Dam

The leakage of crude oil from the Batman-Yumutalik pipeline of the Petroleum Pipeline Corporation (BOTAS) near the Ataturk Dam is among the important accidental soil pollution incidents. In April 2005, about 20-25 thousand barrels of crude oil leaked out and caused to pollution in the bay of Yiginak village and shore of Baglica village and surrounding soil, near Sanlıurfa. While the pipeline was repaired, the dispersion of oil in the lake was prevented and afterward collected with the help of barriers. About 500 tons of contaminated soil were removed and transported to a site of 20 acres. The less contaminated soil was cleaned-up with a bioremediation method at the technically arranged site. The highly polluted soil, on the other hand, was transported together with other oily wastes to IZAYDAS incineration plant.

1999 Kocaeli Earthquake Incident

In August 1999, the earthquake, with a magnitude of 7,4 MW, struck the Kocaeli and Sakarya provinces in north western Turkey. The affected region is one of most economically dynamic regions. Industries, which were damaged, were the petrochemical industry, automotive industry and other industrial facilities like paper mills, steel mills, cement, textile and pharmaceutical factories, etc. Among the state-owned petrochemical complexes the heaviest damage occurred at the TUPRAS refinery itself and associated tank farm with crude oil and product jetties. Six tanks of varying sizes in the tank farm of 112 tanks were damaged due to ground shaking and fire. Considering the IGSAS fertilizer factory, ammonia processing and packing units were partially and the administration building extensively damaged. The PETKIM petrochemical facility had limited damage. Besides structural damages at many facilities and factories, other important ones were the silo collapses at the SEKA paper mill, storage rack collapse, toxic releases from mixing chemicals and damaged piping at the Toprak pharmaceutical firm, damaged tanks at the AKSA chemical installation in Yallova, which was associated with leakage of chemicals. Under such catastrophic conditions the contamination of soil and water resources as a result of leakage from storage tanks, pipelines etc is mostly inevitable (Erdik, 2007).

Fig. 2. A view of the fire in TUPRAS petrochemical complex (Erdik, 2007)

Beykan Oil Field Site

Another petroleum hydrocarbon pollution of soil and water resources occurred at the Beykan Oil Field site, which is enclosed by the watershed of a medium size dam. The reason of pollution was the oil spills at 38 oil producing wells, which are located within the protection zone surrounding the dam's reservoir. Evaluation of available spill data between 1989 and 1995 revealed that a total of 252 recorded spills resulted in a net spill of 395 tons. Major types of oil spills were identified as well heads, return lines and power oil lines. Analyses of 211 soil samples for total petroleum hydrocarbons revealed a concentration range between 600 and 115 500 mg/kg with a mean concentration of 20 300 mg/kg (Ünlü and Demirekler, 2000). In addition to soil and possible reservoir water pollution problems, another primary concern at this site is pollution of the Midyat aquifer due to injection of nearly 20 million m^3 of formation water between the years of 1971 and 1996. Formation water contains high amounts of brine (containing 3 000 mg Cl^-/L and 6 500 mg TDS /L) and some emulsified oil (500 mg/L) (NATO, 1998).

Incirlik PCB Contaminated Soils Site

The contamination at Incirlik was due to PCB oil leaking from storage drums at a military reutilization yard during its operation between 1970 and 1988. In 1991 the contaminated soil was excavated and stored in approximately 300 drums and in a pile. Estimated PCB-contaminated soil volume is 1 600 m^3. PCB concentrations measured in composite soil samples range up to 750 ppm (NATO, 1998). The NATO report indicates that incineration and solidification/stabilization were evaluated for the remediation of contaminated soil; however which one was used is not known.

Chromium Ore Processing Residue Dump Site

Another soil pollution incident reported by NATO (1998) is the chromium ore processing residue dump site. Chromium ore processing residues, which were produced by a chromate production facility, were dumped at a temporary dump site near the factory. Cr (VI) leaching from the residues, containing nearly 25 000 ppm of total chromium, caused to the contamination of soil and groundwater.

Toxic Barrels of Samsun and Sinop

The discovery of 392 barrels at the Black Sea coastline in 1988 is one of the first soil pollution incidents in Turkey. Investigations revealed that the barrels belonged to the Italians. These barrels were thrown into the Black Sea by an Italian Ship, which was carrying about three thousand toxic barrels. About 240 of barrels have been in storage in Alacam near Samsun and the remaining in Soguksu Sinop. Despite the fact that the barrels belonged to Italy, the barrels could not be send back to Italy for 18 years. Finally, it was planned that the toxic barrels are sent to disposal facilities in Germany with a support of the Turkish Cement Manufacturers' Association, which includes Italian companies as well. IZAYDAS, which took the barrels from Samsun and Sinop, transported the barrels to Izmir for further shipment to Europe for their disposal. Unfortunately, about 150 barrels were found empty thus indicating to a contamination of both soil and water resources.

Other Suspected Contaminated Sites

There were some complains about barrels, which were temporary stored by Turkish Petroleum Refineries Corporation (TUPRAS) at a site in Batman for years. These barrels, over hundred pieces, contain chemicals from the petroleum refinery. Competent of TUPRAS stated that the barrels were going to be removed by IZAYDAS. There are many other storage sites, which are accepted as temporary sites, but rather work permanently. According to a hazardous waste management report written by Zanbak and Bayazit Tugal (1997), there is a need for the registration, investigation and rehabilitation of these temporary hazardous wastes disposal sites.

Municipal solid waste dumping sites, especially those used by metropolitan or greater municipalities, cause to significant pollution of soil and groundwater, which is well known. In Turkey, most of the time, commercial and industrial wastes are disposed off together with domestic solid wastes, thus increasing the contamination potential of leachate. Most of the municipal dumping sites do not have a drainage system for leachate collection and a clay layer to prevent leaching into soil and groundwater. When the use of dumping sites is completed, their rehabilitation is carried out. Since measures for the collection and treatment of leachate are generally not taken the contamination of soil and groundwater continues. There are over 3200 municipalities in Turkey, but the number of sanitary landfills and landfills with EIA approval is not over fifty. Under these circumstances it is not difficult to estimate the number of dumping sites suspected as contaminated sites. Some closed or still operating municipal solid waste dumping sites, well known for conditions are Yakacik (with a capacity of 600 000 m^3), Ümraniye (with a capacity of 2 million m^3) and Halkalı (with a capacity of 10 million m^3) dumping sites of Istanbul (closed), Mamak dumping site of Ankara (operating), Cigli, Uzundere, Buca, Isikkent, Güzelbahçe, and Gaziemir dumping sites of Izmir (closed).

5. Conclusion

Since 1990s Turkey has made significant environmental progress with reforms in institutional and legislative frameworks. The Ministry of Environment (now the Ministry of Environment and Forestry) was created and various environmental legislations were adopted. As a part of national development plan, the National Environmental Action Plan was developed. Soil protection and pollution issues were included in this plan and the need for the registration of environmental pollution was emphasized.

Several regulations for the control and prevention of soil pollution were enforced, however there are problems related to their implementation. Despite the enforcement of both Hazardous Waste Control Regulation and Solid Waste Control Regulation, there is still a need for improvements in the waste management system. For the monitoring of waste generation activities and pollution potentials of industries, a complete and correct filling of annual waste declaration form is very important. In addition, the information of the facility activity form required by the Environmental Auditing Regulation will support the development of a waste inventory and the identification of the needs for necessary waste treatment and disposal facilities. The data provided from the implementation of both regulations will also facilitate the identification and registration of contaminated sites.

The Soil Pollution Control Regulation is relatively new and requires a major revision for its practicality on issues like identification, investigation, monitoring and clean-up of contaminated sites. The lists for limiting values of inorganic and organic contaminants need to be extended and indicative values, soil remediation intervention values and target values for all contaminants to be included. Programmes providing a systematic approach for the identification, monitoring and remediation of contaminated sites are needed to be developed.

The Ministry of Environment and Forestry, which is aware on these deficiencies of regulation and current practices of contaminated sites, carries out a project on "the Development of a Management System for Point Source Contaminated Sites," together with the Environmental Engineering Department of Middle East Technical University. The purpose of this project is to determine the necessary points of the Soil Pollution Control Regulation to be amended and publish a conceptually and technically revised Soil Pollution Control Regulation, to adopt specific soil quality standards based on land use, to develop a contaminated sites identification and registration system, a contaminated site evaluation system and a computer software for the contaminated sites information system. It also aims to develop technical guidelines for the investigation, risk evaluation, clean-up and monitoring of contaminated sites. With the accomplishment of the project a contaminated site management system will have been developed, which is of upper most need.

Especially in heavily industrialized regions of Turkey problems with respect to contaminated sites has increased. Main sources of pollution are industrial facilities, disposal of municipal and industrial wastes, mining and energy production activities. Currently, an inventory for contaminated sites does not exist. However, soil pollution incidents, which became public, are mainly illegal dumps and temporary industrial waste disposal sites.

Detailed information on the size of contaminated site, the concentration and characteristics of pollutants and impacts of pollutants on the groundwater and other receptors is mostly not available. Similarly, knowledge about the measures taken and the remediation methods/techniques used is limited to recent contaminated sites. In general, the contaminated soil is removed from the area and either incinerated or controlled landfilled. Soil contaminated by hazardous substances is mostly incinerated at the IZAYDAS incineration plant. For the remediation of petroleum contaminated soil, bioremediation is also considered besides incineration. Municipal disposal sites are mostly rehabilitated after closure. However, any remedial action for the control of underground contaminant transport is generally not carried out.

The importance of soil pollution prevention and remediation of contaminated sites will especially increase in future with the recognition of the groundwater pollution problem. There is an accelerated degradation of groundwater quality and a raised withdrawal in recent years. The use of groundwater resources for domestic water supply purposes, on the other hand, will make the situation more serious. Since contaminants are wider dispersed in groundwater rather than soil, increasing the costs of remediation, the protection of groundwater resources and prevention of pollution will become environmental issues of high priority.

6. References

Commission of the European Communities (2006). *Turkey 2006 Progress Report*, Commission Staff Working Document, pp. 65-68, SEC(2006) 1390, Brussels

Erdik, M. (2007). *Report on 1999 Kocaeli and Düzce (Turkey) Earthquakes*, pp. 30-33, Bogazici University, Department. of Earthquake Engineering, Cengelköy, Istanbul, Turkey, http://www.koeri.boun.edu.tr/depremmuh/eqspecials/kocaeli/Kocaelireport.pdf (last accessed in August 2007)

Keskin, S. (2001). *Gateway to Land and Water Information: Turkey National Report*, pp. 7-8, Head of Data Collection and Analysis, Department of Soil & Water Resources National Information Centre, General Directorate of Rural Services, Turkey, http://www.fao.org/ag/agl/swlwpnr/reports/y_nr/z_tr/tr.htm (last accessed in July 2007)

NATO, North Atlantic Treaty Organization, Committee on the Challenges of Modern Society (2002). *Evaluation of Demonstrated and Emerging Technologies for the Treatment of Contaminated Land and Groundwater (Phase III)*, NATO/CCMS Pilot Study, 2001 Annual Report, Number 250, pp. 251-253, EPA 542-R-02-001.

NATO, North Atlantic Treaty Organization, Committee on the Challenges of Modern Society (1998). *Evaluation of Demonstrated and Emerging Technologies for the Treatment of Contaminated Land and Groundwater (Phase III)*, NATO/CCMS Pilot Study, 1998 Annual Report, Number 228, , pp. 108-113, EPA 542-R-98-002.

OECD Working Party on Environmental Performance (2000), *Environmental Performance Reviews (1st Cycle)*, Conclusions and Recommendations 32 Countries (1993-2000), pp. 245-252, WPEP

Okumuş, K.; (Ed.) (2002). *Turkey's Environment, A Review and Evaluation of Turkey's Environment and its Stakeholders,* The Regional Environmental Center for Central and Eastern Europe, pp. 10-25, ISBN 9639424099, Szentendre, Hungary

Ongan, S. E. (ED) (1997). *Land Use and Management of Coastal Areas*, NEAP, State Planning Organization, ISBN 9751916984, Ankara, Turkey, htpp://ekutup.dpt.gov.tr/cevre/eylempla/arazikul.pdf (last accessed in July 2007) (in turkish)

State Planning Organization (2006a). *Use and Management of Soil and Water Resources*, Ninth National Development Plan, pp. 87-99, DPT, Ankara, Turkey, htpp://plan9.dpt.gov.tr/oik25_topraksu/Rapor.doc#_Toc132780616 (last accessed in July 2007) (in turkish)

State Planning Organization (2006b). *Environment*, Ninth National Development Plan, pp. 20-25, DPT, Ankara, Turkey, htpp://plan9.dpt.gov.tr/oik22%5Fcevre/cevre.pdf (last accessed in July 2007) (in turkish)

State Planning Organization (2001). *Use and Management of Watershed*, Eighth National Development Plan, DPT: 2555, pp. 97-114, ISBN 9751926386, Ankara, Turkey, htpp://ekutup.dpt.gov.tr/suhavza/oik571.pdf (last accessed in July 2007) (in turkish)

State Planning Organization (1998). *National Environmental Action Plan*, DPT.YBM, ISBN 975-19-1995-X, Ankara Turkey, htpp://ekutup.dpt.gov.tr/cevre/eylempla/neap.html (last accessed in July 2007)

Turkish Standard Institute (2006). *Environmental Statistics Compendium of Turkey II*, Environmental Statistics Group, TSI: 3011, pp. 70-83, ISBN 9751938651, Prime Ministry, Ankara, Turkey.

Ünlü, K. (2006). Tour de Table: The Situation of Contaminated Sites in Turkey , *Country Presentation at the NATO/CCMS Pilot Study Workshop Prevention and Remediation in Selected Industrial Sectors: Small Sites in Urban Areas*, 4-7 June 2006, Athens, Greece, organised by the North Atlantic Treaty Organization Committee on the Challenges of Modern Society, pp. 4-5, www.clu-in.org/athens/download/Tour_de_Table/Turkey_Tour_de_Table.pdf

Ünlü, K. & Demirekler, E. (2000). Modeling Water Quality Impacts of Petroleum Contaminated Soils in a Reservoir Catchment, *Water, Air and Soil Pollution*, Vol. 120, No. 1-2, May 2000, 169-193, ISSN 0049-6979

Zanbak, C. & Bayazit Tugal I. (1997). *Hazardous Waste Management*, NEAP, State Planning Organization, pp. 21-38, ISBN 9751916763, Ankara, Turkey, htpp://ekutup.dpt.gov.tr/cevre/eylempla/zanbakc.pdf (last accessed in July 2007) (in turkish)

Zanbak, C. (2006). Industrial Waste Management and Turkey- Problems and Solutions (1996-2006), *Hazardous Wastes: Human and Environmental Health Mini Symposium*, Turkish Society of Toxicology, 30 May 2006, Ankara, Turkey. (in turkish)

ANNEX 1

Natural Resource Management (State Planning Organization, 1998)

Areas of Action	Options
Policies	1. Diffusion and planning of rural development policies 2. Agricultural support policies integrated with efforts to preserve land resources 3. Preparation of action plans for the prevention of soil erosion, desertification and pollution, especially in the GAP region; 4. Using soils according to their suitability classification
Institutional Reform	5. Reorganization of the central and peripheral units of the Ministry of Agriculture and Rural Services and the Ministry of Forestry 6. Making public research agencies and routine analysis laboratories more efficient and effective 7. Ensuring that the service delivery by the General Directorate of Rural Services is carried out by a separate organization than those in charge of research, data collection and training 8. Formation of basin management units and executive boards 9. Rearrangement of the rights, duties and responsibilities of the Union of Chambers of Agriculture 10. Redefinition of the rights, authority and responsibilities of village administrations; strengthening of village unions; democratization of administration 11. Preparation of soil conservation and land improvement action plans at the regional level 12. Establishment of a unit in charge of research, control of techniques and equipment as well as calibration of equipment which are used for agricultural combat.
Legislative Arrangements	13. Amendment of Articles 44 and 45 of the Constitution; 14. Enactment of Soil Conservation and Land Improvement Laws; 15. Enactment of Regulation on the Control of Soil Pollution; 16. Revision of Acts nos. 4070, 4071, and 4072 17. Revision of the Act on Cooperatives; 18. Updating of the Village Act; 19. Revision of the articles of the Land Act and Civil Code that permit the fragmentation and non-agricultural use of agricultural lands; 20. Revising the Act on Mining to ensure that mining enterprises prepare and implement land rehabilitation programs after their extraction activities; 21. Revising the institutional laws of the DSI, GDA, and GDRS to permit integrated work at the catchment basin level

Table 2. Options for the Protection and Management of Land/Soil Resources.

Areas of Action	Options
Policies	22. Diffusion and planning of rural development policies 23. Agricultural support policies integrated with efforts to preserve land resources 24. Preparation of action plans for the prevention of soil erosion, desertification and pollution, especially in the GAP region; 25. Using soils according to their suitability classification
Institutional Reform	26. Reorganization of the central and peripheral units of the Ministry of Agriculture and Rural Services and the Ministry of Forestry 27. Making public research agencies and routine analysis laboratories more efficient and effective 28. Ensuring that the service delivery by the General Directorate of Rural Services is carried out by a separate organization than those in charge of research, data collection and training 29. Formation of basin management units and executive boards 30. Rearrangement of the rights, duties and responsibilities of the Union of Chambers of Agriculture 31. Redefinition of the rights, authority and responsibilities of village administrations; strengthening of village unions; democratization of administration 32. Preparation of soil conservation and land improvement action plans at the regional level 33. Establishment of a unit in charge of research, control of techniques and equipment as well as calibration of equipment which are used for agricultural combat.
Legislative Arrangements	34. Amendment of Articles 44 and 45 of the Constitution; 35. Enactment of Soil Conservation and Land Improvement Laws; 36. Enactment of Regulation on the Control of Soil Pollution; 37. Revision of Acts nos. 4070, 4071, and 4072 38. Revision of the Act on Cooperatives; 39. Updating of the Village Act; 40. Revision of the articles of the Land Act and Civil Code that permit the fragmentation and non-agricultural use of agricultural lands; 41. Revising the Act on Mining to ensure that mining enterprises prepare and implement land rehabilitation programs after their extraction activities; 42. Revising the institutional laws of the DSI, GDA, and GDRS to permit integrated work at the catchment basin level
Economic & Financial Measures	43. Ensuring farmers' contribution to land improvement investments Seeking the contribution of mining industries for land rehabilitation 44. Charging land improvement fees to polluters

Table 2. Options for the Protection and Management of Land/Soil Resources. (continued)

Areas of Action	Options
Economic & Financial Measures	45. Ensuring that monetary incentives introduced by Act no. 4122 are first channeled to soil conservation and land improvement works in water catchments 46. Reconsideration of incentives for pesticide use
Education-Training	47. Development and wider use of soil conservation related in-service training programs in related public organizations and agencies 48. Giving soil conservancy and land improvement training to village group technicians 49. Support to professional and voluntary organizations in their activities for soil conservation and land improvement 50. Diffusion of extension activities in line with the cultural and social conditions of the region concerned; conduct of joint training programs with farmers; implementation of advanced techniques 51. Attachment of extension units to research institutes; coordinated work performance of research-extension-agricultural combat units
Participation	52. Institutionalization of cooperation between the related ministries and professional and voluntary organizations 53. Participation of the farmers of a region in the planning and implementation of integrated eco-basin management plans by exercising specific rights and undertaking duties and responsibilities 54. Strengthening of the committee stipulated by the Regulation on Pesticide Licenses with specialists from different fields
Techniques	55. Establishment and use of remote sensing/geographical information systems; 56. Diffusion of land compacting works; 57. Expansion of the scope of EIA to cover the assessment of the infrastructure investments of public organizations; 58. Monitoring and supervision of land and agricultural input use 59. Expansion of the scope of EIA to cover the issue of non-agricultural use of agricultural lands 60. Limitation of area under tobacco cultivation; 61. Control of the use of phosphorous fertilizers; 62. Preparation and implementation of programs designed for the encouragement of integrated pest management; establishment of early warning schemes; 63. Inspection of agricultural pesticide producers and dealers; 64. Expansion of fodder crop culture;

Table 2. Options for the Protection and Management of Land/Soil Resources. (continued)

Areas of Action	Options
Techniques	65. Control of the importation and use of hormones and carcinogenic substances; 66. Abandonment of the practice of follow with the exception of localities where this practice is scientifically proven as necessary 67. Elimination of differing practices in the field of chemical residue limits
R & D	68. Updating land capacity classifications; 69. Identification of tilling techniques and the methods of agricultural inputs (water, fertilizers and pesticides) with respect to land characteristics and local levels not to harm soils and development of environmentally friendly technologies; 70. Supervision of the producers of agricultural pesticides and equipment 71. Identification of fertilizer compositions suitable to local conditions and environmental impacts of fertilizers specific to locations 72. Development of techniques in order to use drainage water for irrigation purposes.

Table 2. Options for the Protection and Management of Land/Soil Resources. (continued)

ANNEX 2

The Urban Environment (State Planning Organization, 1998)

Areas of Action	Options
Policies	1. Emphasis on minimizing, recycling and recovering wastes; 2. Rehabilitation of old waste dumps; 3. Emphasis on sanitary landfills & hazardous waste treatment.
Institutional Reform	4. Establishment of "Waste Exchange" at national and regional levels; 5. Formation of "Hazardous Waste Management" units at regional level; 6. Establishment of "Emergency Management Centers"; 7. Establishment of "Waste Management Units" with special budgets within the boundaries of greater municipalities.
Legislative Arrangements	8. Enactment of new regulations on industrial accidents, cases requiring urgent intervention, and hazardous waste transportation; 9. Rearrangement of authority and responsibility sharing among relevant organizations in a coordinated manner, and introduction of a functional division of labor; 10. Development of guidelines for the storage of hazardous wastes; 11. Elimination of inconsistent terms used in relevant legislation; 12. Legal arrangements for introducing eco-labeling; 13. Adherence to international agreements on hazardous waste management; 14. Amend Municipal Revenues Law (No. 2464) to increase cost-sharing for infrastructure investment; 15. Revise legislation to allow for more private sector participation in solid and hazardous waste management; 16. Amend Regulation on Solid Waste Control to increase flexibility in lining standards at sanitary landfills where there is no risk of leaching; 17. Allow municipalities to enforce Hazardous Waste Control Regulation.
Economic & Financial Measures	18. Land allocation and credit for regional resource recovery and waste disposal facilities; 19. Replace Environment Cleansing Tax with inflation-adjusted charges based on the quantity and type of waste generated; 20. Incentives for waste reduction at source; 21. Applying hazardous waste disposal charges to original manufacturers as well as users.

Table 3. Options for Solid and Hazardous Waste Management.

Areas of Action	Options
Education -Training	22. Implementation of training programs on hazardous waste management; 23. Training of hospital personnel at all levels; 24. Certification for the information and skill levels of personnel employed in hazardous and medical waste management; 25. Institutionalizing the information flow to the public; 26. Implementation of training programs to encourage the wider use of "eco-packaging".
Participation	27. Support to voluntary organizations in their monitoring and project based activities; 28. Conduct of regional referendums on the collection, dumping and elimination of hazardous wastes; 29. Encouragement of persons and organizations who reduce waste at the source.
Techniques	30. Establishment of waste processing and disposal facilities equipped with appropriate technologies; 31. Support for waste producers to introduce waste minimization (including packaging), recycling and disposal systems; 32. Enlarging the scope of waste inventories and ensuring their sustenance; 33. Supporting the establishment of integrated (reuse+elimination) waste processing facilities; 34. Preparation of programs for the improvement of polluted areas; 35. Updating of topographic maps and aerial photographs; 36. Developing standards for recycled products; 37. Inclusion of hazardous waste management in urban development plans; 38. Wider observance in public and private industries of the principles of ISO 14000, Ecotex, and Triple Responsibility; 39. Encouragement of "environment friendly" production and consumption; 40. Preparation of "Waste Management Plans" by producers in sensitive receiving environments; 41. Support infrastructure for the collection, transport and elimination of medical waste; 42. Preparation of standards for collection tanks, transport vehicles, transfer stations, dumping sites and elimination facilities 43. Development of standards for transportation and temporary storage of hazardous wastes.

Table 3. Options for Solid and Hazardous Waste Management (continued)

Areas of Action	Options
R & D	44. Further enrichment of waste processing options; 45. Establishment of a "Processing Development Center" for reducing wastes; 46. Wider support to the R & D work of industrial enterprises in the field of waste management; 47. Inventory of areas polluted by wastes; 48. Identification of waste dumping sites fit for urban settlements; 49. Development of technologies for the dumping, transport and elimination of medical wastes; 50. Identification of ecosystems affected by dumpsite leachate; 51. Preparation of a "Glossary of Waste Management Terms"; 52. Develop transparent criteria for sanitary landfill site selection.

Table 3. Options for Solid and Hazardous Waste Management. (continued)

9

Distribution of Trace and Major Elements in Lignite and Products of Its Combustion-Leaching Experiments and Cluster Analysis

Aleksandar Popovic , Dragana Djordjevic
University of Belgrade
Serbia

1. Introduction

One of the strategic issues for contemporary manhood is production of a sufficient amount of energy for further technological development. Despite numerous attempts to use new, practically inexhaustible energy sources, such as solar energy (Li, 2002), wind energy (Belanger & Gagnon, 2002) and high and low tides (Omer, 2002), for example, instead of conventional sources, i.e., coal, petrol or gas, the latter ones, nevertheless, remain unavoidable factors in balancing the energy demands of most countries, including Serbia, at the beginning of the 21st century.
On the basis of investigations of coal genesis, its composition, as well as general characteristics of coal deposits, coal can be defined as a combustible sedimentary rock, originating mainly (some coals are algal) from residues of terrestrial and aquatic plants, and of minerals (< 50 %) (Wood et al., 1983). Chemical and physical characteristics of coal are predetermined by the nature of precursor plants, the amount of inorganic material, and by the nature, intensity and duration of biochemical and geochemical processes that are responsible for coal formation. Almost all natural elements have been found in coal (Finkelman, 1993), in variety of forms (Eskenazy & Valceva, 2003; Yudovich & Kertis, 2005), which are responsible for the coal's technological, economical but also ecological impact. Elements can be variously associated (Font et al., 2005). Further, association types may vary within the same deposit (Kuehn & Kurzbach, 1992; Vyazova & Kryukova, 1997).
During coal combustion in power plants, practically only the organic part of coal burns, producing carbon dioxide and water, i.e. its vapor. The inorganic components of coal mainly do not burn, but remain in the ash, which is a by-product of combustion. Since coal, by definition, contains more than 50 % of organic substances, which are not found in the ash, or are found in the very low concentrations, it is clear that the main part of the inorganic components of coal, including trace elements, is concentrated in the ash. During combustion different processes occur, which influence the non-uniform distribution of the elements and phases in different fractions of the combustion products (Xu et al., 2004; Goodarzi, 2006; Ward & French, 2006). One of the basic mechanisms which are present is "evaporation-condensation" (Schultz et al., 1973). Namely, during coal combustion at high temperatures certain elements evaporate. At the lower temperatures, which are typical for the higher parts of the reactor in which the combustion occurs, part of the substances which

had evaporated condenses, and, as a rule, this occurs on the smaller particles which are present in the higher parts of the reactor. Therefore, the fractions which contain the smallest particles of ash are, in the number of cases, enriched with elements the compounds of which are the most volatile. Normally, parts of these most volatile compounds do not condense but are released into the atmosphere, together with the gases which are formed during the combustion (Pacyna & Pacyna 2001; Pavlish et al., 2003).

The chemical composition of ash obtained by coal combustion can be related to the conditions in the original paleo-environment, to geochemical processes occurring during and after the formation of the coal deposit, but also to the technical conditions during coal combustion. It is well documented that coal fly ash characteristics vary with different feed coals, utility systems, and operating parameters; however, coal fly ash characteristics can be expected to be consistent within a range from a single electric generation facility given a consistent feed coal. It often happens that after coal has been combusted in the power plant, ash with a completely different distribution of microelements has been obtained.

Since only a marginal portion of the ash is being used world-wide (Jones et al., 2006; Yunusa et al., 2006), most of it is usually stored in huge coal ash dumps, either in the dry or wet state (Sushil & Batra 2006). Storage of wet coal ash usually prevents the wind from dispersing it, but increases the time necessary for the leaching of various elements (Iturbe et al., 1996). Coal ash in the environment (on the dump; airborne particles deposited on the soil surface, etc.) is exposed to various influences, and solubilization processes occur, sometimes transforming almost insoluble elements to soluble species thus leading to pollution of water and soil (Gupta, 1999).

The quality of waste waters from landfills depends on the quality of the coal, the methods of combustion, the quality and quantity of the "input" water, the quality of the process which is used for the removal of fly ash from the gases resulting from coal combustion (Chu *et al.*, 1978), and on the manner of ash transportation. (Rokita, 1990). As can be seen, the qualities of waste waters are not only dependent on the chemical composition of the coal, which determines the chemical composition of the resulting ash. In order to predict pollution from ash landfills, computer programs have been written which simulate the release of potentially harmful substances under different conditions (Tucek & Konecny, 1975), as well as the content of surface water in the neighborhood of the landfill (Donslund & Eriksen, 1990). Different tracers, i.e. isotopes of strontium and lead (Hurst *et al.*, 1993), sodium bromide, i.e. bromide ion, and lithium were used for the determination of the dispersion of pollution. Distribution of trace and major elements, especially in more mobile and accessible fractions, in ash and slag during coal combustion, is one of the major factors influencing their ability to pollute surrounding waters (Marquenie and Simmers, 1988; Jankowski et al., 2006).

In order to examine and compare distribution of some trace and major elements in lignite and products of its combustion as well as influence of high temperature process on distribution of elements in "Nikola Tesla A" power plant, we performed sequential extraction of lignite as well as ash and slag obtained by its combustion in situ. Sequential extraction is a powerful tool for predicting the behavior of trace elements under various environmental conditions (changes of pH, redox-potential etc.). The use of extraction (Piekos & Paslawska, 1998; Mukherjee & Borthakur, 2004; Senior et al., 2004) and sequential extraction (Fernandez-Turiel et al., 1994; Goodarzi, 1994; Hlavay et al., 1995; Querol et al., 1996) for analysis of coals and coal combustion by-products is being widely used on samples from all over the world including samples of coal, fly and dump ash from Serbia (Popovic et al., 2000; Popovic et al., 2001; Polic et al., 2005; Popovic & Djordjevic, 2005).

2. Materials and Methods

Eight samples of milled lignite, ash and slag obtained by lignite combustion were subjected to a five-step sequential extraction procedure. The procedure itself comprised of the following sequence of extractants:
Phase I: Distilled water (weakest extractant - the amount of elements extracted in this step is the least that can be expected, regardless of the type of water used for ash transport to the dump);
Phase II: 1 M ammonium acetate (this extractant should dissolve the fractions of elements that are bound to the particle surface with stronger bonds than the fraction washed with distilled water);
Phase III: 0.2 M ammonium oxalate / 0.2 M oxalic acid (imitation of reductive dissolution of iron and manganese oxides, which are important microelement scavengers);
Phase IV: 30 % H_2O_2, 0.01 M HNO_3 on a water bath at 85^{0}C, in order to dissolve organic/sulfide matter, with subsequent desorption of eventually re-adsorbed elements, by application a 3.2 M ammonium acetate solution washing for 30 min.
Phase V: Digestion with 6 M HCl bath at 85^{0}C, introduced in order to obtain additional information on the crystalline iron oxides and silicates.
The elected sequence of extractants was, with some alterations, mostly used for examination of coal and combustion product samples (Gupta, 1999; Senior et al., 2004; Jankowski et al., 2006), and it is remarkably similar to the sequence of extractants used for sediments and soils.
After washing, the residuals were treated with the next extractant. The combined extracts and washings after each extraction step were concentrated and the trace elements were analyzed by a "SpectrAA-20+ Varian" atomic absorption spectrometer.
The available data sets were analyzed using the SPSS 10.0 statistical program (Descriptive Statistics and Classify Hierarchical Cluster).

3. Results

Average concentrations and standard deviations of concentrations of extracted major and trace elements from lignite, ash and slag obtained by lignite combustion in "Nikola Tesla A" power plant are shown in Tables 1-3.

4. Discussion

4.1 Lignite

The macro and microelements contained in coal are differently distributed (Tables 1-3, Figure 1). It should be mentioned that during applied sequential extraction process the solid phase is not totally destroyed. As opposed to the organic component dissolved in the fourth phase of extraction, the total mineral component is not dissolved, which means that all the macro elements and some of the microelements may have an additional mineral component, besides the one determined by the extraction. However, it is obvious from the results obtained by extraction that most elements are contained in the inorganic phase of coal, even inside the extractable part.
The investigated macro elements can be separated on the basis of their fraction in the organic phase into elements which are dominantly extracted from the organic phase (aluminium 58.56

± 11.31 % and silicon 58.36 ± 11.66 %) and into elements which are dominantly extracted from the inorganic component, even in an extracted fraction of coal (iron 60.76 % of the totally extracted elements is released in the third phase of the extraction). It is interesting to note that two macro elements, magnesium and potassium, can not be precisely categorized. Actually, a larger amount of potassium and approximately the same amount of magnesium were released in the fourth phase of the sequential extraction, compared with the sum of the amounts extracted in the third and the fifth phase, however, more than 50 % of the extracted amounts are released in the first two phases of the sequential extraction. Since it is not possible to determine the origin of adsorbed and ion-exchangeable fraction using the applied technique, i.e. it is not possible to say whether the adsorbed and ion-exchangeable potassium and magnesium are connected with the organic or with the inorganic fraction of coal, there is no precise conclusion about that whether these two elements in the extractable fraction of coal are distributed dominantly in (or with) the organic or in the mineral phase.

Mg	Phase I		Phase II		Phase III		Phase IV		Phase V		Total
	ppm	%	ppm	%	ppm	%	ppm	%	ppm	%	ppm
Average	2.35	0.08	373.9	12.23	32.91	1.08	314.6	10.29	2333	76.32	3057
St.deviation	0.50	0.02	91.9	3.01	7.50	0.25	50.4	1.65	1082	35.41	1233
Average	163.6	19.27	321.0	37.80	43.68	5.14	184.0	21.67	136.9	16.12	849.2
St.deviation	43.7	5.14	3.4	0.40	2.52	0.30	92.7	10.91	58.1	6.84	200.4
Average	125.0	11.17	285.2	25.49	38.21	3.42	119.0	10.64	551.4	49.29	1119
St.deviation	4.7	0.42	4.2	0.37	21.33	1.91	22.4	2.00	12.9	1.15	66
Al	Phase I		Phase II		Phase III		Phase IV		Phase V		Total
	ppm	%	ppm	%	ppm	%	ppm	%	ppm	%	ppm
Average	103.9	1.30	205.2	2.56	2699	33.71	449.2	5.61	4548	56.81	8005
St.deviation	69.7	0.87	130.7	1.63	757	9.46	209.8	2.62	674	8.42	1842
Average	49.49	2.99	0.05	0.00	431.9	26.09	969.4	58.56	204.5	12.35	1655
St.deviation	35.17	2.12	0.10	0.01	18.7	1.13	187.2	11.31	76.5	4.62	318
Average	3.65	0.15	0.85	0.04	1154	48.20	805.1	33.63	430.6	17.99	2394
St.deviation	3.46	0.14	0.49	0.02	1632	68.16	53.6	2.24	17.8	0.74	1707
Si	Phase I		Phase II		Phase III		Phase IV		Phase V		Total
	ppm	%	ppm	%	ppm	%	ppm	%	ppm	%	ppm
Average	66.3	4.71	399.3	28.38	151.2	10.75	369.6	26.27	420.4	29.88	1407
St.deviation	31.7	2.25	62.5	4.44	35.0	2.49	64.48	4.58	19.5	1.39	213
Average	254.7	5.92	124.2	2.89	229.6	5.33	2512	58.36	1184	27.50	4304
St.deviation	67.8	1.58	20.4	0.47	31.3	0.73	502	11.66	469	10.90	1091
Average	114.3	2.51	174.2	3.83	104.6	2.30	1124	24.71	3032	66.65	4548
St.deviation	5.9	0.13	10.2	0.23	18.9	0.42	45	0.99	268	5.89	348
K	Phase I		Phase II		Phase III		Phase IV		Phase V		Total
	ppm	%	ppm	%	ppm	%	ppm	%	ppm	%	ppm
Average	37.49	2.12	20.24	1.14	560.7	31.70	130.2	7.36	1020	57.68	1769
St.deviation	5.94	0.34	3.47	0.20	56.68	3.20	27.5	1.55	456	25.78	550
Average	33.13	20.01	59.30	35.81	16.32	9.85	45.27	27.34	11.57	6.99	165.6
St.deviation	3.72	2.25	30.81	18.61	2.55	1.54	23.10	13.95	2.87	1.73	63.0
Average	27.54	6.54	91.58	21.76	178.2	42.33	0.00	0.00	123.6	29.37	420.9
St.deviation	0.18	0.04	6.92	1.64	235.5	55.95	0.00	0.00	16.83	4.00	259.4
Fe	Phase I		Phase II		Phase III		Phase IV		Phase V		Total
	ppm	%	ppm	%	ppm	%	ppm	%	ppm	%	ppm
Average	0.33	0.01	177.8	2.90	2832	46.15	272.9	4.45	2853	46.50	6135

St.deviation	0.10	0.00	47.7	0.78	54	0.87	150.4	2.45	1166	19.01	1418
Average	8.81	0.30	0.08	0.00	1761	60.76	746.5	25.76	382.0	13.18	2898
St.deviation	4.47	0.15	0.10	0.00	17	0.57	284.4	9.81	154.0	5.31	460
Average	0.49	0.03	0.00	0.00	942.0	49.70	325.5	17.17	627.5	33.11	1895
St.deviation	0.57	0.03	0.00	0.00	1331	70.21	9.9	0.52	116.4	6.14	1458
Cr	Phase I		Phase II		Phase III		Phase IV		Phase V		Total
	ppm	%	ppm	%	ppm	%	ppm	%	ppm	%	ppm
Average	0.97	1.98	2.85	5.81	5.21	10.62	2.11	4.30	37.91	77.29	49.05
St.deviation	0.35	0.71	1.01	2.05	3.83	7.81	1.26	2.56	23.87	48.66	30.31
Average	0.11	0.92	0.37	3.20	1.01	8.86	7.84	68.61	2.10	18.41	11.42
St.deviation	0.12	1.02	0.68	5.98	0.09	0.75	1.51	13.23	1.52	13.32	3.92
Average	0.64	8.34	0.04	0.52	1.20	15.65	3.58	46.61	2.22	28.88	7.67
St.deviation	0.89	11.62	0.01	0.18	1.64	21.39	1.04	13.55	0.15	1.94	3.73

Table 1. Average concentrations and standard deviations of extracted magnesium, aluminium, silicon, potassium, iron and chromium (ppm) from samples of lignite (first two rows), ash (second two rows), and slag (third two rows) obtained by lignite combustion in "Nikola Tesla A" power plant

Mn	Phase I		Phase II		Phase III		Phase IV		Phase V		Total
	ppm	%	ppm	%	ppm	%	ppm	%	ppm	%	ppm
Average	3.16	1.31	10.55	4.39	8.32	3.46	24.77	10.30	193.7	80.54	240.5
St.deviation	0.76	0.32	3.47	1.44	4.75	1.98	10.41	4.33	80.7	33.56	100.1
Average	1.54	0.86	11.12	6.20	72.00	40.11	14.87	8.28	80.0	44.55	179.5
St.deviation	0.34	0.19	3.40	1.89	25.08	13.97	8.70	4.85	28.0	15.62	65.6
Average	0.07	0.04	4.25	2.65	53.58	33.37	11.89	7.40	90.78	56.54	160.6
St.deviation	0.01	0.01	2.05	1.28	67.35	41.95	0.62	0.38	10.34	6.44	80.4
Ni	Phase I		Phase II		Phase III		Phase IV		Phase V		Total
	ppm	%	ppm	%	ppm	%	ppm	%	ppm	%	ppm
Average	0.13	0.19	1.54	2.23	11.91	17.28	9.18	13.33	46.15	66.98	68.90
St.deviation	0.03	0.04	0.17	0.25	1.16	1.68	1.93	2.80	5.22	7.57	8.51
Average	0.14	0.49	0.06	0.23	15.57	56.80	8.44	30.80	3.20	11.67	27.41
St.deviation	0.03	0.10	0.02	0.05	1.07	3.92	2.79	10.19	1.52	5.56	5.43
Average	0.02	0.07	0.05	0.23	7.84	36.23	5.30	24.51	8.43	38.96	21.63
St.deviation	0.01	0.03	0.00	0.00	11.04	51.04	0.17	0.78	0.97	4.48	12.18
Cu	Phase I		Phase II		Phase III		Phase IV		Phase V		Total
	ppm	%	ppm	%	ppm	%	ppm	%	ppm	%	ppm
Average	0.06	0.20	2.18	7.31	6.49	21.80	3.12	10.47	17.93	60.22	29.78
St.deviation	0.03	0.09	0.44	1.46	1.76	5.91	1.30	4.35	3.94	13.23	7.46
Average	0.08	0.58	0.02	0.15	8.57	65.97	1.44	11.09	2.89	22.21	12.99
St.deviation	0.01	0.10	0.01	0.06	0.97	7.49	0.36	2.80	1.44	11.10	2.80
Average	0.04	0.36	0.02	0.21	1.47	15.21	1.25	12.98	6.86	71.24	9.63
St.deviation	0.01	0.07	0.00	0.00	2.06	21.37	0.10	1.03	0.20	2.06	2.36
Zn	Phase I		Phase II		Phase III		Phase IV		Phase V		Total
	ppm	%	ppm	%	ppm	%	ppm	%	ppm	%	ppm
Average	0.13	0.49	1.44	5.48	0.81	3.09	3.10	11.79	20.81	79.15	26.29
St.deviation	0.08	0.29	0.10	0.38	0.09	0.35	1.63	6.21	4.42	16.83	6.33
Average	3.80	11.91	1.70	5.33	13.24	41.51	2.98	9.33	10.19	31.93	31.90
St.deviation	2.66	8.32	2.72	8.51	2.93	9.18	0.96	3.00	4.69	14.69	13.94
Average	0.42	2.19	0.23	1.21	4.27	22.52	12.38	65.28	1.67	8.81	18.97

St.deviation	0.19	1.01	0.13	0.67	5.87	30.95	1.26	6.64	0.78	4.10	8.22
As	Phase I		Phase II		Phase III		Phase IV		Phase V		Total
	ppm	%	ppm	%	ppm	%	ppm	%	ppm	%	ppm
Average	1.27	8.64	0.00	0.00	0.14	0.97	0.00	0.00	13.24	90.39	14.64
St.deviation	0.70	4.79	0.00	0.00	0.25	1.71	0.00	0.00	2.98	20.38	3.94
Average	1.58	16.98	0.00	0.03	0.89	9.51	5.10	54.85	1.73	18.62	9.30
St.deviation	0.89	9.56	0.01	0.05	0.59	6.40	1.98	21.25	1.27	13.64	4.74
Average	1.43	29.92	0.01	0.10	0.00	0.00	0.00	0.00	3.35	69.98	4.78
St.deviation	0.35	7.40	0.01	0.15	0.00	0.00	0.00	0.00	0.16	3.40	0.52
Cd	Phase I		Phase II		Phase III		Phase IV		Phase V		Total
	ppm	%	ppm	%	ppm	%	ppm	%	ppm	%	ppm
Average	0.03	20.20	0.03	19.19	0.00	0.00	0.00	0.00	0.10	60.61	0.17
St.deviation	0.01	3.13	0.01	7.09	0.00	0.00	0.00	0.00	0.08	46.95	0.09
Average	0.00	0.00	0.01	4.76	0.00	0.00	0.00	0.00	0.15	95.24	0.16
St.deviation	0.00	0.00	0.01	3.17	0.00	0.00	0.00	0.00	0.00	0.00	0.01
Average	0.00	0.00	0.01	100.0	0.00	0.00	0.00	0.00	0.00	0.00	0.01
St.deviation	0.00	0.00	0.01	141.4	0.00	0.00	0.00	0.00	0.00	0.00	0.01

Table 2. Average concentrations and standard deviations of extracted manganese, nickel, copper, zinc, arsenic and cadmium (ppm) from samples of lignite (first two rows), ash (second two rows), and slag (third two rows) obtained by lignite combustion in "Nikola Tesla A" power plant

Concerning the microelements, they can be classified into three groups. The elements having a predominant part of their extractable component associated with the organic fraction of coal are in the first group, i.e., chromium (68.61 ± 13.23 %) and arsenic (54.85 ± 21.25 %). The elements having a significant (although less than 50 %) amount of their extractable component connected with the organic phase are in the second group, the main representative being nickel (30.80 ± 10.19 %). The organic component of the other investigated elements, which are in the third group, are present in only a small amounts in the organic fraction, i.e., manganese (8.28 ± 4.85 %), copper (11.09 ± 2.80 %, zinc (10.19 ± 4.69 %), cadmium (0.00 ± 0.00 %) and lead (10.38 ± 3.10 %) of the total extracted amount. The fact that most of the elements are indisputably connected with the inorganic fraction of coal is not unexpected taking into consideration the large amount of ash remaining after the combustion of coal. The elements which are prevalently connected with the inorganic component, can be divided into groups based on the difference of their extractability in the third and fifth phase of the sequential extraction of the coal. The first group contains elements the extractable fraction of which is connected with the inorganic component and is concentrated in the third phase of the sequential extraction. The second group contains elements the extractable fractions of which, except the part which is organically associated, are mostly present in the fifth phase of the sequential extraction, which is the most difficult to dissolve. Nickel (56.80 ± 3.92 % extracted in the third and 11.67 ± 5.56 % extracted in the fifth phase of the sequential extraction), copper (65.97 ± 7.49 % extracted in the third and 22.21 ± 11.10 % extracted in the fifth phase of the sequential extraction) and iron (60.76 ± 0.57 % extracted in the third and 13.18 ± 5.31 % extracted in the fifth phase of the sequential extraction) belong to the first group. Cadmium (0.00 ± 0.00 % extracted in the third and 95.24 ± 0.00 % extracted in the fifth phase of the sequential extraction) and lead (1.26 ± 2.52 % extracted in the third, and 88.36 ± 44.81 % extracted in the fifth phase of the sequential extraction) belong to the second group, while zinc (41.51 ± 9.18 % extracted in the third and 31.93 ± 14.69 % extracted in the fifth phase of the sequential extraction) and manganese (40.11 ± 13.97 % extracted in the third and 31.93 ± 14.69 % extracted in the fifth phase of the sequential extraction) lie between these two groups.

Pb	Phase I		Phase II		Phase III		Phase IV		Phase V		Total
	ppm	%	ppm	%	ppm	%	ppm	%	ppm	%	ppm
Average	0.21	1.39	0.15	0.98	0.00	0.00	0.03	0.20	14.91	97.43	15.31
St.deviation	0.06	0.38	0.13	0.82	0.00	0.00	0.02	0.11	5.40	35.28	5.60
Average	0.00	0.00	0.00	0.00	0.10	1.26	0.83	10.38	7.03	88.36	7.95
St.deviation	0.00	0.00	0.00	0.00	0.20	2.52	0.25	3.10	3.56	44.81	4.01
Average	0.00	0.00	0.00	0.00	0.00	0.00	2.32	22.73	7.89	77.27	10.21
St.deviation	0.00	0.00	0.00	0.00	0.00	0.00	0.13	1.25	0.32	3.12	0.45

Table 3. Average concentrations and standard deviations of extracted lead (ppm) from samples of lignite (first two rows), ash (second two rows), and slag (third two rows) obtained by lignite combustion in "Nikola Tesla A" power plant

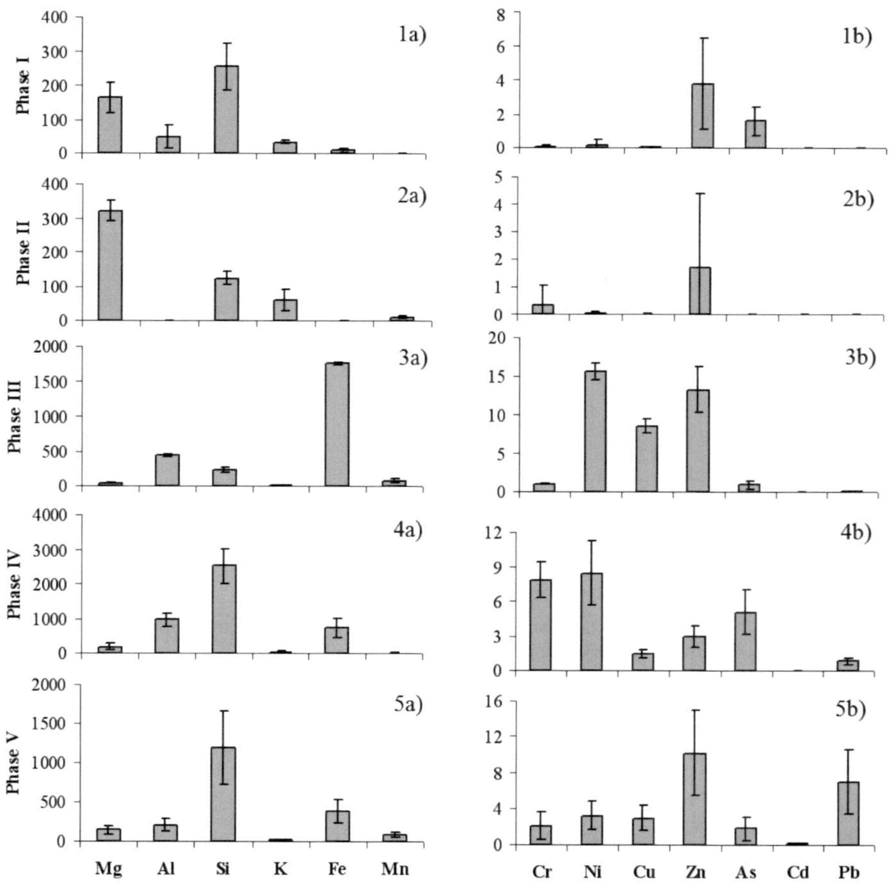

Fig. 1. Average concentrations and standard deviations of a) macro elements and b) microelements (ppm) extracted from samples of lignite

The question arises as to the origin of the adsorbed and ion-exchangeable fraction in the investigated samples of coal. These two fractions of different elements certainly existed in the original samples of coals but it is questionable whether the adsorbed and ion-exchangeable elements, which were present in the original coal, were also present in the coal used in the thermo-electric power plant, or whether these fractions present during the combustion of the coal were of a more recent origin. Taking into consideration that the coal had been exposed to rain after its mining, i.e., in stock piles and during transportation, it is probable that most of the elements which were present in the original "easily extractable" fractions of coal were rinsed out long before the combustion process, and that the adsorbed and ion-exchangeable elements, which were found in the experiment, were there as a consequence of sorption occurring during transportation, storage and grinding of the coal.

4.2 Filter Ash

Fly ash is the most important product obtained by coal combustion. High temperature during combustion in power plants enables melting, crystallizing as well as sintering process of alumosilicates and generation a new structures that are procuring their chemical property. Since only a small part of organic substances remain in the ash, it would not be realistic to expect fly ash to contain elements the dominant fractions of which are associated with organic component of coal. Thus, it is logical that the highest concentrations of all the investigated elements exist in the extraction phases in which the inorganic matrix of the ash particles dissolve. On the basis of their extraction characteristics, the elements present in the ash (Tables 1-3, Figure 2) can be divided into those which are to a significant degree still associated with the organic component of the ash (silicon - 26.27 ± 4.58 %), those which are to a small extent bonded to the organic component (magnesium - 10.29 ± 1.65 %; manganese - 10.30 ± 4.33; nickel - 13.33 ± 2.80 %; copper - 10.47 ± 4.35 %; zinc - 11.79 ± 6.21 %), those which are to minimal extent associated with the organic component of the ash (aluminum - 5.61 ± 2.62 %; potassium 7.36 ± 1.55 %; iron - 4.45 ± 2.45 % and chromium 4.30 ± 2.56 %) and those which, practically, do not exist in a form bonded to the organic component of the ash (arsenic, cadmium and lead).

When the amounts extracted in the third and fifth phase are compared, larger amounts of all the investigated either macro- or microelements were extracted in the last phase, and only for five elements (aluminum, silicon, potassium, iron and copper) is the ratio of these amounts less than 3:1 in favor of the concentration dissolved in the fifth phase of the sequential extraction. It should be noted that higher concentrations of silicon, manganese and zinc were extracted in the fourth than in the third phase of the sequential extraction. This was also true for magnesium but there is a possibility that the results of the extraction of this element in the third phase were influenced more by the applied extraction mean (extraction with oxalate will yield non-soluble magnesium oxalate) than by the nature of the substrate by itself.

4.3 Slag

The arrangement of the elements in the slag, shown in Tables 1-3 and Figure 3 is the most irregular and it is difficult to give any precise classification of the investigated elements, such as the ones made in the case of coal and fly ash.

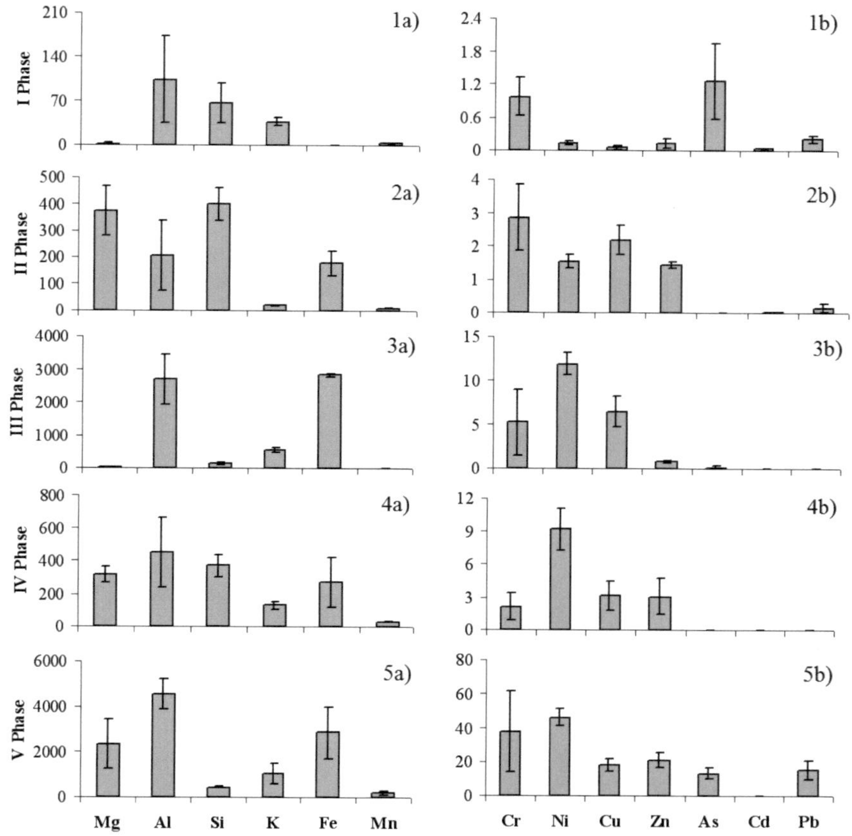

Fig. 2. Average concentrations and standard deviations of a) macro elements and b) microelements (ppm) extracted from samples of filter ash

The absolutely dominant bonding to the organic phase of slag (in relation to the totaly extracted, and not absolute amount) is by zinc (65.28 ± 6.64%) and relatively dominant by chromium (46.61 ± 13.55%). Three elements, potassium, arsenic and cadmium, are not bound to the organic phase of the slag, while different amount of the other elements are associated with the organic phase. There are still relatively high amounts (compared with the average amount of organically bound elements in fly ash) of aluminum (33.63 ± 2.92 %), silicon (24.71 ± 0.91 %), iron (17.17 ± 0.52 %), nickel (24.51 ± 0.78 %) and lead (22.73 ± 1.25 %) in the organic fraction. Concerning the inorganic component, higher amounts of four elements, aluminum, potassium, iron and zinc, are extracted in the third than in the fifth phase of the sequential extraction.

4.4 Cluster Analysis of Elements Content in the Lignite and in the High Temperature Product

In order to identify the main associations of elements in the coal, the most common subset of clustering methods that is generally referred to an agglomerative hierarchical method was used. This method is the most appropriate to evidence correlations between variables. The series of stages is summarized in the dendrogram (Figure 4).

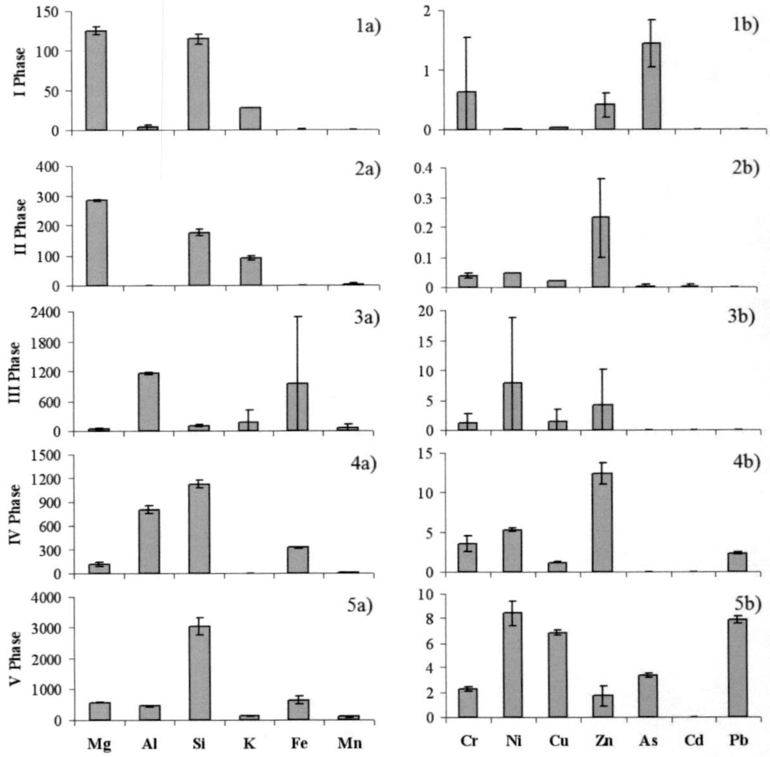

Fig. 3. Average concentrations and standard deviations of a) macro elements and b) microelements (ppm) extracted from samples of filter ash

The lower value is on the axis, the more significant are the associations (Facchinelli et al., 2001; Djordjevic et al., 2004; Relic et al., 2005), i.e. groups of elements in the cluster that are correlated with high significance. As it can be seen, strong associations between the concentrations of different elements leached from lignite and ash are different in their nature, and clearly witness that lignite combustion results in distribution of elements in resulting ash that is different from distribution present in lignite. There is strong association of aluminium, potassium and silicon in the first phase of extraction of lignite (Figure 4-1a), while in same phase of extraction of ash silicon has strong association with arsenic and cadmium (Figure 4-1b). Cluster analysis performed on concentrations leached in the third phase of extraction shows that there is a strong association of aluminium and silicon with magnesium in lignite (Figure 4-3a), while in the ash obtained from lignite aluminium and silicon are associated with potassium (Figure 4-3b). The change in distribution of elements caused by combustion process is also revealed by cluster analysis of associations of elements leached in fourth (e.g. manganese, associated with magnesium and iron in lignite (Figure 4-4a), but with iron and aluminium in ash (Figure 4-4b)) and fifth phase of sequential extraction (for example, magnesium, associated with iron and copper in lignite (Figure 4-5a), but with chromium and arsenic in ash (Figure 4-5b). Cluster analysis thus demonstrates an influence of high temperature process of coal combustion on elements associations.

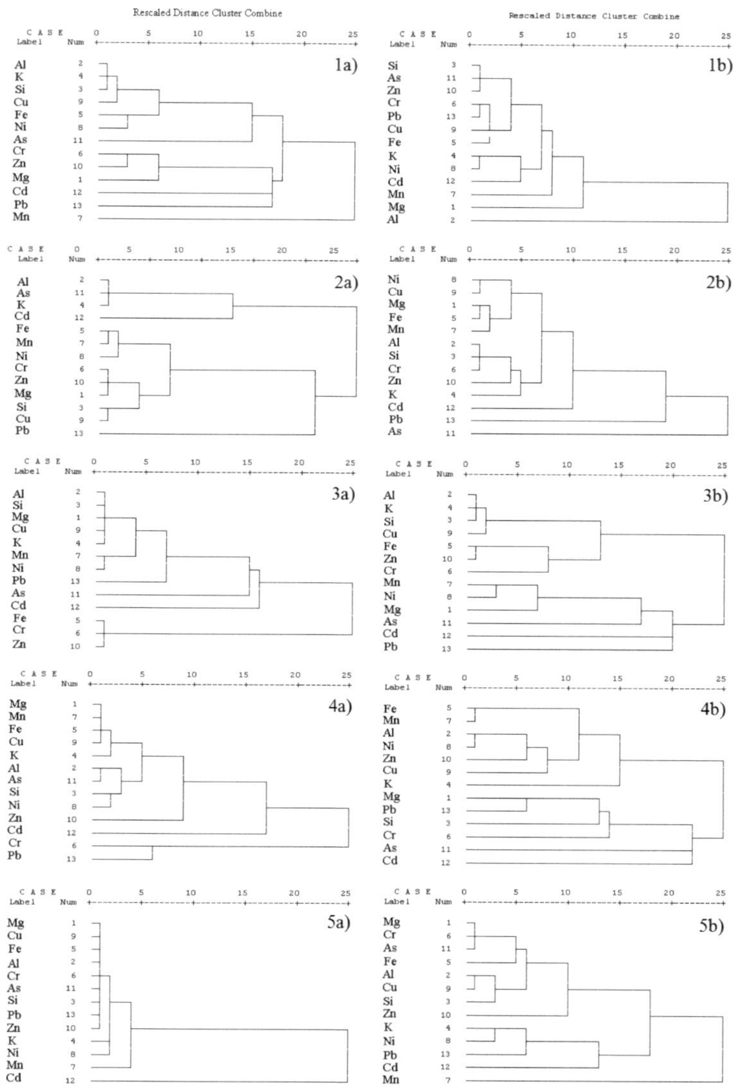

Fig. 4. Dendograms of cluster analysis for a) lignite and b) fly ash; 1) first phase, 2) second phase, 3) third phase, 4) fourth phase and 5) fifth phase

5. Conclusions

First, it must be emphasized that a larger amount of totally extractable components are present in the ash than in the slag for all the examined elements, except for silicon. It can only be speculated about the reasons for this phenomenon: the cause can be the higher amount of a probably relatively pure organic fraction in the slag, but also the higher content of smaller particles that are carriers of large amount is soluble fractions of element in the ash.

Comparing the characteristics of the extraction of coal, ash and slag, it can be concluded that:

- arsenic (or at least the part which does not leave the system with the combustion gases) during combustion of coal completely transforms to silicates, and this occurs in both slag and ash. The fraction of arsenic which is extracted in the third and fourth phase of sequential extraction of coal disappears during combustion;
- cadmium, which in coal is present in the last extraction phase, is also found in the same phase in ash, while it is not present in slag.
- lead, which was present in coal in the organic phase and in the last phase of the extraction, is present exclusively in the last phase in ash, while lead associated with the organic part is registered in the slag;
- organically bound chromium transfers during combustion mainly to the slag, while chromium, which is bound to the least soluble fraction, goes mainly to the ash;
- both the organic fraction and fraction dissolved in the fifth fraction of extraction are almost equally present in the both lignite combustion products in case of manganese, while the third phase, which is present in coal, transfers almost completely to the slag;
- the fly ash and slag have a similar arrangement of copper (with a somewhat larger amount of the element being extracted from ash in the fifth phase);
- the fraction of nickel soluble in the fifth phase of extraction of coal is almost completely transferred to the ash during combustion;
- organically bound zinc which is left after combustion, and zinc which is bound to the inorganic compounds which are soluble in the third fraction are mainly found in the slag.

Cluster analysis confirmed the difference in associations of trace and major elements in lignite and ash that are caused by transformation occurring during combustion.

6. Acknowledgement

This work was supported by Ministry of Science, Republic of Serbia (Grant # 146008).

7. References

Belanger, C. & Gagnon, L. (2002). Adding wind energy to hydropower. *Energy Policy* 30, 14, 1279-1284, ISSN 0301-4215

Chu, T-Y. J.; Ruane, R.J. & Krenkel, P.A. (1978). Characterization and reuse of ash pond effluents in coal-fired power plants. *Journal of the Water Pollution Control Federation* 50, 11, 2494-2508, ISSN 1047-7624

Djordjevic, D.; Radmanovic, D.; Mihajlidi-Zelic, A.; Ilic, M.; Pfendt, P.; Vukmirovic, Z. & Polic, P. (2004). Principal associations of trace elements in airborne particulate matter at the South Adriatic Coast. *Environmental Chemistry Letters*, 2, 1, 147-150, ISSN 1610-3653

Donslund, B. & Eriksen, P.B. (1990). Disposal of residues from electricity production. Practical application of a mathematical model to predict the impacts on ground water quality. In: *Developments in Environmental Modelling*, Fenhann, J.; Larsen, H.; Mackenzie, G.A. & Rasmussen, B. (Ed.), 321-331, Elsevier, ISBN 0-444-88609-5, Amsterdam

Eskenazy, G.M. & Valceva, S.P. (2003). Geochemistry of beryllium in the Mariza-east lignite deposit (Bulgaria). *International Journal of Coal Geology*, 55, 1, 47-58, ISSN 0166-5162

Facchinelli, A.; Sacchi, E. & Mallen, L. (2001). Multivariate statistical and GIS-based approach to identify heavy metal sources in soils. *Environmental Pollution*, 114, 3, 313-324, ISSN 0269-7491

Fernandez-Turiel, J.-L.; De Carvalho, W.; Cabans, M.; Querol, X. & Lopez-Soler, A. (1994). Mobility of heavy metals from coal fly ash. *Environmental Geology (Berlin)*, 23, 4, 264-270, ISSN 0943-0105

Finkelman, R.B. (1993). Trace and minor elements in coal, In: *Organic Geochemistry*, Engel, M.H. & Macko S.A. (Ed.), 593-607, Plenum Press, ISBN 0-306-44378-3, New York

Font, O.; Querol, X.; Huggins, F.E.; Chimenos, J.M.; Fernandez, A.I.; Burgos, S. & Pena, F.G. (2005). Speciation of major and selected trace elements in IGCC fly ash. *Fuel*, 84, 11, 1364-1371, ISSN 0016-2361

Goodarzi, F. (1994). Inorganic constituents of coal and their impact on coal quality. *CIM Bulletin*, 87, 983, 47-56, ISSN 0317-0926

Goodarzi, F. (2006). Characteristics and composition of fly ash from Canadian coal-fired power plants. *Fuel*, 85, 10-11, 1418-1427, ISSN 0016-2361

Gupta, D.C. (1999). Environmental aspects of selected trace elements associated with coal and natural waters of Pench Valley coalfield of India and their impact on human health. *International Journal of Coal Geology*, 40, 2-3, 133-149, ISSN 0166-5162

Hlavay, J.; Polyak, K.; Bogod, I. & Csok, Z. (1995). Importance and determination of chemical species in solid samples. *Microchemical Journal*, 51, 1-2, 53-63, ISSN 0026-265X

Hurst, R.W.; Davis, T.E.; Elseewi, A.A. & Page, A.L. (1993). Strontium and lead isotopes as monitors of fossil fuel dispersion. In: *Trace Elements in Coal and Coal Combustion Residues*, Keefer, R.F. & Sajwah, K.S. (Ed.), 99-118, Lewis, ISBN 0-87371-890-9, Boca Raton

Iturbe, R.; Cruickshank, C.; Vega, E. & Silva, A.E. (1996). Solubility and Transport of Arsenic in Coal Ash, *Proceedings of the 3rd International Conference on Tailings and Mine Waste*, pp. 305-317, ISBN 9-05410-594-1, Fort Collins, Colorado, January 1996, Balkema Publishers, Rotterdam

Jankowski, J.; Ward, C.R.; French, D. & Groves, S. (2006). Mobility of trace elements from selected Australian fly ashes and its potential impact on aquatic ecosystems. *Fuel*, 85, 2, 243-256, ISSN 0016-2361

Jones M.R.; McCarthy, A. & Booth, A.P.P.G. (2006). Characteristics of the ultrafine component of fly ash. *Fuel*, 85, 16, 2250-2259, ISSN 0016-2361

Kuehn, W. & Kurzbach, H. (1992). The chemical composition of the coal ash in the White Elster basin with special emphasis on the titanium content. *Braunkohle (Duesseldorf)*, 44, 1, 27-32, ISSN 0341-1060

Li, F. (2002). Hydropower in China. *Energy Policy*, 30, 14, 1241-1249, ISSN 0301-4215

Marquenie, J.M. & Simmers, J.W. (1988) A method to assess potential bioavailability of contaminants. In: *Earthworms in Waste and Environmental Management*, Edwards, C.A. & Neuhauser, E.F. (Ed.), 367-375, SPB Academic Publishing, ISBN 9-05103-017-7, The Hague.

Mukherjee, S. & Borthakur, P.C. (2004). Effects of alkali treatment on ash and sulphur removal from Assam coal. *Fuel Processing Technology*, 85, 2-3, 93-101, ISSN 0378-3820

Omer, A.M. (2002). Overview of renewable energy sources in the Republic of the Sudan. *Energy*, 27, 6, 523-547, ISSN 0360-5442

Pacyna, J.M. & Pacyna, E.G. (2001). An assessment of global and regional emissions of trace metals to the atmosphere from anthropogenic sources worldwide. *Enviromental Reviews*, 9, 4, 269-298, ISSN 1208-6053

Pavlish, J.H.; Sondreal, E.A.; Mann, M.D.; Oslon, E.S.; Galbreath, K.C.; Laudal, D.L. & Benson, S.A. (2003). Status review of mercury control options for coal-fired power plants. *Fuel Processing Technology*, 82, 2-3, 89-165, ISSN 0378-3820

Piekos, R. & Paslawska, S. (1998). Leaching of assimilable silicon spieces from fly ash. *Fuel Processing Technology*, 56, 3, 201-213, ISSN 0378-3820

Polic, P.S.; Ilic, M.R. & Popovic, A.R. (2005). Assessment of lignite fly ash and its utilization products as recycled hazardous wastes on surface and ground water quality, In: *Environmental Impact Assessment of Recycled Wastes on Surface and Ground Waters-Chemodynamics, Toxicology and Modeling*, Aboul-Kassim, T.A.T. & Williamson K.J. (Eds.), 461-510, Springer-Verlag, ISBN 3-54000-268-5, Berlin.

Popovic, A. & Djordjevic, D. (2005). Speciation of Selected Trace and Major Elements in Lignite Used in "Nikola Tesla A" Power Plant (Obrenovac, Serbia). *Journal of Serbian Chemical Society*, 70, 12, 1497-1514, ISSN 0352-5139

Popovic, A.; Djordjevic, D. & Polic, P. (2001). Trace and Major Element Pollution Originating from Coal Ash Suspension and Transport Processes. *Environment International*, 26, 4, 251-255, ISSN 0160-4120

Popovic, A.; Djordjevic, D. & Polic, P. (2000). Leaching of Trace and Major Elements in Coal Ash Dumps. *Toxicological and Environmental Chemistry*, 75, 141-150, ISSN 0277-2248

Querol, X.; Juan, R.; Lopez-Soler, A.; Fernandez-Turiel, J.-L. & Ruiz, C.R. (1996). Mobility of trace elements from coal and combustion wastes. *Fuel*, 75, 7, 821-838, ISSN 0016-2361

Relic, D.; Djordjevic, D.; Popovic, A. & Blagojevic, T. (2005). Speciations of trace metals in the Danube alluvial sediments within an oil refinery. *Environment International*, 31, 5, 661-669, ISSN 0160-4120

Rokita, J. (1990). Energy-saving hydraulic system for removing of fly ash and slag. *Polish Technical Review*, 6, 7-8, ISSN 0032-3012

Schultz, H.; Hattman, I.A. & Bocher, W.B. (1973). Fate of some trace elements during coal pretreatment and combustion, *Proceedings of 18 Annual Meeting of Division of Fuel Chemistry of the American Chemical Society*, pp. 108-113, Chicago, September 1973, American Chemical Society, Washington https://www.anl.gov/PCS/acsfuel/preprint%20archive/Files/Volumes/Vol18-4.pdf

Senior, C.S.; Bustard, C.J.; Durham, M.; Baldrey, K. & Michaud, D. (2004). Characterization of fly ash from full-scale demonstration of sorbent injection for mercury control on coal-fired power plants. *Fuel Processing Technology*, 85, 6-7, 601-612, ISSN 0378-3820

Sushil S. & Batra, V.S. (2006). Analysis of fly ash heavy metal content and disposal in three thermal power plants in India. *Fuel*, 85, 17-18, 2676-2679, ISSN 0016-2361

Tucek, J. & Konecny, D. (1979). Treatment of wastewater containing dispersed mechanical contaminants. *Czech Patent* 177679, 2 pp.

Vyazova, N.G. & Kryukova, V.N. (1997). Regularities of distribution of some toxic elements of the coals of East Siberia. *Fuel and Energy Abstracts*, 38, 3, 189-189, ISSN 0140-6701

Ward C.R. & French, D. (2006). Determination of glass content and estimation of glass composition in fly ash using quantitative X-ray diffractometry. *Fuel*, 85, 16, 2268-2277, ISSN 0016-2361

Wood, G.H.; Culbertson, W.C.; Kehn, T.M. & Carter, M.D. (1983). Coal Resource Classification System of the U.S. Geological Survey. U.S. Geological Survey Circulation 891, http://pubs.usgs.gov/circ/c891/index.htm

Xu, M.; Yan, R.; Zheng, C.; Qiao, Y.; Han, J. & Sheng, C. (2004). Status of trace element emission in a coal combustion process: a review. *Fuel Processing Technology*, 85, 2-3, 215-237, ISSN 0378-3820

Yudovich, Ya.E. & Ketris, M.P. (2005). Arsenic in coal: a review. International Journal of Coal Geology, 61, 3-4, 141-196, ISSN 0166-5162

Yunusa, I.A.M.; Eamus, D.; DeSilva, D.L.; Murray, B.R.; Buchet, M.D.; Skilbeck, G.C. & Heidrich C. (2006). Fly-ash: An exploitable resource for management of Australian agricultural soils. *Fuel*, 85, 16, 2337-2344, ISSN 0016-2361

10

Air Radioactivity Monitoring in Serbia

Dragana Popovic, Dragana Todorovic*,
Vesna Spasic Jokic* and Gordana Djuric
University of Belgrade,
Serbia,
**Institute for Nuclear Sciences Vinca, Belgrade, Serbia*

1. Introduction

Today radioactivity monitoring in the environment is required not only around sites where the significant amounts of radioactive material are used or stored, but in a number of other processes as land remediation or decommissioning of nuclear plants. Since the end of the 20th century there has been an increasing awareness of the so-called technologically enhanced natural radioactivity and the need for its monitoring in a wide range of non-nuclear industries including oil and gas extraction, ceramic and glass industries, production of phosphoric acid, production of different phosphoric fertilizers and other chemical industries. Coal electric power plants as significant sources of ashes and other contaminants are included, too. Contrary to nuclear facilities or radioactive waste control these processes require measuring radionuclides at very low levels of radioactivity. This demands specific methodology and instrumentation and responding to the very real economic importance of classifying waste products from decommissioning and remediation processes. However, natural or man made, waste or industrial, environmental radioactivity monitoring requires adequate equipment calibrated in the full range of radionuclides likely to be encountered, as well as trained and experienced personnel. It is the task of national metrological institutions to provide radioactive reference materials and standards for calibration as pure isotopes, mixtures of isotopes or reference materials for particular applications and to supervise the processes of quality control and quality assurance. Besides primary standards and reference materials national metrological institutions should provide laboratory proficiency testing exercises to support radioactivity measurements within academic, industrial or research laboratories (Spasic, 1984; Spasic, 1985; Spasic, 1987a; Spasic et al. 1987).

2. Environmental Radioactivity Monitoring

To protect the general public, national and international regulations require that levels of radioactivity in the environment and in foodstuffs are regularly monitored. This include first, the primary standards for radionuclides to be measured and secondly, a unique range of very low activity standardized solutions (e.g. about 1 Bq per ml) based on the primary standards to be provided (Fisenne, 1993; CCEMRI, 1996). The main advantage of the solutions is that they are 'ready-to-use', thus improving the accuracy of calibration and reducing the risk of contaminating expensive analytical instrumentation. Also,

environmental monitoring laboratories are obliged to demonstrate the accuracy of their measurements, for compliance with quality assurance management systems such as ISO 17025 (Spasic-Jokic, 1998; Spasic-Jokic et al., 2002; Todorovic et al., 2003; Spasic-Jokic et al., 2006).
Methods used in the environment radioactivity monitoring include a number of highly sophisticated nuclear techniques and instrumentation such as 4¶ beta-gamma counting, electron capture gamma counting, alpha-gamma and X ray-gamma coincidence counting, tracer-efficiency counting, liquid scintillation counting, high resolution gamma ray spectrometry, re-entrant ionization chambers, gas counting and total X ray counting (CCEMRI, 1994; Quin, 1996; Spasic-Jokic, 1998; Spasic-Jokic et al., 1998).

2.1 Quality Assurance (QA) and Quality Control (QC) in Air Radioactivity Monitoring

To assure the quality of the environmental radioactivity measurements it is necessary to provide optimum site location and sampling procedure, appropriate and high quality instrumentation, best operating practices, calibration procedures and frequency, appropriate data and documentation handling, statistical procedures and compatibility with national network standards. All measurements undertaken have to be underpinned by accurate calibration techniques with traceability to international standards. Thus it is necessary to design and undertake industrial and urban air and water pollution surveys to meet European Union Directive requirements, to develop and implement new measurement protocols, to carry out site and operation audits, to advise on CEN (Comité Européen de Normalisation, European Committee for Standardization) standards requirements and to research into novel techniques and instrumentation for pollution monitoring. Also, it is imperative to train local staff to carry out day-to-day monitoring functions, to design and optimize fixed air and water-quality monitoring networks, and to negotiate with local authorities and State Environment Agency (Fisenne, 1993).
Quality Assurance and Quality Control programs in radionuclide metrology deal with primary reference materials, standards, techniques and systems, secondary standards and systems, validation and demonstration of equivalence for measurement standards, infrastructures for environmental monitoring, nuclear data for applications of ionizing radiation and last but not least, the transfer of knowledge (Spasic, 1982; Spasic-Jokic et al., 2002).

2.2 Traceable Radionuclide Standards in Environmental Radioactivity Monitoring

National regulations and legislation demand the highest standards for calibration and traceability to National Measurement Standards in radioactivity measurements. Nuclear site licensing, environmental impact assessments, national and international accreditation and ISO 9000 certification increasingly rely on the use of traceable low level activity radionuclide standards for spectrometers calibration and radiochemical assay validation. Low activity mixed nuclide gamma ray and single gamma ray emitters, alpha and beta particle emitters as environmental radionuclide standards are now in routine laboratory use by nuclear fuel cycle facilities, governmental regulatory agencies and reactor research centers. Wide range of low level activity mixed nuclide gamma-ray emitters, single nuclide gamma ray emitters, alpha particle and beta particle emitter radionuclide solutions, radon gas and spiked standards have been specifically developed to meet the needs of radioactivity environmental measurements. Thus, single gamma ray

radionuclide solutions and milk powder are suitable for calibration of spectrometers with either Ge or NaI detectors as they efficiently overcome problems associated with cascade summing and sample density effects. Alpha and beta particle emitter solutions are widely used as chemical yield tracers, in isotope dilution techniques, as internal standards for radiochemical assays as well as in calibration of alpha spectrometers. These solutions are also suitable for calibration of proportional counters, Geiger-Müller tubes and liquid scintillation counters. All low activity environmental radionuclide standards have to be directly traceable to national absolute primary radioactivity standards and further to meet the validation and intercomparison with other national standards laboratories organized under the auspices of the International Bureau for Weights and Measures (BIPM) in Sevre, France (Spasic, 1986; Spasic-Jokic, 1998; Quin, 1996; BNM, 1995; CCEMRI, 1991; CCEMRI, 1994; CCRMRI, 1996).
For efficiency calibration of hyper-pure Germanium gamma ray detectors mixed gamma-radionuclide standard are used, with typical activity values in Bq/g (in brackets): ^{241}Am (30), ^{139}Ce (9), ^{60}Co (50), ^{109}Cd (250), ^{57}Co (7), ^{137}Cs (30), ^{203}Hg (30), ^{113}Sn (30), ^{85}Sr (50) and ^{88}Y (160), and the uncertainties in the range ±0.5% to ±4% (at 95% confidence level). The most frequently used environmental radioactivity standards are: Strontium-90, Caesium-137, Caesium-134, Mixed Nuclides, Europium-152, Technetium-99, Americium-241, Iodine-129, Plutonium–242, Americium–243, Carbon-14, Uranium-232, Neptunium-237, Lead-210, Plutonium-239, Curium-244, Thorium-229, Uranium-238, Uranium-238, Tritium and Organically Bound Tritium (all in activity range 0.3 Bq/g - 20 kBq/g, mass 1 - 10 g nominal) (Spasic et al, 1987; Spasic-Jokic, 1998; CCEMRI, 1996).The methods and instrumentation most frequently used for verification of radioactive sources as certified reference materials are Ge spectrometers, proportional counters, digital coincidence counting (DCC), triple-to-double coincidence ratio (TDCR) and selective sampling (Fisenne, 1993; Spasic-Jokic, 1998; Spasic et al., 1987).

2.3 Metrological Legislative in Serbia

Officially, the system of metrological assurance in the field of ionizing radiation in Serbia was initiated in 1981 when Advisory Commission for Ionizing Radiation was formed within Federal Bureau of Measures and Precious Metals, in Belgrade. The very first task of the Commission was to prepare various acts of metrological legislation as a base of metrological assurance. That became highly urgent during and after the Chernobyl nuclear plant accident in 1986. Since than more than 60 regulatory papers, guidelines and classification schemes were prescribed by the Commission and became legally obliged (Spasic, 1984;, Spasic & Bek-Uzarov, 1986; Spasic-Jokic, 1998; Spasic, 1987b; Spasic, 1987c).
Among 64 regulatory papers 28 of them are dedicated to radionuclide metrology and to metrological assurance of the environmental radioactivity measurements. These include:

1) 6 classification schemes (hierarchy of standard instruments and sources) for dissemination of unit for activity of radionuclides alpha, beta and gamma emitters, issued until 1998;
2) 12 regulatory papers dedicated to metrological conditions for various instruments as are: semiconductor and scintillation gamma spectrometers and counters, semiconductor and scintillation alpha spectrometers and counters, GM counters,

proportional counters in various geometries, plastic and liquid scintillation counters and various ionization chambers; and

3) 12 guidelines for verification of instruments as in line 2).

2.4 Instrumentation and Methods in Radioactivity Measurements: State of Art in Serbia

Since 1991 the Commission for Ionizing Radiation in the Federal Bureau of Measures and Precious Metals in Belgrade has been elaborating the data base on human, equipment and laboratory resources in the environment radioactivity research area. The Working Group for Radioactivity Measurements prepared and distributed the Questioner on Instrumentation and Methods in Radioactivity Measurements which provided data on instrumentation, methods, personnel, sampling procedures and general working conditions in the institutions involved in the environmental radioactivity measurements. The Questioner was aimed to improve the Quality Assurance and Quality Control procedures. It was prepared according to Environmental Measuring Laboratory Procedures Manual "Quality Control and Quality Assurance in Radioactivity Measurements" HASL-300 (US Energy Department. New York, 1990/92) and was distributed to 15 laboratories and institutions involved in the environmental radioactivity monitoring in the country. The principal users of the results were and still are Ministries for health, science and environmental protection as well as Military, Police and other governmental offices (DPL, 1998).

The data from the Questiner on the laboratory space showed significant differences among laboratories in total working area (30 m^2 - 400 m^2) and in number of available premises (2 – 21). All laboratories were air-conditioned and with appropriate detectors protection but only two could be considered as low-background laboratories. Staff and experts, mostly physicists, electrical engineers and chemists with MSc and PhD degree were well trained, but there was a significant shortage of well trained technicians. The number of personal differed significantly among the laboratories, too (4 - 16 per laboratory) (Spasic-Jokic et al., 1998).

Average number of samples measured per laboratory varied between 50 and 1000 (maximum 6000). As sampling network was very loosely defined, location and collecting frequency became the subject of the national Low of Radiation Protection, issued in 1994 and of other regulatory papers (Spasic-Jokic et al., 1998).

In the time when Questioner was sent the laboratories were equipped mostly with HPGe gamma spectrometers, also with NaI (Tl) and Ge(Li) detectors, liquid scintillation detectors, some flow proportional alpha-beta anti-coincidence counters and alpha semiconductor spectrometers and some nationally developed radiometric laboratories (LARA). Most of the instruments were purchased after the Chernobyl accident, in 1986. Instruments are of standard construction, with commercial lead shielding (5-10 cm). Depending of the shielding, the background counts were between 0.8 cps and 7 cps for semiconductor gamma detectors, 0.1 cps to 1.2 cps for beta and alpha radiation detectors and up to 20 cps for Na(I) detectors. To note, during the Chernobyl accident there was only one laboratory for low level activity measurements in the environmental samples, with background count of 1 cps (100 – 2000 keV) at the Department of Radiology, Faculty of Veterinary Medicine, Belgrade. Over 90 % laboratories were monitoring gamma activity, only a few of them were equipped for alpha and beta emitters monitoring.

Among natural radionuclides most commonly determined were ^{235}U, ^{238}U, ^{226}Ra, ^{232}Th, ^{40}K, ^{220}Rn, ^{7}Be, ^{3}H, ^{210}Pb, ^{210}Po and among fission products those were ^{134}Cs, ^{137}Cs, ^{90}Sr and ^{60}Co (Spasic-Jokic et al., 1998).
Calibrations were performed by standard certified reference materials generally in geometry of point sources and/or Marinelli bakers. Radionuclides in mixed or pure sources were usually ^{152}Eu, ^{133}Ba, ^{137}Cs, ^{241}Am + ^{239}Pu + ^{244}Cm, ^{60}Co, ^{57} Co, ^{144}Ce, ^{54}Mn, ^{22}Na, ^{85}Sr, ^{88}Y, ^{203}Hg, ^{40}K, but some laboratories also used standard reference samples of sludge, soil, sand and phosphates. All laboratories use commercial sources provided from International Agency of Atomic Energy (IAEA), Hungarian National Office of Measurement (OMH), National Institute of Standards and Technology (NIST) and Amersham Co., with extended measuring uncertainty up to 5 %. Most of the laboratories participated in the first national intercomparison organized and coordinated by Federal Bureau of Measures and Precious Metals in 1992. The intercomparison pointed to some problems in sampling procedures, inappropriate calibration, and shortage of legally verification culture and non-existence of well defined and applied QC and QA programs and methods. Those problems were over-bridged by numerous intercomparisons organized at national, regional and international level and closer connection with international bodies (IAEA) and organizations in the field (Kandic et al., 1993; Spasic-Jokic et al., 1998).

3. Air Radioactivity Monitoring

On global scale air radioactivity monitoring provides important data for comparing environmental impact of radioactivity from anthropogenic sources to natural ones and for studying global atmospheric processes (Baeza et al., 1996; Gaggelar, 1995; Gustafson et al., 1981; Arimoto et al., 1999; Djuric & Popovic, 1994). Most of air-born radionuclides are attached to aerosols and their concentration in ground level air is manly due to behavior of aerosols in the atmosphere. As aerosols play an important role in the atmospheric processes as carriers of atmospheric materials and available surfaces for different chemical processes, radionuclides can be used to trace their transport, as well as transport of ozone in the atmosphere (Gaffney et al, 1994; Gaffney et al., 2004; Groundel & Papastefanou, 2004). Among naturally occurring radionuclides in air, radon (^{222}Rn) and its progenies (^{210}Pb, ^{210}Bi, ^{210}Po) are most significant, while ^{131}I, ^{137}Cs and ^{90}Sr are of major interest among the fission products. Most commonly radionuclides monitored in ground level air are cosmogenic ^{7}Be, natural ^{210}Pb and anthropogenic ^{137}Cs.
Berillium-7 (half life 53.28 days) is produced by cosmic rays in spallation processes with light elements (N,O,C) in the upper troposphere and lower stratosphere and its annual mean concentrations are good indicator of the changes in the atmospheric production rate due to cosmic ray intensity and are correlated to stratosphere-to-troposphere exchange processes (Gaggelar, 1995; Gaffney et al., 2004). Concentrations of ^{7}Be in air at midlatitudes exhibit maxima in spring/summers periods attributed to seasonal thinning of tropopause that allows stratospheric masses rich with ^{7}Be to enter the troposphere (Agelaio et al., 1984; Bettoli et al., 1998; Cannizzaro et al., 1995; Hirose et al., 2004; Ioannidou & Papastefanou, 1997; Todorovic, 1997). Concentration of ^{7}Be in air is effected by air exchange between stratosphere and troposphere, vertical mixing

within the troposphere and the washout effects (Todorovic, 1997; Gerasopoulos et al., 2003).

^{210}Pb (half life 22.3 years) is produced in the decay of ^{238}U into gaseous ^{222}Rn (half life 3.8 days) that escapes from soils into the atmosphere where one of its short lived daughters (^{218}Po) decays into ^{210}Pb (Gaggelar, 1995; Gaffney et al., 2004; Groundel & Postendourfen, 2004; Todorovic et al., 2000). In small amounts radon and thus ^{210}Pb are produced in coal combustion processes and nuclear explosions (Todorovic et al., 2007). Deposition of ^{210}Pb varies with season and geographic position and variations in ^{210}Pb concentration in ground level air are used to trace continental emission and to investigate aerosols transport (Peters et al., 1997; Tokieda et al., 1996). ^{210}Pb concentrations exhibit maxima in autumns that may be attributed to increased radon emanation (Hirose et al., 2004).

Long lived ^{137}Cs (half life 30 years) is most significant among fission products and a reliable indicator of anthropogenic pollution in the environment due to nuclear weapon atmospheric tests and nuclear power plant accidents. After the moratorium on the atmospheric nuclear weapon tests in the 1970ties and the decrease in concentrations of ceasium released in the atmosphere during the Chernobyl nuclear plant accident in 1986 due to its migration into deeper layers of soils, concentrations of ^{137}Cs in ground level air in the 1990ties were of order of $\mu Bq/m^3$. The concentration pattern was exhibiting one or two maxima in summers and winters, the summer maxima due to stratosphere to troposphere air exchange and the winter one due to soil dust resuspension in air (Cannizzaro et al., 1995; Larsen et al., 1995; Todorovic et al., 1997).

4. Air Radioactivity Monitoring In the Institute Vinca, Belgrade

Air radioactivity control in the Institute for Nuclear Sciences Vinca started in the late 1950ties, after an accident at the experimental reactor. It comprised continuous monitoring of ground level air radioactivity in several sites in the immediate vicinity of the Institute, chosen to enable the control of the most probable trajectories of radioactive matter possibly relised from the reactor. Two of the four control stations, at the distance of 2 and 7 km were in the direction of the dominant wind blowing towards the city of Belgrade. Air radioactivity monitoring program included continuous control of short and long lived radioactive aerosols, background spectra, radioactive gasses in the free atmosphere, wet and dry deposition, as well as logistic meteorological measurements (Todorovic et al., 1996; Todorovic et al., 2007).

4.1 Materials and Methods

At the very start of the monitoring program only 4 automatic devices for aerosols total beta activity with movable filter belt and continuous counting (Frieseke-Hoepfner 59A) were used. The devices enabled measuring activity of the radionuclides in 2 different delay periods compared to the time of sampling, as 2 scintillation detectors were positioned along the belt. Sampling period was 228 minutes, while delay periods were 78 minutes and 60 hours. Thus the first measurement provided the data on the total short and long lived radionuclides in air, and the second one provided the data on the activity of long lived radionuclides only. Data on short lived radionuclides in air were obtained by simply contracting the results of the measurement.

Under normal conditions, the rate of natural radionuclides originated from radon and thoron is dominant in the measurements after 78 minutes, showing pronounced daily variations: maxima in early mornings and minimums in the afternoons. This daily pattern is closely connected with meteorological conditions in the ground level air, and the best early indication for increased level of anthropogenic radionuclides in air.

Measurements of total beta activity in air in 24- hour's aerosols samples were introduced in the late 1960ties and they completely replaced the automatic devices in 1976. The activity were determined by 2π proportional alpha/beta counters in air voluminous samples (600 m^3), taken on filters with effective diameter of 13 cm. The samples were measured 5 hours and 5 days after the sampling. Under no anthropogenic pollution, first measurement provided data on the level of radionuclide thorium-B and its daughters, and the other one on the activity of long lived radionuclides in the aerosols, originated from the nuclear tests or the reactor emission. As alpha activity data are very significant (fission products are in general beta/gamma emitters) so when increased beta activity after 5 hours is followed by an adequate increasement of alpha activity, it could be attributed to natural radionuclides. On the contrary, non proportionality of alpha and beta activity in air would undoubtifully pointed to anthropogenic origin of radioactive pollution.

Monitoring of gamma emitting radionuclides in aerosols by gamma spectrometry started in the mid 1960ties, first with an Na(Tl) detector (well-type, "3x3" crystal with 100 channel amplitude analyzer). Today three high computerized gamma spectrometric systems with HP Ge detectors are in use (see 5.1). Under normal conditions composite monthly filter samples previously ashed up to 370°C are analyzed, while when radioactivity is increased, filters could be analyzed without previous preparation.

To determine plutonium in air a method of radiochemical separation of plutonium from aerosols was developed. The process included air filters ashing at 400°C, dissolving in nitric and cholorohydorgenic acid, separation on anions resins and finally electro-deposition of plutonium on steel plates. Plutonium was than determined by alpha spectrometry. Efficiency of the method was obtained by adding a standard ^{236}Pu solution (Harwel ,UKAEA) (Todorovic et al., 1996).

Concerning different psychical-chemical forms of iodine that could be present in the atmosphere during episodes of iodine pollution, a special sampling procedure was developed in 1969. A complex sampler type May-pack was constructed, containing an aerosol filter for sampling iodine attached to aerosols, two coal impregnated filters for sampling large organic compounds and 6 layers of active coal for sampling lighter iodine organic compounds. The system has an efficiency of nearly 100% for all iodine compounds, and proved highly efficient during the nuclear plant accident at Chernobyl in 1986 (Todorovic et al., 1996).

4.2 Air Radioactivity Monitoring at the Institute Vinca up to 1991

The data of total beta activity in air measured at the Institute Vinca from 1961-1968 are presented as average monthly values in Fig.1. As the results are affected by the time delay between the measuring and the sampling periods the results for the 5 days delay period and for the 60 h delay period for years 1967 and 1968 are presented on the figure, too. The effects of the level of natural radionuclides (ThB) on the total beta activity measured with the time delay of 60 hours are obvious and therefore only the results obtained by measuring 24 sampling hours' aerosols with time delay of 5 days should be considered as reliable. Daily sampling and determining total beta activity in air with a 5 days delay period over the long time period enabled forming a comprehensive data base giving a reliable insight into the

fluctuations of the activities of radionuclides in air and a possibility to define a reference level for further investigations and intervention (Todorovic et al., 1996).
The results of long time trend of total beta activity in air in 1967-1991 measured in the Institute Vinca are presented in Fig. 2.
The seasonal maxima (fig. 2.) observed in spring (April, May, June) are due to the transport of air from higher to lower heights while other maxima are due to nuclear weapon tests performed up to 1980. A sharp increase in May 1986 is due to the nuclear plant accident at Chernobyl that resulted in a general increase in the level of the radionuclides activities in air in the next months, as well as to incidental increases due to soil resuspension in January 1987 (Todorovic et al., 1996; Todorovic, 1997; Todorovic et al., 1997; DM Report 96-01; Popovic & Spasic-Jokic, 2006).
Generally, the level of natural radionuclides in air is about two orders of magnitude higher that the level of anthropogenic radionuclides. Daily fluctuations in the level of natural radionuclides are pronounced: during temperature inversion in early mornings the level of natural radionuclides could be even three orders of magnitudes higher than in the afternoons. Natural radioactivity is much affected by microclimate, too, presenting a seasonal pattern with maxima in late summers/early autumn's period (August - October) due to the increased radon emanation from soils (Popovic & Todorovic, 2006).
Fission product ceasium-137 was detected in the atmosphere first after the nuclear bomb attacks at Hiroshima and Nagasaki, in 1945. In the period of intensive nuclear weapon tests in 1966 and 1967 the level of ^{137}Cs in air measured at Vinca Institute in Belgrade was in the range of 0.1 - 0.8 mBq/m^3. In the late 1970ties and early 1980ties ceasium was on the limit of detection until the nuclear plant accident at Chernobyl, in April-May, 1986. Activities of ^{137}Cs, ^{134}Cs, ^{103}Ru and ^{131}J in daily air samples during the nuclear plant accident in Chernobyl are presented in Table 1., together with total beta activity in air. Since 1991, continuous gamma spectrometry of aerosols has been performed. Ceasium-137 had specific seasonal pattern following total beta activity seasonal variations (Fig.3.) (Smiljanic et al., 1989; Todorovic et al., 1996; Todorovic et al., 1997).

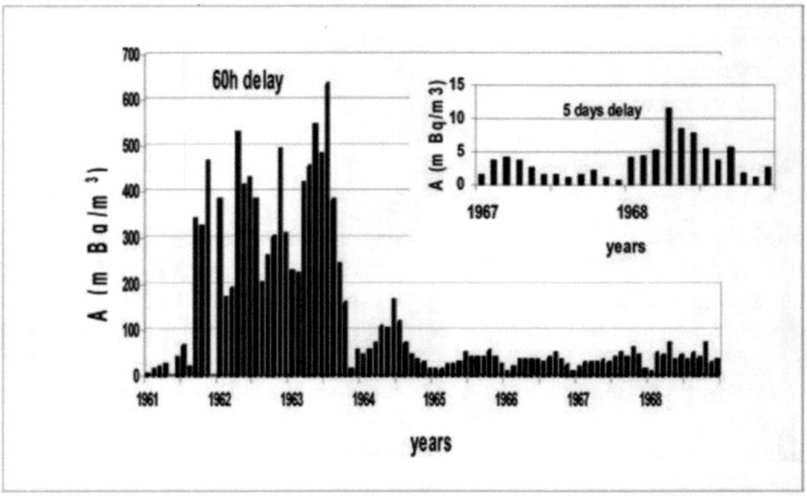

Fig. 1. Total beta activity in air 1961-1968, Institute Vinca, Belgrade (Todorovic et al., 1996)

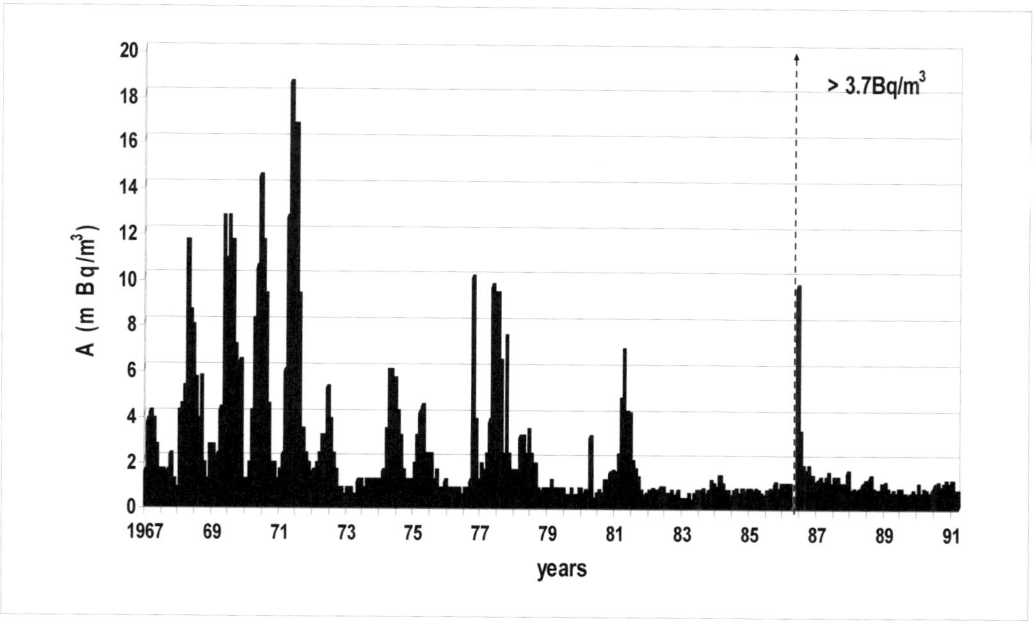

Fig. 2. Total beta activity in air, Institute Vinca 1967 -1991 (Todorovic et al., 1996)

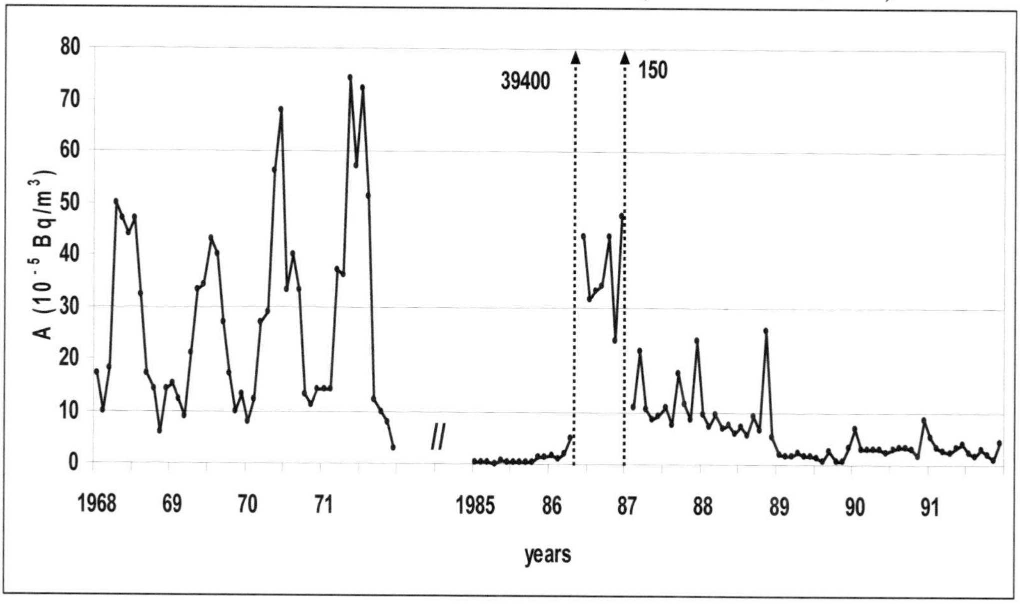

Fig. 3. ^{137}Cs in ground level air, Institute Vinca up to 1991 (Todorovic et al., 1996)

Date	Total β activity	Activity (Bq/m³)			
May 1986.	(Bq/m³)	^{137}Cs	^{134}Cs	^{131}J$^{(1.*)}$	^{103}Ru$^{(*)}$
01.	18.9 10^{-3}	22.4 10^{-3}	<10^{-3}	>38.6$^{(2)}$	9.6 10^{-3}
02.	33.4	2.6	1.3	>20.8$^{(2)}$	9.2
03.	28.5	2.9	1.4	>4.1$^{(2)}$	8.1
04.	6.2	0.5	0.3	>15.0$^{(2)}$	1.4
05.	2.6	0.3	0.1	7.3	0.73
06.	17.7	1.9	1.0	8.2	9.1
07.	12.9	1.5	0.8	3.8	7.7
08.	6.3	0.6	0.3	4.6	4.5
09.	4.2	0.4	0.2	1.5	3.4
10.	1.3	4.9 10^{-3}	<10^{-3}	1.0	20.3 10^{-3}
11.	23.4 10^{-3}	5.9 10^{-3}	3.2 10^{-3}	0.5	41.5 10^{-3}
12.	28.0 10^{-3}	2.7 10^{-3}	<10^{-3}	0.2	15.3 10^{-3}
13.	10.4 10^{-3}	1.2 10^{-3}	<10^{-3}	0.1	16.3 10^{-3}
14.	11.1 10^{-3}	0.5 10^{-3}	<10^{-3}	0.3	17.6 10^{-3}
15.	20.9 10^{-3}	7.7 10^{-3}	<10^{-3}	0.2	51.1 10^{-3}
16.	29.5 10^{-3}	5.8 10^{-3}	1.9 10^{-3}	0.1	59.6 10^{-3}
17.	13.6 10^{-3}	0.7 10^{-3}	<10^{-3}	0.1	11.8 10^{-3}
18.	7.7 10^{-3}	0.8 10^{-3}	<10^{-3}	0.1	9.6 10^{-3}
19.	7.0 10^{-3}	1.2 10^{-3}	<10^{-3}	0.1	30.6 10^{-3}
20.	9.7 10^{-3}	3.5 10^{-3}	<10^{-3}	0.1	39.8 10^{-3}
21.	17.8 10^{-3}	<10^{-3}	<10^{-3}	<0.1	17.5 10^{-3}
22.	14.3 10^{-3}	1.2 10^{-3}	<10^{-3}	<0.1	15.3 10^{-3}

(1) maximum daily activities (30 minutes sampling period), (2) sampling on coal filter
* NaJ(Tl) detector, well-type.
Table 1. Average daily air radioactivity in May 1986, Institute Vinca (Smiljanic et al., 1989)

In May 1986, only about 30% of the total iodine released in the atmosphere could be detected in the aerosols filters. Other 36% of elemental iodine and less active organic

compounds was detected by coal impregnated filter and the rest 34% of active organic compounds by active coal (Todorovic et al., 1996; Smiljanic et al., 1989; Andrasi, 1986).
The data on the activity of plutonium in air in the period of intensive nuclear weapon tests in 1965-1969 are presented in Table 2. The results are in agreement with the reported data on plutonium in air on the same latitudes (Todorovic et al., 1996).

Year	Month	$^{239,240}Pu$	^{238}Pu
1965	March	13.1 ± 1.3*	(0.5)***
	April	9.7 ± 1.3	n.d.**
	April	7.6 ± 0.6	(0.2)***
	May	16.7 ± 1.1	n.d.**
1967	July	2.1 ± 0.3	1.0 ± 0.2
	August	1.6 ± 0.4	0.9 ± 0.4
	October	0.7 ± 0.2	(0.3)***
	October	0.8 ± 0.7	(0.4)***
	November	(0.3)***	(0.1)***
	November	(0.4)***	(0.2)***
	December	0.5 ± 0.3	(0.3)***
	Jan.Feb.Mar.	(1.5)***	n.d**
	Apr.May.June	2.9± 0.6	1.0± 0.14
	July.Aug.Sep.	2.2 ± 0.3	1.1 ± 0.2
	Oct.Nov.Dec.	1.5 ± 0.4	0.7 ± 0.4
1969	Apr.May June	1.5 ± 0.4	0.6 ± 0.4
	Oct.Nov.Dec.	0.8 ± 0.2	(0.1)***

* 95% uncertainty level, ** non detected, *** values with low uncertainty level due to short counting period about 24 hours or less

Table 2. Activity of $^{239,240}Pu$ (mBq/m^3) and ^{238}Pu (mBq/m^3) in air 1965-1969, Institute Vinca, Belgrade (Todorovic et al., 1996)

Daily activity of plutonium in air during nuclear accident at Chernobyl, from May, 1st -15th, 1986 were in the range of 0.4 – 10.6 x 10^{-6} Bq/m^3 for $^{239,240}Pu$ and 0.3 – 5.7 x10^{-6} Bq/m^3 for ^{238}Pu, with a maximum on May the 5th (Todorovic et al., 1996).

4.3 Air radioactivity monitoring at the Institute Vinca in 1991-2006

Monitoring of ^{7}Be in ground level air in the Institute Vinca started in 1991. Average monthly concentrations of ^{7}Be in air up to 2006 were in the range of 2.0-7.0 mBq/m^3 and exhibited one/two maxima in summer/early autumn and a minimum in winter (Fig.4), corresponding to the values measured in Europe and elsewhere (Gaffney et al., 1994; Groundsel & Postendurfen, 2004; Hernandez et al., 2005; Ioanidou & Papastefanou, 1997; Iskihawa et al., 1995). The maxima were correlated with the increment of temperature, while the minimums were linearly correlated with precipitation. Sharp increase of ^{7}Be concentrations in air in 2001 and 2003 was probably due to increased stratosphere-to-troposphere exchange (Hernadez et al., 2005; Todorovic et al., 1997; Todorovic et al., 2000; Todorovic et al., 2005). Concentrations of ^{137}Cs in air in the same period were in the range of 0.1 - 8.5 x 10^{-5} Bq/m^3 with a maximum in spring/summers and one in winter due to local resuspension effects (Fig.5). We should note that since 1989, ^{137}Cs concentrations were decreasing and in 1998 obtained the level before the Chernobyl accident (Todorovic et al., 2005; Todorovic et al., 2007). Activity of ^{210}Pb in ground level air has been measured since 1985 and was in the range of 0.1 – 31.7x10^{-4} Bq/m^3, with a maximum in early/late autumns (Fig.6) (Todorovic et al., 1997; Todorovic et al., 1999; Todorovic et al., 2000; Todorovic et al., 2002; Todorovic et al., 2005). This corresponds to the values reported by other authors (Arimoto et al., 1999; Duenas et al., 2004; Gaffney et al., 1994; Ionadiou & Papastefanou, 1997). Higher values of ^{210}Pb measured in Belgrade air in some periods are probably due to the anthropogenic sources (heavy traffic run by leaded gasoline and city heating plants run by crude oil and coal). The maxima are due to increased radon emanation from soils (Todorovic et al, 2005).

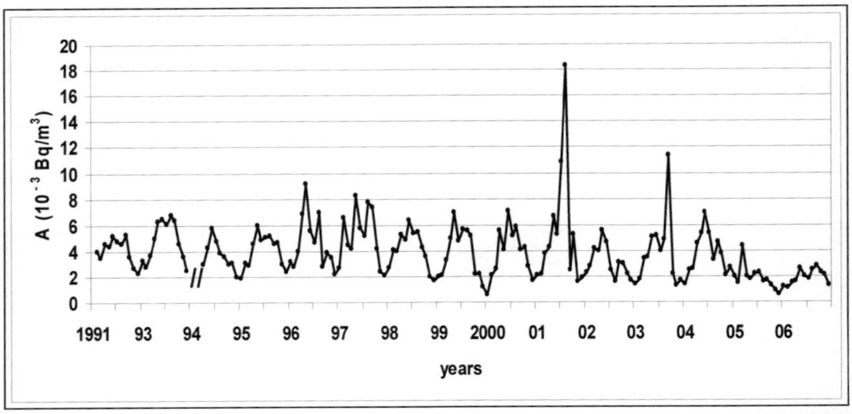

Fig. 4. Average monthly concentrations of ^{7}Be in air (Institute Vinca, 1991-2006)

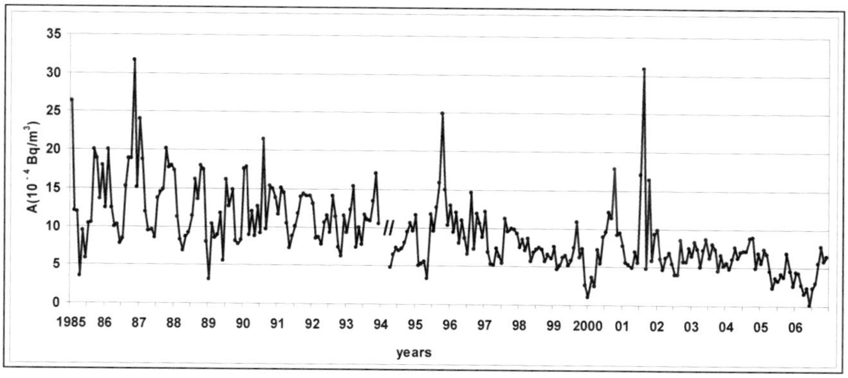

Fig. 5. Average monthly concentrations of ^{210}Pb in air (Institute Vinca, 1985-2006)

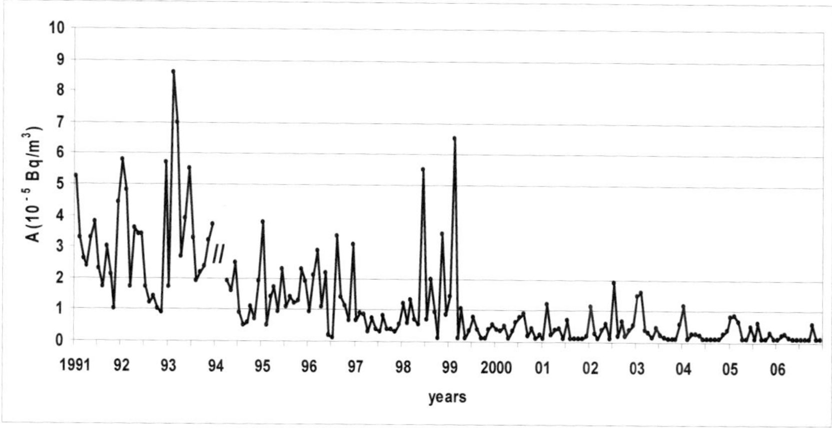

Fig. 6. Average monthly concentrations of ^{137}Cs in air (Institute Vinca, 1991 -2006)

5. Air Radioactivity Monitoring Program in Belgrade Central City Area

Due to high population density, heavy traffic and industrial plants located in the outskirts urban areas especially central city areas are exposed to severe air pollution. In the last decade, air radioactivity monitoring in urban areas is a part of the pollution monitoring program in most of the European countries. In Serbia, there are many studies on the natural and anthropogenic radinuclides in ground level air but still there is no comprehensive database on the air radioactivity in the central city areas and the monitoring program started only recently (Manic et al., 2006; Popovic et al., 1996; Popovic et al., 1996a; Popovic et al., 2000; Todorovic et al., 2007; Popovic et al., 2000a).

5.1 Site, Materials and Methods

The city of Belgrade (44°47′N, 20°32′E; 205 m a.s.l) is located in West-Central Serbia on the junction of the rivers Sava and Danube and has about 2 millions inhabitants. The climate is moderate continental. In the last decades, severe air pollution in the form of aerosol smog occurred in the city center during winters due to increasingly heavy traffic and household heating. Leaded gasoline is still in use and city heating plants and domestic heating plants use natural gas, crude oil and coal.

Aerosol for radionuclides study were collected on the representative "black spots" in very center of the city. Samples were collected on filter papers (FILTRAK/Whatman 41/DDR, 15 cm diameter, relative efficiency for deposited dust 80%) by constant flow rate samplers (average air flow 20 m^3/h, average daily volume 600 m^3), ashed at temperatures below 400°C and a monthly composite sample containing 30-31 daily filters was formed (average volume $15x10^3$ m^3). The samples were measured in small metalic containers.

The activity of the radionuclides was determined on 3 HPGe detectors (Canberra, Ortec relative efficiency 23%, 20%, 18% respectively, resolution 1.89 keV at 1332 keV) by standard gamma spectrometry. The two detectors (vertical, coaxial) were placed in a shielding cage of Pb bricks (width 10 cm) with layers of Cu (3 mm) and Fe (6 mm - 10mm). The third detector has the commercial shielding; it is a reverse-electrode Ge, with a thin Be window (Todorovic et al., 1994).

Energy calibration (100 keV-2000 keV) was performed with a set of standard point sources (Coffret d'etalon gamma ECGS-2, Sacle, France) containing ^{133}Ba, $^{57,60}Co$ and ^{137}Cs (10^3 Bq - 10^4 Bq, 25.11.1987). Geometric efficiency was determined with IAEA-083 (AIR4) stimulated air filter (spiked with solution of ^{60}Co: 2160 Bq/filter, ^{133}Ba: 846 Bq/filter, ^{137}Cs: 1182 Bq/filter and ^{210}Pb: 151 Bq/filter, uncertainty 5%, 1.1.86). Another reference radioactive material was also used: an aerosol powder matrix in plastic box (11 g, with ^{137}Cs: 207 Bq, ^{54}Mn: 14 Bq, ^{65}Zn: 16 Bq, ^{57}Co: 5 Bq, ^{40}K: 7 Bq: 09.03.1988) (ZND89). These are standard calibration procedures applied in air radioactivity monitoring in the Vinca Institute.

Radionuclides were determined at gamma energies 661.6 keV for ^{137}Cs, 477 keV for ^{7}Be and 46 keV for ^{210}Pb. Minimum detectable concentrations, were derived from the lower limit of detection as LLD = $k^2 \pm 2LC$, where *k* is the coefficient of normal distribution corresponding to the confidence level of 95%, while *LC* is the critical level depending on background photo peak counts. Minimum detectable concentrations (MDC) of the radionuclides in air were: 1.0 $\mu Bq/m^3$ for ^{137}Cs, 10 $\mu Bq/m^3$ for ^{7}Be and 20 $\mu Bq/m^3$ for ^{210}Pb.

Counting time intervals were from 150 ks – 250 ks. Background spectrum integral mean count was 1.7 cps. Data were statistically analyzed on IBM/PS2 by SPECTRAN-AT and Genie 2000 programe. Total standard error of the method (including relative errors in geometric efficiency estimation, photopeak counts estimation, sample volume determination, etc.) was estimated below 25%.

Precipitation data for the period were obtained from the Department of Meteorology, Institute of Nuclear Sciences Vinca (DM Report 96-01).

5.2 Monitoring of ^{137}Cs, ^{7}Be and ^{210}Pb in Air in Belgrade Central City Area

Monitoring program on radionuclides in ground level air in Belgrade central city area started in 2002. The concentrations of radionuclides in air on different sites within the city differed significantly (45% for ^{7}Be and ^{210}Pb and about 80% for ^{137}Cs), but the overall seasonal variations pattern was similar to one obtained at Institute Vinca, outside the city. The average concentration values were (0.4 – 13.5) $x10^{-3}$ Bq/m^3 for ^{7}Be, (0.1 – 1.50) $x10^{-5}$ Bq/m^3 for ^{137}Cs and (0.1 - 11.0)$x10^{-4}$ Bq/m3 for ^{210}Pb. The lowest concentrations were measured at sites exposed to good ventilation, while higher ones were measured at sites with dense vegetation and poor ventilation. Since 2004, monitoring of air radioactivity continued on a site with heavy traffic and one of the most polluted areas in the city (Popovic

et al.,2000a; Todorovic et al., 2002; Todorovic et al., 2005; Todorovic et al., 2007; Todorovic et al. 2005a).
The average monthly concentrations of ^{7}Be and ^{210}Pb in air measured at the site in central city area, in 2004-2006 are presented in Fig.7 and Fig.8. The results for ^{137}Cs concentrations in air are not presented, as more than 80% of the values were below the lower limit of detection that was $1x10^{-6}$ Bq/m^3 (Todorovic et al., 2007).

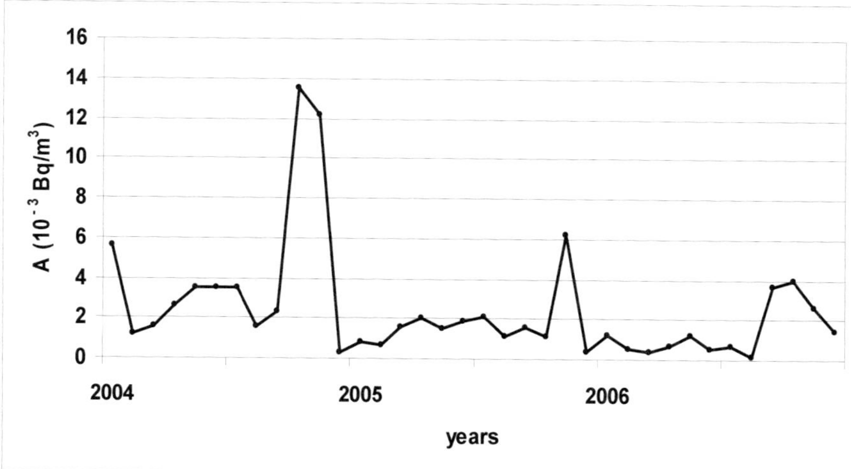

Fig. 7. Average monthly concentrations of ^{7}Be in Belgrade air (2004-2006) (Todorovic et al.,2007)

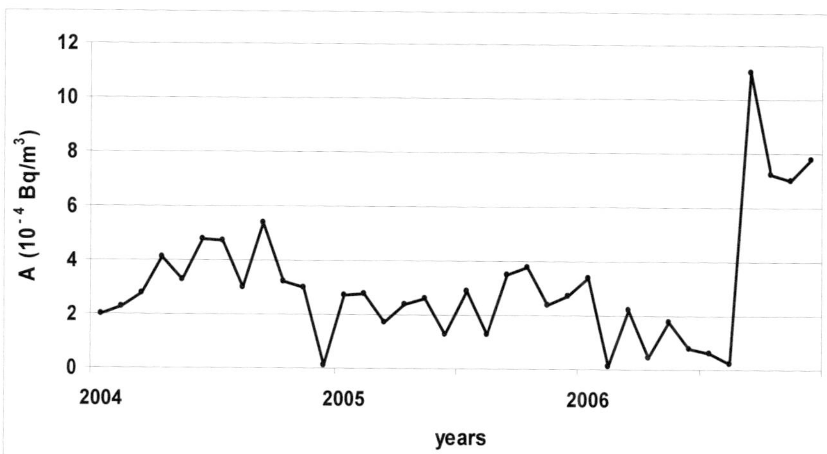

Fig. 8. Average monthly concentrations of ^{210}Pb in Belgrade air (2004-2006) (Todorovic et al., 2007)

The concentrations of ^{7}Be and ^{210}Pb measured in air in 2004- 2006 in Belgrade center city area were in the range of values measured in the same period at Institute Vinca, but the pattern of seasonal variations was different. This could be attributed to local climate conditions and short sampling period. Both ^{7}Be and ^{210}Pb exhibited seasonal variations with maxima in (late) autumns. The pattern of ^{7}Be variations is more pronounced than the one of ^{210}Pb variations, what is probably due to the vicinity of the highway and, apart from the different origin of the radionuclides, to the short period of sampling. Concentrations of both radionuclides are very low in winters, as the result of preticipation and show coverage effects (Popovic et al., 2000; Todorovic et al., 2007). Extremely low values of radionuclides concentrations in air are often due to technical problems (deposition of dust within the pumping system, air flow saffocation, etc), but in general sampling procedure is the main couse of errors in air radioactivity monitoring and control.

6. Conclusion

It can be concluded that monitoring natural and anthropogenic radionuclides in ground level air provide important information on the content of radionuclides due to their origin, weather and climate conditions (rate of precipitation, washout effects, temperature differences and inverstion, wind direction and intensity, troposphere-to-stratosphere transport and exchange, etc). The local topology, as well as anthropogenic factors such as the effects of traffic and heating plants are highly important, especially when monitoring radionuclides in air in urban areas. Sampling procedures, adequate measuring techniques and calibration are essential for providing reliable results and data. National and international standards regarding sampling sites and measuring laboratories network are to be precisely applied so as to provide high quality assurance and quality control in the air radioactivity monitoring system.

7. References

Agelaio,G., Cannizzaro, F., Greco, G., Rizzo, S. & Spitale, M.S. (1984). Sampling and concentration measurements of ^{7}Be and ^{137}Cs in ground level air at Palermo. *Health Physics* 47, pp. 96 –101

Andrasi, A. (1986). Monitoring the Radiation Consequences due to the Disaster at the Chernobyl Nuclear Facility from April 28 to June 12,1986, Prec.rep.,KFKI-1986-49/K

Arimoto,R., Snow, J.A., Graustein, W.C., Moody, J.L., Ray,B.J., Duce, R.A., Turekian, K.K. & Maring, H.B. (1999). Influences of atmospheric transport pathways on radionuclide activities in aerosol particles from over the North Atlantic. *J.of Geophyic .Research* 104 (D17) , pp. 301-321

Baeza, A., Delrio, L.M., Jimenez, A., Miro, C., Paniagua, J.M. &Rufo, M. (1996). Analysis of the temporal evolution of atmospheric berilium as a vector of the behavior of other radionuclides in the atmosphere. *J.of Radioanalytical and Nuclear Chemistry* 207 (2), pp. 331-344

Bettoli, M., Bonasoni, P., Calzolari, F., Colombo, T., Ravanelli, M., Tositti, L. & Tubertini, O.(1998). Determination of Atmospheric Radionuclides at Mt.Cimone Observatory. *Annali di Chimica* 88, pp. 769-770

BNM (1995). Bulletin du Bureau National de Metrologie: Situation de la métrologie de la radioactivité et des rayonnements ionisants: le Laboratorie primaire des rayonnements ionisants (LPRI), No.100, Paris

Cannizzaro,F., Greco,G., Raneli, M. Spitale, M. & Tomarchio, E. (1995). Behavior of ^{7}Be air concentrations observed during a period of thirteen years and comparision with sun activity. *Nuclear Geophysics* 9, pp. 597-607

CCEMRI 1991. Comité Consultatif pour les Étalons de Mesure des Rayonnements Ionisants, Rapport de la 12e session, BIPM.1991, Sèvres, France

CCEMRI 1994. Comité Consultatif pour les Étalons de Mesure des Rayonnements Ionisants, Rapport de la 13e session, BIPM.1994, Sèvres, France

CCEMRI 1996. Comité Consultatif pour les Étalons de Mesure des Rayonnements Ionisants, Rapport de la 14e session, BIPM.1996, Sèvres, France

Djuric,G. & Popovic, D. (1994) .Radioactive contamination in the environment. *Ecologica I/2*, pp. 19-24

DM Report 96-01. Department of Meteorology - Institute of Nuclear Sciences Vinca Annual Reports, Belgrade, 1996-2001.

DPL98 :Directive on permissable levels, measuring methodology and criteria for immission monitoring sites and collecting data, Službeni Glasnik Republic of Serbia 8.08.1998, No. 54, 1998.

Duenas, C.,Fernandez, M.C., Carretero,J., Liger, E. & Canete, S. (2004). Long-term variations of the concentrations of long-lived radionuclides descendents and cosmogenic 7Be and determination of the MRT of aerosols. *Atmospheric Environment* 38, pp. 1291 - 1301

EML Procedures Manual (1990/92). Quality Control and Quality Assurance in Radioactivity Measurements, HASL-300, US Energy Department. New York

Fisenne, I.M. (1993). Long lived radionuclides in the Environment, in Food and Human Beings, In: *CEC Report on Natural Radiation Environment, EUR 14411,* EN, Luxemburg

Gaggelar,H.W. (1995). Radioactivity in the Atmosphere. *Radiochimica Acta* 70/71, pp. 345-353

Gaffney, J.S., Orlandini, K.A. Marley,N.A. & Popp,C.J. (1994) Measurement of 7Be and 210Pb in rain, snow and hail. *J.Of Appleid Meteorology* 33, pp. 869-873

Gaffney, J.S., Marley, N.A. & Cunningham,M.M. (2004). Natural radionuclides in fine aerosols in the Pittsburgh area. *Atmospheric Environment* 38 , pp. 3191-3200

Gerasopoulos,E., Zerefos,C.S., Papastefanou,C., Zanis,P. & O,Brien,K. (2003). Low frequency variability of Berillium-7 surface concentrations over the Eastern mediterranean. *Atmospheric Environment*, 37, pp. 1745-1756

Groundel,M. & Postendourfen,J. (2004). Difference between the activity size distributions of the different natural radionuclides aerosols in outdoor air. *Atmospheric Environment* 38, pp. 3723-3728

Gustafson, P.F., Kerrigan, M.A.& Brar, S.S.(1981). Comparison of ^{7}Be and ^{137}Cs in ground level air. *Nature* 191, pp. 454-456

Hirose,K., Honda,T., Yagishita, S., Igarashi,Y. & Aoyama,M. (2004) Deposition behaviors of 210Pb, 7Be and thorium isotopes observed in Tsukuba and Nagasakim Japan. *Atmospheric Environmnet* 38, pp. 6601-6608

Hernandez,F., Hernandez-Armas, J., Catalau, A., Fernandez-Aldecoa,J.C. & Karlsson, L. (2005). Gross alpha, gross betha activities and gamma emitting radionuclides composition of airbotne particulate samples in an oceanic island. *Atmospheric Environment* 39, pp. 4057-4066

Ioannidou,A. & Papastefanou,C. (1997). 7Be,22Na and 210Pb in the Atmosphere At Mid Latitude 40oN, *The Nucleus* 34, 1-2, pp. 111-115

Ishikawa,Y., Murakami,H., Sekine,T. & Yoshihara,K. (1995). Presipitation scavenging studies of radionuclides in air using cosmogenic 7Be. *J.of Environmental Radioactivity* 26(1), pp. 19-36

Kandic, A., Spasic – Jokic, V., Todorovic, D. et al (1993), Intercomparison of semiconductor HPGe detectors, *Proceedings of full papers, XVII Conference of Yugoslav Society of Radiation Protection* pp. 195-198, Vinca, May 1993, Yugoslav Society of Radiation Protection, Beograd-Vinca (in Serbian)

Larsen, R.J., Sanderson,C. & Kada, J.(1995). EML Surface Air Sampling Programme, EML/572. US Energy Dept., New York

Manic, G, Petrovic, S., Manic, V., Popovic, D. & Todorovic, D.(2006) Radon concentrations in a spa in Serbia, *Environment International* 32, pp. 533-537

Peters, A., J., Gregor, D., J., Wilkinson, P. & Spenser, C.(1997). Deposition of ^{210}Pb in the Agassiz Ice Cap, Canada. *J. of Geophysical Research* 102(5), pp. 5971-5978

Popovic, D.,Djuric, G. &Todorovic, D. (1996). Radinuclides in building materials and radon indoor concentrations. *Radiation Protection Dosimetry* 63/3, pp. 223-225

Popovic, D., Djuric, G. &Todorovic, D. (1996a). Chernobyl fallout radionuclides in soil, plants and honey of a mountain region. *IAEA Tech.Reports* No.964, Vol.II, 432-437, Vienna

Popovic, D., Todorovic, D., Radenkovic, M. & Djuric, G. (2000). Natural and man-made radionuclides in ground level air in Belgrade area. *J.of Environment Protection & Ecology* 3, **S** pp. 124-130

Popovic, D., Djuric, G., Todorovic, D. & Spasic-Jokic, V. (2000) Radionuclides in building materials and radon concentrations in Belgrade dwellings. *Central European J.of Environmental And Occupational Medicine*, 6 (2/3), pp. 129-134

Popovic, D., Todorovic, D., Radenkovic, M. &DJuric, G. (2000a). Natural and man made radionuclides in ground level air in Belgrade area. *J.of Environm. Protection and Ecology*, I, S134-130

Popovic,D & Todorovic D. (2006).Radon indoor concentrations and activity of radionuclides in builiding materials in Serbia, *Facta Universitatis - Series: Physics, Chemistry and Technology*, 4/1, pp. 11-20

Popovic, D. & Spasic-Jokic, V. (2006). Consequences of the Chernobyl nuclear plant accident in Serbia. *Vojnosaniteski pregled* 63/5, pp. 481-487

Quin, T.J. (1996). Results of recent international comparisons of national measurement standards carried out by the BIPM. *Metrologia*, Vol.33, pp. 271-287

Smiljanic, R., Novkovic, D., Paligoric, D., Milosevic, Z, & Zaric, M. (1989). Determination of ^{137}Cs and ^{134}Cs in Air Samples Collected Near Belgrade /Yugoslavie/Following the Chernobyl Accident, *J.Radionalytical and Nuclear.Chemistry Letters*, 136/6, pp. 437

Spasic, V. (1982). Accreditation and supervision of institution for QC of measuring instruments and equipment, *Proceedings of full papers, ATESTIRANJE 82*, pp. A7.1-A7.8, Svetozarevo, Jun 1982, Federal Bureau for Standardization, Belgrade (in Serbian)

Spasic, V. (1984). Ionizing radiation metrology from the standpoint of Federal Bureau of Measure and Precious Metals, *Proceedings of full papers, XI JUKEM*, pp.137.1-137.8, Novi Sad, October 1984, JUKEM, Beograd (in Serbian)

Spasic, V. (1985). Metrological assurance of ionizing radiation quantities, *Proceedings of full papers, XIII Conference of Yug. Society of Radiation Protection,* pp 625-628, Pula, June 1985, Yugoslav Society of Radiation Protection, Zagreb (in Serbian)

Spasic, V. &Bek-Uzarov, Dj. (1986). Actual problems of legal metrology through the activity of Commission of Ionizing Radiation Metrology, *Proceedings of full papers, II Symposium of Yugoslav Society of Radiation Protection,* pp 165-168, Kragujevac, October 1986, Meeting of Yugoslav Society of Radiation Protection, Beograd (in Serbian)

Spasic, V. (1986). Background influences on low environmental activity measurements, *Proceedings of full papers, XII JUKEM,* pp. 522-526, Beograd, October 1986, JUKEM, Beograd (in Serbian)

Spasić, V. (1987a). Affirmation of measurement uniformity in the field of ionizing radiation in Yugoslavia, *Proceedings of full papers, 13. International Seminar ISEMEC on Ionizing radiation,* pp. 15-29, 13. Ljubljana, October 1987, ISEMEC *87:1*, Ljubljana

Spasic, V.(1987b). Officially verification and regulatory papers for semiconductor gamma ray spectrometers, *Proceedings of full papers, XIV Regional Congress IRPA,* pp. 425-428, Kupari, October 1987, IRPA, Beograd

Spasic, V. (1987c). Standard instruments in Ionizing radiation metrolgy, *Proceedings of fill papers, Workshop of ionizing radiation-measurement and protection,* pp. 61-88, Beograd, October 1987, JUKEM, Beograd (in Serbian)

Spasic, V., Paligoric, D.& Novkovic, D. (1987), Metrological assurance in radionuclide metrology, *Proceedings of full papers, XIV Conference of Yugoslav Society of Radiation Protection,* pp.425-428, Novi Sad, June 1987, Yugoslav Society of Radiation Protection, Beograd (in Serbian)

Spasic -Jokic, V. (1998). Ionizing Radiation Legal metrology, *Proceedings of full papers, Conference of Yugoslav Metrology, Academy of Science,* pp.321-332, Belgrade, May 1998, Federal Bureau of Measures and Precious Metals and Academy of Science, Beograd (in Serbian)

Spasic - Jokic, V., Popovic, D., &Djuric, G. (1998), Radionuclide activity measurements in environment: questioner data analysis, *Proceedings of full papers, Symposium of measurements and measuring techniques* pp. 735-741, Beograd, October 1998, Federal Bureau of Measures and Precious Metals, Beograd (in Serbian)

Spasic-Jokic, V., Popovic, D. &Djuric, G. (2002). Quality Assurance and quality control in environmental radiation protection metrological and legislation system in Yugoslavia, *Proceedings of full papers, 5-th International Conference of Balkan Environmental Association BENA,* pp.299, November 2002, BENA, Beograd

Spasic- Jokic, V., Orlic, M. & Djurovic, B. (2006). Risk of potential radiation accidental situations as TESLA Accelerator Installation, *Proceedings of full papers, Second European IRPA Congress on Radiation Protection: From knowledge to action,* CD Rom, P4111-17, Paris, May 2006, Paris, French Society of Radiation Protection, Paris

Todorovic, D., Djuric, G. & Popovic, D. (1994). Influence on the quality of shielding on the evaluation of the contenmts of natural radionuclides in the environmental samples. In *»Strahlenschutz: Physic und Messtechnik«* Eds. W. Koelrer and R. Manshart, pp. 331-335, IRPA Publ.Ser.on Radiation Protection, FS 94-71T, Munich

Todorovic, D., Smiljanic, R. & Manic, S. (1996). Thirty Years of Air Radioactivity Monitoring in Vinca Institute, *Ecologica* 10(2), pp. 33-40 (In Serbian)

Todorovic, D. (1997). The effect of tropopause height on the content of radioactive debris in surface atmopshere, *Environmental International* 23(6), pp. 815-818

Todorovic, D., Popovic, D. & Djuric, G. (1997). Activity of Cs-137 in air before and after the nuclear plant accident at Chernobyl. *Vinca Bulletin* **2**(1), pp. 635-638

Todorovic,D., Popovic,D., Radenkovic, M. & Djuric,G. (1999). Concentration Measurements of ^{7}Be and ^{137}Cs in Grund Level Air in Belgrade City Area, *Environmental International* 25(1) pp. 59-66

Todorovic,D., Popovic, D., Djuric, G.& Radenkovic, M. (2000). 210Pb in ground level air in Belgrade city area, *Atmospheric Environment* 34 (19) pp. 3245-3248

Todorovic,D., Popovic, D., Radenkovic M. & Djuric,G. (2002). Concentrations of 7Be, 137Cs and 210Pb in ground level air in Belgrade area from 1985 to 2001. In *Applied Physcs in Serbia,* SANU, XVCIII (2/1), Eds. S.Koicki, N.Konjevic, Z.Petrovic, Dj.Bek Uzarov, pp. 63-66

Todorovic, D., Radenkovic, M., Popovic, D. & Spasic-Jokic, V. (2003). Environmental and radiation protection laboratories, *Proceedings of full papers, BPU 5. Fifth General Conference of the Balkan Physical Union*, pp SP12-019, Vrnjacka Banja, August 2003, Balkan Physical Union, Beograd

Todorovic, D., Radenkovic M., Popovic, D. & Tasic S. (2005). Ground level air radioactivity monitoring in urban areas, In: *Recent Advances in Applied Physics*, Ed. A.Mendey-Vilas, Elsevier, Amsterdam, pp. 479 - 483

Todorovic, D., Popovic, D., Djuric, G. & Radenkovic, M. (2005a). ^{7}Be to ^{210}Pb concentration ratio in ground level air in Belgrade area. *J.of Environmental Radioactivity* 79 (2005), pp 297-307

Todorovic, D.; Popovic, D.; Rajsic, S.& Spasic,M. (2007). Radionuclides and Particulate Matter in Belgrade Air. In: *Environmental Research Trends*, Ed.M.A.Cato, Nova Science Publishers, Inc., New York, pp 271-301

Tokieda, T., Yamanaka, K., Harada, K. & Tsunogai, S. (1996). Seasonal variations of residence time and upper atmospheric contribution of aerosols studied ^{210}Pb, ^{210}Bi, ^{210}Po and ^{7}Be. In: *Tellus 48B*, pp. 690-702

ZND89. Protocol on the coordination on a project of atmospheric radioactivity investigation. SEV Intercomparision Programme. Sofia, SEV, 1989.

11

Improving the Grapevine Technology by Optimising the Utilisation of the Environmenthal Resources in the Murfatlar Vineyard

Aurora Ranca
Research Station for Viticulture and Oenology Murfatlar, Romania

1. Abstract

Optimizing the vine breeding technology in water stress conditions is useful to the wine producing companies from the Dobrogea area (an area with a pronounced dry climate).

This research was financed by the Research and Education Ministry through a special programme named AGRAL. It took place in 2001-2006 interval on Columna variety on a 3 ha plot.

The main objective of the research is the improving the actual vine breeding technologies in the Murfatlar vineyard by better use of the environmental resources in the condition of climate changing.

The conversion of intensive technology into a new balanced exploitation of the vine ecosystem was taken into consideration.

In the Murfatlar vineyard the climatic conditions of the latest six years has a different evolution in comparison with the multiyear mean, causing several specific features in the vine growing process. The climatic changes - the global warming processes affected also the quality of harvest.

Following these causes it must be adapted the breeding technologies in order to better use the environmental resources and to protect the vineyard ecosystem.

The main problem is to keep the healthy status of the grapes with a specific range of bio-selective pesticides and to choose the most efficient irrigation system.

2. Introduction

The integrated production concept is used in the agricultural system which obtain high quality food which utilizes resources and mechanisms of natural control aiming at replacing the interventions which are harmful for the environment and which assure a sustainable agriculture on the long term.

The *objective of the integrated production* (Bull. OILB, 1993, 2000) are:

- It is a holistic concept which assures the preserving of the ecosystem's equilibrium by biodiversity and natural resources conservation

- Undesirable side-effects are reduced (ex: the contamination of potable water with pesticides and nitrites) (Baicu, 1986,1996)
- It is a systemic approach centered on the whole farm as a basis unity, important strategies such as equilibrium of the nutrients circuit and optimum utilization of environmental resources can be applied
- Man is the key factor in the realization of the integrated production: his perspicacity, motivation, need to be prepared professionally are request needed to make a durable agriculture
- The nutrients circuit needs to be balanced and the loss must be minimum
- The soil fertilization need to be preserved and ameliorated
- Taking the decision regarding the protection of plants is done in concordance with the integrated control; integrated protection is applied to the harmful organisms (phytophags, weeds) "the fight" meaning eliminating the part of the pest population which generates effective losses
- Biological diversity needs to be encouraged regarding both genetic level and species from ecosystem, these forming the backbone of the ecosystem's stability, factor of the natural control and the landscape quality
- The product's quality needs to be evaluated regarding both the ecological parameter of the production system and through the classic criteria of internal and external qualities (Tardea, 1995)
- In viticulture obtaining an integrated production requires the preserving and fertilizing of the soil with moderate doses on one hand, and on the other hand the exact determination of the disequilibrium produce by "key" diseases and pests, on which the fate of the harvest relies
- From the conclusion of the research made on this topic in Romania we mention:
- The soil work must be limited, done at the right time and the fresh organic matter must not be buried deep (Dejeu, 1997, Olteanu, 2000)
- The fight against weeds: is recommend on one hand preemptive methods such as removing the dissemination sources, early spring soil ploughing, the soil's covering with dry herbs or other biological materials, on the other hand, the control will be done mechanically, manually and by burning
- The soil's inhering is the ecological and economic alternative for preserving the soil with a "equilibrating" role of the soil components (Condei, 1991)
- Using green fertilizers contribute to the reduction of the soils erasure, ameliorates the structure, being a control means of the chrolosis (Cotea, 1996), mainly through the cultivation of leguminous plants (Davidescu, 1994, Catanescu, 1987)
- The integrated control of the diseases and pests in viticulture relies on warning through the calculation of PED (the economical harm threshold), elaborating control schemes of disease and pests, using biological means (biological insecticides, pheronomal traps and others) (Filip, 1994), elaborating datum bases and foreseeing models (Ranca, 1998)

The Research Station Murfatlar is directly involved in research and extension in vineyards located on southeast of country area of this integrated breeding system.

The Murfatlar vineyard is situated in the Dobrogea region, in the South-East part of Romania, between the Danube river and the Black Sea coast.

The climate and the soil of the region make the quality of the Murfatlar wines.

Vineyards are placed on plateaus with blunt edges or slow slopes, S – SW oriented. The climate is excessive continental with large quantities of solar radiation, rich heliothermic resources and one of the driest climate of Romania. The influence of the sea plays a positive role in autumn, as thermal regulator.

Some *climatic data* (average for the last 40 years) are presented below:

- total annual temperature: 4103 degree C
- total active annual temperature (over 10 degree C): 3657 degree C
- number of possible vegetation days (with temperature higher than 10 degree C): 179
- total sunshine hours (April-October):
  - potential : 2936; real: 1604
- total annual precipitation: 413 mm
- total precipitation from April to October: 195 mm
- mean annual temperature: 11 degree C

The soil is mostly carbonated brown chernozem with 2-4 % humus, 7.5-8.3 pH, total $CaCO3$ - 3-12 % and parental rock- loess.

Main breeds in vineyard are:

- for white wines: Pinot gris, Chardonnay, Sauvignon, Muscat Ottonel, Riesling Italian, Columna.
- for red wines: Pinot noir, Cabernet Sauvignon, Merlot, Feteasca neagra, Mamaia.

Direction of production is superior dry, half dry and sweet; white and red natural wines; special wines on high quality.

Mean production on hectare is 6 to 10 to/ha after the type of breed and the type of wine that will be obtained.

Sugar content at the harvest time is 196 - 260 g/l , depending on breed and wine type to be produced.

3. Material and Methods

The pilot plot is represented by a plantation with the Columna variety (Romanian variety for the white wines, created at Murfatlar), established in 1995.

The vines are grafted on the rootstock Berlandieri x Riparia – Sel. Oppenheim 4, the plantation distance being 2.5 x 1m, the form of leading being Guyot on semi-trunk.

The soil is typical chernozem with a favorable fertility to the growing vine – having middle humus content: 2.7% and small to medium supplier with nutritive substances.

The general slope is 2% with southern exposure.

The following activities have been accomplished:

- The monitoring of the main climatic factors and of the water status content of the soil;
- Establishing and executing of some technological research with minimal inputs – reducing the number of the agro-phytotechnic work, drip irrigation and fertilization
- Establishing a control program of the health status of the plantation accentuation on the biotechnical methods;
- Utilizing and evaluating the efficacity of a drip irrigation system.

4. Results and Discussions

Climate conditions in the 2001-2006 period have had a different evolution in comparison to the multi-annual mean (tables 1 and 2) determining a series of peculiarities in the growth and development processes of vines.

Analyzing the thermal evolution we notice a moving of the maximal values of the annual and real (over the 10 ^{0}C – the biological threshold of the vine development) thermal sums from the summer months towards the spring ones, respective in the March-June interval. During all years the annual temperature mean was higher than the multiyear mean registered between 1960 – 1990 period.

The 2001 year was excessively dry, the precipitation fallen in the vegetative period (89.3mm) represented only 32% of the multi-annual mean during the same interval. The other years are characterized through a non-uniform repartition of the precipitations, thus moths without precipitations (January, February, May) and months with excessive precipitation (March, September, October in 2002, August in 2003 and 2005, September in 2004, July in 2006) being observed.

In this context of hydric deficit in soil the atmospheric dryness has been present with unusual strength in the May – October interval 2002 and in the April-June months in the most of the other years. The drought effect can be observed through the little growth of the shoots and sometimes through the involution of the flowers in tendrils. In august 2001 a falling of the leaves from the shoot top was noticed. This disturbed the physiologic processes of the plants.

The precipitation fallen in the autumn-winter period was insufficient thus assuring a humidity accessible normally only in the surface horizon of the soil (0-20 cm). Alternatively on the 20-100 cm depth, where the maximal spreading of the vine roots was registered at the end of March a deficit of utile water of over 600 mc/ha. The deficit doubled at the end of the summer months.

Month	Multiyerly mean	2001	2002	2003	2004	2005	2006
I	0,5	3,5	0,1	0,8	-2,5	3,3	-3,2
II	1,3	2,7	7,3	-4,3	2,1	-0,7	0,3
III	4,2	9,7	8,5	3,4	8,6	4,5	7,1
IV	10,2	10,8	10,3	9,0	12,3	11,2	11,5
V	16,2	16,1	18,5	21,2	15,8	18,0	17,1
VI	20,4	21,2	23,3	24,2	21,4	21,9	22,7
VII	22,6	27,9	27,2	25,2	23,9	24,3	24,2
VIII	22,6	25,2	23,4	24,7	22,8	24,4	26,3
IX	17,6	20,0	18,1	16,3	18,6	20,5	18,6
X	12,0	13,4	13,2	11,5	15,7	12,1	13,8
XI	7,2	5,6	9,9	6,8	8,8	6,4	7,1
XII	2,3	-2,3	-1,4	2,6	4,1	3,1	4,7
Year mean	***11,0***	***13,0***	***13,3***	***11,8***	***13,8***	***12,4***	***12,5***

Table 1. The average monthly of temperature, Murfatlar, 2001-2006

Year Month	Multiyearly mean	2001 Rains (mm)	2002 Rains (mm)	2003 Rains (mm)	2004 Rains (mm)	2005 Rains (mm)	2006 Rains (mm)
I	31,0	0,0	0,0	35,1	60,5	35,1	17,7
II	33,0	49,0	7,4	30,7	12,0	30,7	35,6
III	21,7	43,2	101,9	19,2	28,2	19,2	76,6
IV	33,5	18,0	18,4	1,4	24,2	1,4	34,1
V	50,2	11,9	7,2	140,8	7,8	140,8	73,7
VI	53,2	24,3	25,9	65,2	22,4	65,2	16,7
VII	35,6	1,8	20,7	31,9	31,4	31,9	130,1
VIII	31,6	2,0	33,9	144,0	16,1	144,0	69,9
IX	41,6	28,5	71,9	11,6	167,4	11,6	86,6
X	30,2	2,8	118,8	8,9	74,1	8,9	4,8
XI	40,4	53,5	27,9	25,0	9,3	104,1	25,8
XII	34,0	37,3	45,7	19,5	93,6	29,9	7,5
Sums	**419,8**	**288,6**	**479,7**	**533,3**	**547,0**	**622,8**	**579,1**

Table 2. Monthly and yearly precipitations, Murfatlar, 2001-2006

Establishing and executing of technological work with minimal input:

- Two variants of pruning where practiced with a 24 and 32 buds on plants
- No chemical treatment was used of the weeds, choosing instead to manually remove them
- The some work was eliminated on the canopy such as tying the shoots on fills, cutting the top of shoots, the removing of the leaves during the vegetation period
- 2 irrigation were applied of 200 mc/ha each and 2 fertilizations (100kg/ha) with a complex fertilizer with microelements (N:P:K - 19:19:19 supplemented with Mg and Mn) at the flourishing and berry growing phenophasis.

Establishing a program of integrated control

A fight integrated program of the main diseases and pests in the pilot plot was established and executed, the highlight being on utilizing biotechnological methods. In the plantation was used sexual pheromones traps called ATRABOT for keeping the pest population of grape mouth (*Lobesia botrana* Den. et Schiff) under control (figure 1).

The integrated control scheme applied in the conditions of 2003 year on the Columna variety at Murfatlar (table 3) assured an identical efficacity with the one of the classical scheme (chemical treatments) at a lower cost. Thus this cost 170 euro/ha versus the 200 euro/ha of the classical scheme.

The number of treatments was smaller by one, both for manna and for odium and the botrytis of the grapes was controlled through 2 partial removing of leaves at the grapes level. The main pest of the Murfatlar vineyard, grape moth, was controlled by using of 3 traps/ha achieving an efficacy close to the level of chemical treatments.

Using and evaluating efficacyty of the drip irrigation system

The un-uniform distribution of precipitation in the vegetative period, the constant water deficit in the soil which registered values between 100 and 1200 mc/ha and the strong winds specific of the area make the using of irrigation in vine plantation in the Murfatlar region necessary.

Fig. 1. ATRABOT traps in Columna experimental plot

The disease or pest controlled	Integrated control scheme				Clasic control scheme			
	Used product and the dose - l or kg/ha	No. of treat.	Efic. comp. with the witness un-treated	Cost Euro /ha	Used product and the dose- l or kg /ha	No. of treat.	Efic. comp. with the witness un-treated	Cost Euro /ha
Manna	Champion -3,0 Ridomil gold -2,5	1 2	87 94	20 70	Mancozeb -2,0 Curzate Manox -2,5 Bordelese -7,5 Champion-3,0	1 1 1 1	81 88 77 87	20 40 10 25
Oydium	Topas 0,2 Wet Sulph -5,0	2 2	96 88	30 20	Sulph dry -20,0 Sulph -5,0 Karathane-0,5	2 2 1	82 90 95	10 20 15
Botrytis	Two partial leaves removing	-	89	20	Konker -1,5	1	95	30
Grape moth	3 pheromonal traps	-	85	10	Decis -0,4 Larvin -1,0	2 2	94 98	10 20
Total			**90**	**170**			**89**	**200**

Table 3. Integrated control scheme for mains vine diseases and pests, Columna variety, 2003 year at Murfatlar

Very important is to choose the most efficient irrigation system taking into consideration that the water quantity used for one vine hectare irrigated by sprinklers (most frequent system in the area) is between 800-1000 mc water.

An optimum solution is to choose the less water consumer system, like drip irrigation.
The efficiency of utilizing water in the case of the sprinkle irrigation system is 50% compared to the 85-90% of the drip irrigation (Ranca, 2005). Once with the local application of the water in reduced quantities we can concomitantly apply the fertilization with soluble fertilizers.

The description of the drip irrigation system existing on the pilot plot

In this step the system serves a single plantation, the Columna one, but allows for the ulterior linkage of other plots. The bringing of water is realized through a PEHD 80D 50mm conduct on the distance of 230 m, fueled from a common underground network of irrigation water supply.

The first filtering is done with a main sand filter that assures a 230 mc/h debit and the second filtering is done with a small plastic filter having a filtrating surface of 527 cm^2, having a debit of 28.8 mc/h. The filtering is preview with two manometers on entrance and exit and evacuating cock.

The fertilization is done through an injector integrated in a circuit with a cock access that assured a medium debit on the injector of 501 /min (3 mc/hour) of fertilized solution injected in the irrigation water.

Distribution is done with tubes PEHD D50 mm respectively 40 mm for the distribution antenna water on the plant rows (figures 2 and 3). After the technological probes and the determining of the functional parameters of the installation will have introduce an automatic programming system.

Fig. 2. Dirip irrigation distribution on the plants row in the Columna experimental plot

The watering system is previewed with plastic tubes with double incorporated drippers situated at 60 cm from the other on the tubes length. Each dripper assures a 3 l/h debit at a supplied water pressure of between 0.5 and 2 atmospheres, having from the construction a pressure regulator

The functional parameters of the installation are: maximum debit assured: 34 mc/h, maximum watering intensity: 26 mc/h/ha. A standard watering of 200 mc/ha can be done in 8-10 hours on normal functioning parameters. From this distribution center the water can be supplied to other parcels, alternatively, with the condition that it must not surpass the distance from the station as to not lose pressure.

The benefit of implementing drip irrigation system in vineyards

There are multiple direct and indirect advantages in using drip irrigation system on the vineyards located in dry or half-dry areas.

The ***direct advantages*** are:

- The ***water quantity is reduced*** mainly in the areas with a continental semiarid climate with lack of water resources
- Need of ***labor force is lowed***
- Small and controlled ***fertilizers can be applied*** (fertigation) increasing the efficiency of these and reducing pollution risks
- The ***intervention of the soil*** level was ***diminished,*** being protected the chemical and physical soil proprieties
- Slope areas can be cultivated without the danger of washing of fertile level of soil
- The risk of phreatic water contamination with fertilizers is extremely reduced.

Fig. 3. The watering surface in drip irrigation system

Between the ***indirect advantages*** are:

- Unfavourish the weeds growth, the interval between plants is dry
- The cryptogrammic diseases risk is lower and the costs of treatments is diminished
- There is the possibility for underground irrigation by using herbicides in this system

➢ This system permits applying of different breeding technologies, which lead to *increase the grapes quality* (table 4).

The wine quality was studied in comparison with a witness obtained on a classical breeding technology.

The Columna wine has a constant quality on the experimental period assured by the new technology: water supply on necessary time and of appropriate quantity, a rational fertilization, preserving a good status health of plant, respectively grapes. In opposition the wines obtained on a witness Columna plot has hard influences due to the lack of water at the principal developing moments of plants (flourishing, shoots growing, berry forming) and of the main physiological processes (photosynthesis rate, transpiration, stomata conductance). So, the witness vines are in all this years (exception 2006 year), below the quality of witnes obtained by the integrated technology (figure 4).

Un-irrigated plot			Sprinkler irrigation			Drip irrigation		
Tones/ ha	Sugar g/l	Acidit.g/l H_2SO_4	Tones /ha	Sugar g/l	Acidit. g/l H_2SO_4	Tones /ha	Sugar g/l	Acidit. g/l H_2SO_4
6,4	183,6	4,1	8,2	181,6	5,2	8,0	190,08	4,8

Table 4. Characterisation of harvest - Columna – average data 2001-2006

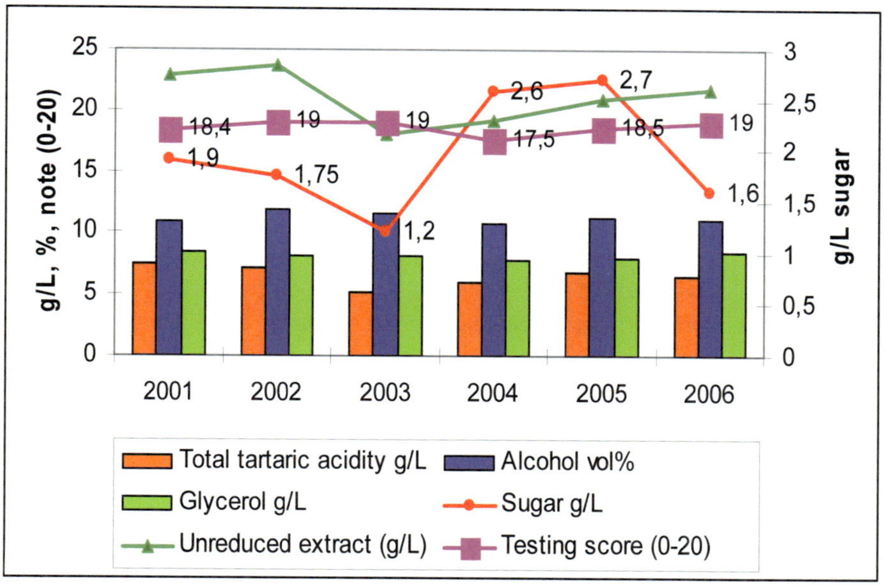

Fig. 4. The oenological potential of the wines obtained from Columna variety

5. Conclusions

The yearly climatic condition represented mainly by thermal, lightening and hydro parameters specific for the vineyard biotope are directly related with growing and developing rate of plants, changing the active period of the these influencing the achievement of a constant and qualitative harvest (Jones, 2005, Stock, 2005).

The improved technology applied in the Columna plantation at Murfatlar proves to be not only ecologic, diminishing the number of interventions, putting the accent on a integrate scheme of the control of the health plants but also is economically profitable, the total annual costs being with 15-20% smaller.

Drip irrigation system is usefull mostly in the spring and early summer months for to supply the water needs of the vine plants in the intensive growth processes florish, berry formation and growing).

6. References

Baicu T., Savescu A. – *Sisteme de combatere integrata a bolilor si daunatorilor pe culturi.* Ed. Ceres, 1986.

Baicu T. – *Principles of intedrated pest and disease management.* Ed. Ceres, Bucuresti, 1996.

Condei G., Ciolacu, M., et.all. – L'approche ecologique du systeme entegre d'entretien du sol en plantations viticoles intensives. *III Symp. intern. sur la non-culture de la vigne et les autres techniques d'entretien des dols viticoles,* Montpellier, 1991, 289-296.

Cotea V, Cotea V. – *Viticultura, ampelografie si oenologie.* Ed. Did. si Ped. Bucuresti, 1996.

Davidescu D., Davidescu V. – *Agricultura biologica – o varianta pentru exploatatiile mici si mijlocii.* Ed. Ceres, 1994.

Dejeu L., Petrescu C., Chira A. – *Horticultura si protectia mediului,* Ed. Didactiac si Pedagogica Bucuresti, 1997.

Filip I. – Realizari ale combaterii moliei strugurilor in acord cu natura. *Combaterea integrata a buruienilor,* p/ 270-277, Constanta, 1994.

Jones G. V., With M. A., Cooper O.R., Storchmann K., 2005 – Climate change and global wine quality. *Climatic Change* 73, 319-343.

Olteanu I. si colab. – Influence of the integrated technological system under ecological control on some physiological and biochemical processes with impact on the viticultura yield. *Analele Univ. Bucuresti,* 2000.

Ranca A. – Studii preliminare in modelarea dezvoltarii moliei strugurilor la Murfatlar, *Analele ICVV,* vol. XVI, 1998.

Ranca A. - Key concerning vine breeding in the dryness conditions at the Murfatlar vineyard, *Analele Univ. Ovidius Constanta Seria Geografie Vol. II,* Univ. Ovidius Press., 2005

Stock M., Gherstengarbe F. W., Karstschall T., Werner P. C., 2005 – Reliability of ClimateChange Impact Assessments for Viticulture. *Acta Horticulture,* 689, ISHS, 29-39.

Tardea C, Dejeu L., 1995, *Viticultura,* Ed. Didactica si Pedagogica Bucuresti, 120-126.

XXX, Production integree. *Bull. OILB,* 16, p.41-61, 1993.

XXX, *Bull. IOBC* 23, 2000

12

Integrated Sustainable Fisheries Management for Pearl Mullet of Lake Van, Turkey

Mustafa SARI
Fisheries Department of Agriculture Faculty of the Yuzuncu Yil University Turkey

1. Introduction

The pearl mullet is the sole endemic fish species that can survive in the salty and alkaline waters of Lake Van. Covering an area of 3712 km^2, with mean depth of 171 m, maximum depth of 451 m, and 1648 m above sea level, this is Turkey's largest lake. Its waters are extremely alkaline and salty. Due to the nature of these waters, it is characterized as a "soda lake". Researchers have reported a pH level of approximately 9.8, with saltiness at 0.19% (Kempe *et al.*, 1978). The lake's biological diversity is significantly different from both fresh and salt waters. Its phytoplankton resources consist of 103 species belonging to the *Diatome, Bacteriophyta, Cynophyta, Chlorophyta, Flagellate* and *Phaeophyta* groups, and its zooplankton resources include 36 species from the *Rotatoria, Cladocera* and *Copepod* groups (Selcuk 1992). The pearl mullet (*Chalcalburnus tarichi*, Pallas 1811), a species belonging to the *Cyprinidae* family, is the only fish that can survive in Lake Van. The pearl mullet is a migrating species. Although it generally lives in the lake, it immigrates to the surrounding freshwater rivers for reproduction purposes and returns to Lake Van after the reproduction period.

The earliest available information on the fishery practices related to the pearl mullet comes from travelers that visited the Lake Van region. Among them, Evliya Celebi (17th century) provides the most detailed account. Having arrived to Van with a general, Melik Ahmet Pasha, Celebi speaks of the lake as the "Van Sea", stating that a certain species of fish exist in these waters that migrate annually to the Bendi Mahi River creek for a month as a flock. He goes on to explain that the fish are captured on their way back to the lake by the provincial treasurer, salted and then taken to the Iran, Nagorno Karabakh, and Azerbaijan area to be sold. He says 900 loads of silver coins are thus collected for the treasury each year and this income is then distributed among the soldiers serving in the fortress around the lake (E. Celebi, 17th Century). The first detailed study on the biology of the pearl mullet was carried out by Akgul (1980); further research was conducted by Danulat&Selcuk (1992) and Danulat & Kempe (1992) revealing new parameters about some physiological and biological characteristics not articulated in the previous study. Other studies on the pearl mullet include: Arabaci's (1995) research on physiological changes that take place during reproductive migration; Cetinkaya et al's (1995) on some selectivity characteristics of nets used for pearl mullet fishing; Sari's (1997a) research which presents the basics of current fishery management practices and proposes an alternative model based on estimated pearl mullet stock quantities; Sari's study (1997b) on mesh selectivity for pearl mullet fishery; Sari and Tokac's research (1997) on the technical features of Lake Van fishing boats; Sari & Ipek's study (1997) specifying new fishing areas around Lake Van, drawing on satellite imageries; and Sari & Tokac's research (2000) about the productivity of nets used for fishing the pearl mullet.

As of 1997, technological approaches have prevailed in fishery management at Lake Van. With the establishment of a center at the Yuzuncu Yil University focusing specifically on this field, these studies have acquired an institutional identity (Sari 2000a). Sari (2000b) has declared that new fishing grounds have been successfully determined with the aid of the remote sensing and geographic information system. Using the daily AVHRR satellite images taken from the NOAA HRPT ground station situated at the remote sensing center to determine the surface temperature map (SST), in conjunction with depth and flow data, this study has verified that the pearl mullet's distribution area can be successfully established. The effects of these applications, which will ease the transition from the erroneously administered spawning period fishing practices to professional fishing, have been clarified in detail by Sari (2001).

In the event of a failure to prevent spawning period fishing, the new technology will be an added fishery pressure on the pearl mullet population; in light of this possibility, the need for cooperation among relevant public authorities, and local, national, and international non-governmental organizations is clearly an imperative. This paper is a detailed examination of the transition process to sustainable fishery of the pearl mullet, the role of relevant governmental and non-governmental bodies, security forces and universities within this process.

2. The Biology of the Pearl Mullet (*Chalcalburnus tarichi*)

A member of the Carp family *(Cyprinidae*), the pearl mullet is a fish species that only inhabits the Lake Van Basin. Generally of bright-silver color, its back is grayish green, and the abdominal region is silver (Figure 1). Its body is covered with small scales, and its eyes are large (Kuru 1987, Geldiay & Balik 1988). On the average, it measures around 19.5 cm in length, and has an average weight of approximately 80 g. It feeds on phyto and zooplanktons. Its average life span is around 7 years, and the fish reaches reproductive maturity at 3 years old. The reproduction period, at which time the fish immigrate to freshwater rivers in flocks, starts in early April and lasts through the end of July. As they cannot make the move from salty-alkaline water to freshwater directly during the migration, the fish have a brief waiting period at the river mouths for osmotic adaptation, both on their way to the area where they will lay their eggs and back. As the river water temperature reaches approximately 12 °C, the pearl mullets go into the freshwater and start to lay their eggs. Having laid the eggs in small pebbly, sandy areas where the river tends to sprawl out and the speed of the flow is lessened, the adults return to the lake.

Their youngsters start on the journey towards the lake within a week or two after the eggs are hatched. The young pearl mullets wander and feed in flocks around the nutritiously rich shore areas of the lake throughout the summer. Pearl mullets tend to disperse all over the lake in the summer months, but avoid areas where the water is deeper than 25 m. In the wintertime, the fish move within parts of the lake that have a maximum depth of 60 m (Sari 2001, Sari 2003).

Fig. 1. Pearl Mullet (*Chalcalburnus tarichi*)

Fig. 2. Lake Van, important settlement areas surrounding

3. Historical Development of Pearl Mullet Fishery and its Management

The pearl mullet is captured by two different fishing methods, in two different seasons. The fish migrates to freshwaters in order to breed between the months of April and July. However it is not able to have direct access from the salty-alkaline waters of the lake to the freshwater rivers. Due to its biological constitution, the fish must undergo an osmotic adaptation process whereby it needs to wait for a certain time at the river mouths known as "mansap". During this "waiting" period, large flocks gather within the river mouths. The first fishing method involves the capture of the pearl mullets, by casting beach seined nets along the shore (beach seining), from small fishing boats as they are waiting to continue on their reproductive migration by the river mouth. Simple traps laid along the rivers are also employed to capture the fish during this period. Whereas 90% of the total pearl mullet fishing was done by this method in 1996, by 2006 around 40% was done during the breeding period. The local population has been employing this traditional method, capturing the fish during its migration period since ancient times (Sari 1997b, Sari 2001).

The second method is "winter or professional fishing", whereby the fish is captured at Lake Van, which is its main habitat, between September and April. During this period, fishermen use 8-16 m boats with trammelnets that mesh sizes of 20-22 mm. They start fishing at depths of around 15-20 m in September, proceed to 50-60 m as the weather gets colder, and go back to 20-30 m depths as spring approaches and temperatures get higher. This fishing technique was started during the 1970s and is becoming increasingly more widespread. Nevertheless, at present only 60% of pearl mullet fishing is being carried out in the winter months (Sari, 2006).

Until the 1960's, pearl mullet fishing was at a minimum due to several reasons such as lack of adequate fishing gear and equipment, and the fact that fresh fish consumption culture had not yet developed among the local populace. In those years, the fish could not find a place to migrate for breeding purposes, and sometimes went as far as the irrigation canals at the surrounding fields. As a result, piles of dead fish would be seen for days on end by the river shores in the springtime (Sari 1997b). Albeit in small amounts, all fishing activity was carried out only during the breeding period. Since it affected a minimal part of stock distribution within the lake, there was no need for fishing regulations. However after the 1950s, as fishing activities that took place during the breeding season took on a commercial characteristic and as "winter or professional fishing" also started to become more prevalent in the 1970s, pressures increased pearl mullet fishery. Following the '70s, it became imperative to take some administrative measures.

The first regulatory measure related to pearl mullet fishery was the "closed season", put into application during the reproductive migration period even if it was for a short duration. The dates through which fishing was prohibited during the reproduction period (closed season), and hence the number of days when the lake was off-limits for pearl mullet fishers, kept changing each year due to political or social conditions. However in the 1980's, the ban became more or less consistent and started to be applied between the dates of May 15 and July 1st. In the years that followed, while there was noticeable negligence in monitoring illegal fishing activities, it was also observed that for some reasons, the dates of the closed season on Lake Van was haphazardly designated in different regions and the fishing ban, devised as precautionary management strategy, was abolished in practice, even if not legally. In the 1980s, along with the above mentioned closed season application, a limit was also set on the mesh size to be used, both for winter and reproduction period fishing.

However, as in the case of the closed season application, since the limitations on trammel net mesh sizes were based on the views and demands of the fishermen rather than scientific data, this resolution too failed to be effective (Sari, 1997b). In addition, fishing during the reproduction season was encouraged in a way, because the National Real Estate Organization rented out the river mouths exactly at this period. Unfortunately, this practice was continued until the lastet few years.

As a result of these management strategies, the 600-ton pearl mullet yield in 1967 showed an increasing trend, which can be seen when production is analyzed in 10-year periods as follows: 4000 tons in 1977 (approximate increase of 700%), 10000 tons in 1987 (250% increase), 21000 tons in 1997 (200% increase) and has reached the highest yield within the inland fish production category, with 15654 tons in 2000. Whereas the ratio of pearl mullet production to total inland fish production varied between 5-9% in the initial years, recently this figure has gone up to 36%. This increasing trend in pearl mullet yield has shown a decline in the last years due to overfishing. Sari (2001) has pointed out that pearl mullet fishery has taken a course that corresponds to the "the theory of development of uncontrolled fishing", stating that preventive measures must be taken.

4. Transition to Sustainable Fishery

Basically, transition to sustainable fishery of the pearl mullet means that fishing during the reproductive period must be prevented, and should only be practiced at other times. In essence, this is the main factor that threatens the species. This approach is prevalent in the new management model recently developed by Sari (1997b). Effective prevention of fishing during the reproductive period will not only result in increased revenues, but will also require less labor and input, as well as ensure the preservation of the species. Efforts to achieve this goal can be evaluated in three phases.

4.1 Phase 1: Defining the Existing Situation

Despite the fact that the pearl mullet presently constitutes 36% of inland fish production in Turkey, unfortunately it is a species that has not been researched sufficiently. As a result, studies on the general biology, stock quantity and fishery of this fish were practically non-existent. As emphasized above, the existing studies were realized with inadequate number of samples and within limited time restrictions. Therefore, revealing the species' reproductive, growth, recruitment and other characteristics had to be the first priority. Studies conducted between the years 1993-1996 focused on the parameters of reproduction, growth, and recruitment, thereby enabling an assessment of stock sizes and the determination of fisheries management basis (Sari, 1997b). The same researcher also stated that calculations related to fishing practices had revealed a exploitation ratio (E) mean of 0.684, and reported that existing fishing practices led to over-exploitation of the pearl mullet population, resulting in a decrease in average length of the species, and a decline in unit and total catch yield. A series of proposals were set forth, with a core approach that can be summarized by the following suggestions: minimum mesh size must be 20 mm in order to prevent over-exploitation; each boat must hold no more than 5000 m of trammel net; the dates of the fishing ban to be applied during the species' spawning season must be re-adjusted according to the reproductive migration patterns of the fish; a separate unit must be established for centralized fisheries management. At that time the sole restrictions on pearl mullet fishing were the unmonitored "closed season" whereby the dates of the fishing ban were determined according to the fishermen's requests, and a limitation of minimum mesh sizes of "16-18 mm" for the beach seine nets- an apparently random figure for which there is no scientific basis. The "closed season" application for the lake started at two different dates, but ended on the same date. As a result, the fishing ban was applied between May15-July 1st for Van (central province) and the river mouths, rivers or streams within the boundaries of the districts of Gevaş, Edremit and Muradiye, whereas the ban was put into effect between the dates of May 25-July 1st for the district of Erciş and within the boundaries of the province of Bitlis. Yet, the distance between Bendi Mahi River (within the boundaries of the district of Muradiye) and Delicay or

Haydarbey Stream (within the boundaries of the district of Erciş) consisted of 17 km only.

As the first step for the realization of these proposals, local and federal authorities responsible of fisheries management were informed about the situation, and were asked to make the necessary regulatory adjustments. However, coming from a region that had not requested any change for years on end, this appeal for change - which could easily be considered "radical" for that period- did not immediately elicit any response. Following lengthy disputes with local and national fisheries management agencies, a decision was reached to unite the dates for the "closed season" application. As of the year 1996, the fishing ban would be applied throughout the entire lake between the dates of May 10-July 1st, in accordance with the recommendation of the university. Thus, as the ban was put into effect synchronously, it was also pulled forward, even if only by 5 days. On the other hand, the subject was brought to the attention of the media, and the dangers of incorrect fishing were explicated.

This transformation was initially met by the negative reaction of the villagers, who regarded spawning-period fishing as their "irrefutable birthright". The previously established dialogue was not enough to convince them. Unfortunately, NOT A SINGLE OUTCOME could be obtained from the measures taken to prevent fishing during the reproductive phase. Spawning–period fishing continued full blast, and there were no controls whatsoever. The local public agency responsible of monitoring fishery activities recommended that the security forces should solve the problem. The security forces responded by emphasizing that basically, this was not their responsibility.

4.2 Phase 2: Developing New Approaches and Experimenting with These Methods

An evaluation of the results (failure) of the first phase led to the conclusion that a "state-centered" approach to fisheries management was not applicable for Lake Van, at least at that time and under those circumstances. Therefore, non-centralist new methods were required. As of 1996, efforts were focused in this direction. First, the correlation between the amount of fishing and prices was clearly explained, based on observations from the previous period. Emphasis was put on the fact that intense fishing activity during the spawning-season was leading to decreased earnings for the fishermen. Having gained an insight about this relationship, the fishermen were inclined to find some solutions by themselves. These fishers were brought together with the sellers, and were encouraged to agree on a quota system, which they, themselves would monitor. The most important aspect of this strategy based on the quota system, was that it would bring a novel approach to the monitoring process that was somehow never accomplished by the state authorities, by drawing on the supply and demand balance as a brand new leverage element, while at the same time preventing erroneous fishing practices through the active involvement of voluntary inspectors. The fishermen and sellers, under the supervision of lead researcher, reached a consensus on this issue and signed a page-long mutual agreement document. The agreement stipulated that the "closed season" application would start on June 1st, rather than May 10th of each year; the fishermen would not catch more than 3 tons of fish daily, and the sellers would not buy below a price specified in the document. This management strategy would be in practice for 2 years, by the end of which, spawning-period fishers would have saved some capital. At the end of the 2-year period, they would all join forces to support the prevention of spawning-season fishing.

The official authorities would not be involved in any part of this process, except for making the necessary adjustments in the circular (Sari, 1998). However, the local fisheries management authority opposed this proposal on the grounds that it would lead to a "fish massacre" in spite of the fact that they had never taken action to enforce control mechanism. So this novel approach, the result of nearly a year's work and built on hard-earned trust, never came to realization. Nevertheless, many national NGOs became aware of the issue during the preparation of this proposal. The errors of the methods used for the capture of pearl mullets took place in the media time and again. National NGOs declared to the relevant ministries and fishers that they would keep on pursuing this subject. This cooperation, while perceived as a hard-to-beat union by the fishermen, also enabled the pearl mullet to become a national issue rather than a local matter. On the other hand, the General Directorate of Fisheries Circular Number 33/1, put into application in the years 1999-2000 required that the above-mentioned dates stay the same, but also stipulated a limit on mesh size and on the number of nets that can be used for pearl mullet fishing. In effect, the circular enforces a 20 mm mesh size and does not allow more than 5000m netting per boat. Throughout this process, the dialogue that had established with the fishermen developed even further. Those who were convinced that spawning-period fishing is the wrong method were brought together, and training programs that explicated the damage caused by erroneous fishing practices were carried out in villages where spawning-period fishing prevailed. These efforts, encouraging spawning-period fishers to convert to winter-fishing instead, were not viewed positively by those fishers who were already fishing during the winter season. They were uneasy about the possibility of newcomers exploiting the few fishing grounds they had found through years of experience, and further decreasing the already lowered yield. In order to determine new fishing areas, a "Remote Sensing Center" was established within the Yuzuncu Yil University structure. New fishing grounds have started to be located using the satellite imageries obtained daily from this center.

4.2.1 Remote Sensing Applications

Remote sensing is broadly defined as collecting and interpreting information about a target without being in physical contact with object. The term remote sensing is commonly restricted to methods that employ electromagnetic energy (such as light, heat, and radio waves) as the means of detecting and measuring target characteristics (Priede 1983, Sabins 1986).

Satellites of NOAA series are scientifically designed and supply two types of imaging products. One of them is Automatic Picture Transmission (APT) and the other is High-Resolution Picture Transmission (HRPT). *Advanced Very High-Resolution Radiometer* (AVHRR) sensor produces 1.1-km resolution images, which are suitable for terrestrial, aquatic, and climatic studies. NOAA AVHRR images have 5 channels and cover visible and infrared range of the electromagnetic spectrum (Sari et al., 2000). In a study conducted by Sari (2000), 12 identical fishing net was used; during the study 6 locations became the chief focus of the investigation. Three of these locations were old, which fishermen have been using for fishing for a long time. The other three were newly-determined- locations being based on the maps showing water surface temperatures and the currents in the lake. Throughout the course of the study it was determined that the average catch obtained from old fishing locations was found between 4.8 to 22.3 kg, while the average catch in newly determined locations was between 15.4 and 52.3 kg. This translates to 2 to 5 times more product in new fishing locations than old ones (Sari, 2000). The difference between old and new locations possibly stems from the direct relation between the distribution of pearl mullet and high water of the temperature of the lake in the winter.

Smith & Baker (1978) determined the higher primary productivity areas using remote sensing techniques and then tracked the locations of sardine shoals in the seas. Kemmerer & Benigno (1973) successfully detected the locations of fish herds in the open seas using sea surface temperature maps. On the other hand, Narayana et al., (1995) produced sea surface temperature maps by using remote sensing techniques which are beneficial for determining potential fishing locations

In the light of all these findings it might be inferred that temperature plays a determining role in finding new locations for fishing through the use of NOAA AVHRR images; but it would be misleading. Because temperature can not be considered as the single factor determining these results. Biological characteristic of pearl mullet, the currents and bathymetric properties of the lake are considered as the other factors causing this condition. Simpson (1994) pointed out that the most important event of remote sensing in fishing was the determination of potential fishing areas; and then he reported water temperature, coastal and bathymetric borders, currents and salinity as natural factors, which affect the distribution of fish.

When the maps charted and measured by Sari (2000) related with surface temperatures of Lake Van are overlapped with bathymetric and current maps measured by satellite images. When analyzing maps, temperature changes are observed to take place in locations in which currents were much intense while currents show changes depending on the depts of lake. In Van Bay in which bathymetric values change rapidly and irregularly temperature changes are also observed to be higher. If such a relationship (fish distribution, surface temperature of lake, depth and currents) can be put into expression through a mathematical model, daily-obtained- satellite images of fish distribution could be observed quite effectively and fishermen could be directed to much productive areas for fishing. Sari (2000) pointed out that the algorithm compiled through the limited data showing the relationship between surface temperature, currents, depth and distribution of pearl mullet in Lake Van would be insufficient and that much more comprehensive information of every season of the year should be collected in order to compose much valid algorithm to be used every time. When primary productivity values directly affecting fish distribution are added to this relationship, a perfect algorithm could be formulized and put into expression in observing potential areas of fishing through satellite data.

In this way the technical and economical aspect of fisheries management will come to the fore and survey studies of fish lasting for years in order to determine fish distribution areas will be conducted in a relatively shorter period of time and much effectively. The duration for fish finding directly effecting the cost of fishing will be shortened and the expenditures caused by the fuel floating the boats will also be reduced.

Besides this positive contributions to fisheries, remote sensing could also turn out to be negative when it is misused in every branch of technologies. If the fishermen have not enough professional discipline and moral conduct of their works, and have a lack of supervision and control they will exploit the fish stocks easily determined by satellites images in a short time. For this reason, it would be of greater importance to fill the gaps and irregularities wrought by lack of education and awareness for this profession before this new type of fishing is introduced to communities in the region.

4.2.2 GIS Applications

Geographic Information Systems (GIS) is a computer-based tool for mapping and analyzing things that exist and events that happen on earth. GIS in fisheries are being effectively used in fisheries management, in tracking of spawning grounds of some pelagic fish, in classification of habitats according to environmental factors, in monitoring and conservation of spawning ground of the anadrom species, in coastal and river basin management (Sari et al., 1998).

Lake Van Fisheries Geographic Information System (LVFGIS) has been designed for managing fisheries in an effective way and determining the technical, economical, social, cultural and demographical aspects related with fisheries in Lake Van as well as for compiling data in ArcView software environment. . With this study which encompassed all the communities dealing with fisheries around the Lake Van, LVFGIS facilitated to effective decision making and interpretations of the impacts of every kind of change in the means and modes of this profession.

In the previous chapter, we stated that pearl mullet was the only species adopting itself to live in Lake Van due to extreme values in the water quality in it. For this reason, in this study of LVFGIS only the fisheries management within in the context of pearl mullet has been investigated and information provided here could only be available for such management. We overlapped on LANDSAT MSS images with the borders of the lake, streams, settlements in which fisheries are carried out. Topological pinpoints pertained to the determined settlements were setup and then two categories were composed by defining two different times of fisheries as winter fishery and reproductive period fishery. Then the attribute table related with the sub-branches of every category was composed and linked with layers.

In the category of winter fishing such topics as the name of settlement dealing with winter fishing according to years, its location, the name of cooperative, if existed, the number of boats, the number of nets, number of fishing days, total catch, unit catch, total population of the settlement, people working in fisheries, demographical structure of the population dealing with fishing, mean family members and educational status of fishermen. In the category of fishing in reproductive period, such topics as the name of the settlement dealing with fishing according to years, their locations, number of beach seine net, the number of fishing days, the demographical structure of fishermen, mean family members and educational status of fishermen were investigated.

All of the data compiled through LVFGIS from 1994 on were brought together within the context of GIS. In case of any queries concerning the access of information, decision making and applications made by the authorities LVFGIS can be easily used. For instance, when we want to view the entire domains of the settlement in which the number of the boats are more than 10 it is possible to access information in seconds on the query screen; on the other hand if we want to see the relationship between the education status of fishermen, demographical condition and catch yield, it is possible to evaluate the results by overlapping to different layers in seconds by this system.

Fisheries management is a complicated and difficult task. Especially in such countries as Turkey in which fishermen have no access to scientific data, the fishermen are quite liable to violate the laws and rules of fishing in a way to persuade the authorities for exploiting the fish stock in an irregular and immoral way of fishing. Such fishermen usually come to the authorities and pledge that fishing is the only way of source of their financial and economical well-being; they also threat the authorities that they would begin strikes and uprisings by migrating to the city center. However the functionality of such system as LVFGIS will never allow the ill intentions and misuse of such fishermen for the fact that the pretexts put forward by them will quite easily be refuted through the scientific data provided by such system since this system will monitor all the information about fisheries, total populations, financial sources, the situation of settlement; and thus a much more sustainable and reliable environment for fishing will be established. Fishermen are also having the other demands such as benefiting from subventions and subsidies as well as support from the government on the pretext that their fish were exploited by the other fishermen of the near-by communities and a decrease in catch yield occurred, as well as replacements which they claim to have taking place in their fish stocks. Such cases show us that the applications of LVFGIS are of vital prominence in filling these gaps and regulating such chaotic environment of fishermen (Sari&Guven, 2004).

4.3 Phase 3: Putting New Approaches into Practice

Even though the first and second phase of this study did not produce tangible outcomes, they did enlighten our path to find the true solution to the problem. Having concluded by the end of the first phase that a "state-centered" approach was not applicable, and by the end of the second phase that relying on a "fisher-centered" strategy also did not lead to a solution, the third phase was designed so that the NGOs would have a pivotal responsibility in solving the issue, with the support of "public administrative bodies" and the "fishers". At the outset, the initiative involved touring the fishing villages at regular intervals explaining the damages incurred through the use of improper fishing methods. The local authorities and security forces were also visited at regular intervals and the ecological, economical, social and cultural significance of the pearl mullet was explicated. Brochures and posters addressing the fishing community, local authorities, security forces and consumers were prepared and distributed. Flow of information was established with the Ministry of Agriculture and Rural Affairs, the Ministry of Environment and Forestry and similar federal organizations responsible of fisheries management, in order to ensure that the subject was accurately perceived. Consequently, the gendarmerie in rural areas, and security forces at provinces increased their monitoring activities. Whereas no action had been taken against a single fisherman for violation of the law since 20 years, as a direct result of improved controls, now there were times when action was taken against more than 20 fishermen per day. Whereas 15 truckloads of fish was being sold right across the street from the Governor's office at the center of Van City during the "closed season", it was now difficult to find pearl mullets even at the smallest vendors at remote neighborhoods. However, the spawning-season fishers felt that they had been pushed into a corner and convinced the General Directorate of State Hydraulic Works authorities to direct all of the water in the Bendi Mahi River into the irrigation canals for agricultural purposes, just at the period when the fish were about to migrate to the streams and rivers to lay their eggs. Despite all efforts, no water was channeled into the streambed until the end of the spawning season. As a result, approximately 1000 tons of fish died in the Bendi Mahi Streambed due to dehydration and the fishers collected truckloads of fish from the streambeds. As of the year 2001, after the issue was taken to court and a dialogue was established with the farmers' and irrigators' unions, a sufficient amount of water was left within the streambeds. An "irrigation training project" was launched in the Muradiye area, so that the farmers would learn proper irrigation techniques and would stop using the wrong methods. Two posters, and 4 different booklets were published for the training project. In parallel with this venture, another project was started to help steer the spawning-season fishers towards professional-fishing, with the support of the United Nations Development Program Global Environment Funds Small Grants Program (UNDP-GEF/SGP). Within the scope of this project, two separate training sets were designed for fishing villages, and were applied throughout all of the villages in the region. In 2002, all of the parties dealing with pearl mullet fishery management were gathered together for the "Sustainable Fisheries Management Workshop". The workshop was concluded with the signing of a proclamation stating all the subjects on which all of the participants had reached an agreement. Thus, even though there had been some arguments, for the first time all the parties had put their signature under a single text. In fact, this agreement was the turning point for sustainable fishing of the pearl mullet. After this stage, every unit began to own up to their institutional responsibilities regarding fishery management. The Preservation and Control General Directorate of the Ministry of Agriculture and Rural Affairs, which had also participated in the shaping of this agreement, in its circular to be effective in the 2001-2002 periods, accepted the joint proposal made by the university and the NGOs to set the dates of the fishing ban from April 15 to June 30. In support of the on-going scientific efforts, the Preservation and Control General Directorate also declared in its circular that the fishing ban would continue to be in effect for the preservation of the pearl mullet and that the same dates (April 15-June 30) would apply for the periods covering 2002-2004 and 2004-2006. This put an end to the legal inconsistencies that had been creating the greatest obstacle to sustainable fisheries management. With the beginning of the fishing ban on April 15, the long held practice of renting the river mouths for fishing purposes in the spawning-season was effectively ended. At present, none of the river mouths around the lake are being rented for spawning-season fishing activities. In fact, although focused specifically on the fishery of Lake Van pearl mullets, these efforts have turned into a model for Turkey's inland fishery management because the problems confronting our lakes are pretty much the same throughout the country, with minute differences. Lake Van is unique in one aspect: the severe pollution seen around the lakes of the Mediterranean and Agean regions is only at the beginning levels in Lake Van. Contrary to the first and second phases, the third phase of this project has provided positive results. The over-exploitation pressures on the pearl mullet stock have been decreased, fishing yield has increased, and profit levels in the fishing sector have risen. The sustainability of these outcomes depends on the dedication of relevant local institutions and organizations, and their determination to maintain these applications. Inevitably, the need has arisen for a local NGO that can ensure coordination between the local institutions and organizations, own up to the heritage of previous efforts, and develop projects geared towards the future. The Association of Nature Observers was established in 2003 as a result of this necessity.

5. Results of the Transition to Sustainable Fishery

The achievement of sustainable pearl mullet fishery is a subject that is of great importance for the 14000 people around the lake who depend on it for their livelihood. Therefore, even though the studies that have been carried out bear ecological implications, in reality, their economic justifications are the main priority. The studies conducted between the years 1993-2006 in three phases as summarized above, provided excellent results, especially with the NGO-centered local fishery management strategies adopted in the third phase; and have become a model for inland fishery management in Turkey. 17 different training and cautionary materials were produced within the span of this 13-year study. Clearly, the endurance of the messages resulting from studies that deal with societal issues depends on the adequacy of educational and cautionary documents.

The results and benefits attained through the 3-phase studies conducted with the support of these training-cautionary materials are presented in Table 1. As can be seen in Table 1, the number of seine nets employed for spawning period fishing has decreased by half, whereas the number of professional-fishers has gone up by the same proportion. This indicates that most of those who gave up spawning season fishing have started to engage in winter fishing (Sari, 2004).

Parameters	**Years**	
	1996	**2006**
Number of seine nets used for spawning season fishery	92	32
Number of river mouths rented for spawning season fishery	12	0
Number of villages entirely opposed to the project (total 15 fishing villages)	9	2
Number of professional-fishers	101	165
Unit catch in professional fishing (kg/100 m/day)	2.622	8.50
Mean fork length (cm)	16.74	19.8
Number of fish per 1 kg (unit)	16-18	10-12
Total fishery revenues (USD)	3659000	7143000

Table 1. Output data obtained during transition to sustainable pearl mullet fishery

Whereas there were only 6 villages out of a total of 15, that supported the efforts to prevent spawning-season fishing at the initial stages of the transition to sustainable fishery management, today this number has gone up to 12 villages. Unit yield obtained through winter fishing has increased by three-fold in comparison to 1996 figures, going up to 8.50 kg. Decrease in size of captured fish due to over-fishing had been observed in over-exploited grounds (Pauly, 1983). Sari (1997b) has stated that this decrease in size of captured pearl mullets is a direct result of over-exploitation and has emphasized that the fish size will increase once over-fishing is successfully prevented. As can be seen in Table 1, the increase in mean size of the pearl mullets is very noticeable. In the meantime, as a result of the decline of spawning season fishery, a supply and demand balance has started to emerge, and the revenues obtained from this fish have increased nearly 100%. .

6. Conclusion and Suggestions for Sustainable Fishery

Although there is a rising trend in pearl mullet fishery towards sustainable fishery practices, it must be clarified that "sustainable fishery management" has not yet been achieved. Considering the fact that tens of species have been lost within our inland waters in the last 50 years, the achievement of this current level of protection for the pearl mullet - the only species that can survive in the extreme ecosystem of Lake Van- definitely cannot be underestimated. On the other hand, neither is it possible to paint a rosy picture, or to claim that every problem has been solved.

Presently, 3 of the 15 fishing villages still insist on fishing during the spawning period. A new project has been launched in these villages, with the support of UNDP-GEF/SGP. This study aims to determine the reason for this difference in attitude by researching the social, cultural, and traditional structure of villages that have abandoned spawning season fishery practices and those that have not, while at the same time trying to change the traditional style of consumption, which encourages fishing in the reproductive period. Also within the context of this project is the assessment of alternative income resources for all the fishing villages, with a special emphasis on those villages that have abandoned spawning period fishing practices. Possible alternative income resources pinpointed during previous studies, such as winter fishing, canned fish and salted fish production workshops, and eco-tourism, are being studied for their adaptability to the local population's socio-cultural and traditional structure. Other alternative income resources generated by the local townspeople will also be appraised.

The first priority for full transition to sustainable fishery management is the acceptance of an NGO-centered local fishery management scheme by the fisheries management authorities. Significant steps have been taken to this effect, demonstrated by the fact that the dates for the fishing ban have been determined as a result of regional meetings, and the efforts that are being made to reach decisions with the highest possible participation in the last few years. However, the same sensitivity has not been observed in the application of the legislature. Presently, the security forces still play a major role in monitoring fishing activities at Lake Van. Yet legally, the security forces are only expected to assist the local branches of the Ministry of Agriculture and Rural Affairs, which are actually responsible of monitoring the lake. As a result, inspections are at times slackened, depending on how the administrators of the security forces interpret their range of responsibility. Institutionalized sustainability can only be attained if the responsible organizations are provided with the personnel and equipment they need, and are enforced to fully take on their monitoring responsibilities. This goal is only achievable through the establishment of a single "local administrative unit" that will be the sole decision-maker with regards to all fisheries management decisions for major inland waters such as Lake Van.

Preservation of the habitat is a basic component of sustainable fishery. Precautionary measures that need to be taken for sustainable fishery management of the pearl mullet and the preservation of its habitat can be listed as below:

- Establishing a local administrative unit that can achieve fishery management single-handedly in special ecosystems like Lake Van;
- Creating alternative employment opportunities that will provide a livelihood for the fishermen who decide to abandon spawning season fishery;
- Instigating amendments in the Fisheries Statute for the confiscation of fishing equipment and transport vehicles in order to prevent poaching, taking the Forest Statute as an example;
- Initiating measures to protect Lake Van, the natural habitat of the pearl mullet, from pollution;
- Preventing the transfer of sand from riverbeds, this damages their spawning areas.
- Mandatory construction of fish passage to ease the migration of the pearl mullets on all kinds of water structures that are to be built on rivers;
- Building boats' shelters at suitable places for effectual management and monitoring of fishery activities;
- Setting up processing and cold storage facilities in order to enhance the value of the fish for marketing purposes;
- Applying ecological water distribution in order to ensure that enough water is left in riverbeds for spawning during the reproduction period;
- Ensuring that the ecological needs of the pearl mullet are recognized as the first priority in all the long term planning related to the lake basin.

7. References

Akgul, M. (1980). Studies on the Bio-ecology of the Pearl Mullet (*Chalcalburnus tarichi*, Pallas 1811) Endemic to Lake Van, Scientific and Technological Research Council od Turkey (TÜBITAK) VII. Scientific Congress, Biology Proceedings, Aydin, pp.533-544.

Arabaci, M. (1995). Study on Some Blood Parameters of the Pearl Mullet (*Chalcalburnus tarichi*, Pallas 1811), Yuzuncu Yil University Science Institute, Graduate Thesis, Van, 52 pp.

Cetinkaya, O., Sari, M., Arabaci, M. (1995). A Preliminary Study of Fishing Yield and Selectivity of Beach Seine Nets Used for Lake Van Pearl Mullet (*Chalcalburnus tarichi*, Pallas 1811) Fishery, Ege University, Faculty of Fisheries, Journal of Fisheries, Vol. 12(1-2): 1-13.

Danulat, E. & Kempe, S., (1992). Nitrogeneus Waste Extrection and Accumulation of Urea and Ammonia in *Chalcalburnus tarichi* endemic to Lake Van, Fish. Physiol. Bioch., 9:377-386.

Danulat, E. & Selcuk, B. (1992). Life History and Enviromental Conditions of the Anadromous *Chalcalburnus tarichi (Cyprinidae)* in the Highly Alkaline Lake Van, Eastern Anatolia,Turkey, Arch. Hidrobiol., 126(1):105-125.

Evliya Celebi (1986). The Travels of Evliya Celebi, Vol.: 3-4, (Ed., Temelkuran, T. and Aktaş, N.) Ücdal Yayinevi, Istanbul, 766 pp.

Geldiay, R. & Balik, S. (1988). Freshwater Fishes of Turkey, Ege University Publication. Izmir, 519p

Kemmerer, A.J. & J.A. Benigno (1973). In "Symposium of Significant Results Obtained from ERTS" Vol. I (P. Brock, Ed.). Goddard Space Flight Center, Greenbelt, Maryland.

Kempe, S, Khoo, F. & Gurleyik, Y. (1978). Hydrography of Lake Van and Its Drainage Area, 30-44 The Geology of Lake Van, Degens, E.T. and Kurtman, F. (Eds.), MTA Pub. No:169, Ankara, 158 pp.

Kuru M. (1987). A Systematic and Zoo-Geograpgic Analysis of Pisces Endemic to Dicle-Firat, Kura-Aras, Lake Van and Black Sea Basin Freshwaters (Unpublished thesis) Ataturk University, Faculty of Science, Erzurum

Paluly, D. (1983). Some Simple Methods for the Assessment of Tropical Fish Stocks, FAO Fish. Tech. Pap., No:234, Rome, 52p.

Pride, I.G. (1983). "Use of Satellites in Marine Biology, In:Experimental Biology at Sea" (Edited by A.G. MacDonald and I.G. Priede) 3-50, Academic Press, London.

Sabins, F.F. (1986). "Remote Sensing Princibles and Interpretation", Second Edition, W.H. Freeman and Company, New York, 449 p.

Sari, M. (1997a). Selectivity of Nets used in Pearl Mullet Fishing (*Chalcalburnus tarichi*, Pallas 1811), Mediterranean Fishery Congress, April 9-11, 1997, Izmir, pp. 93-102

Sari, M. (1997b). Assessment of Stock Sizes of the Lake Van Pearl Mullet (*Chalcalburnus tarichi*, Pallas 1811) and Determination of the Basics of Fisheries Management. Ege University, Science Institute Doctorate Thesis, 150 pp., Izmir 1997

Sari, M. & A. Tokac (1997). Some Technical Features of Lake Van Fishing Boats, *Mediterranean Fishery Congress pp. 693-696*, April 9-11, 1997, Izmir.

Sari, M. & S. I. Ipek (1997). A Preliminary Study on the Determination of New Fishing Grounds Through the Use of Satellite Imary, *IX National Fishery Symposium pp. 673-682*, Sept. 17-19, 1997, Eğirdir.

Sari, M. & M. Arabaci (1997). Reproduction and Fecundity in the Pearl Mullet (*Chalcalburnus tarichi*, Pallas 1811) Population, *Mediterranean Fishery Congress pp. 537-544*, April 9-11, 1997, Izmir.

Sari, M. (1998). Restructuring is needed for effective use of Lake Van. Deniz Postasi Number:4, TÜDAV, Istanbul.

Sari, M., A. Tumay, A.C. Saydam (1998). Geographic Information Systems and Use in Fisheries, *First International Symposium on Fisheries and Ecology Proceeding pp 108-114*, September 2-4 1998, Trabzon.

Sari, M. & A. Tokac (2000). Comparative Study of Net yiled of Two Different Beach Seine Nets Used in Pearl Mullet (*Chalcalburnus tarichi*, Pallas 1811) Fishing, Ege University, Faculty of Fisheries, *Journal of Fisheries, Vol.* 17(3-4): 27-33.

Sari, M. (2000a). Yuzuncu Yil University Remote Sensing Center, Projects in Progress and Selected Abstract of Results, *Eastern Anatolian Region IV. Fishery Symposium Proceedings pp.185-201*, June 28-30, 2000, Erzurum.

Sari, M. (2000b). Identification of New Fishing Grounds by Using RS and GIS Techniques in Lake Van, Eastern Turkey. Agroenviron 2000, *2nd International Symposium on New Technologies for Environmental Monitoring and Agro-Applications Proceedings* pp 47-56, Trabzon.

Sari, M., I. Polat, A.C. Saydam (2000). NOAA AVHRR Uydu Göruntuleri Ile Van Gölu Yuzey Sicakliğinin Izlenmesi, *4. Doğu Anadolu Su Ürunleri Sempozyumu*, 825-842 pp., 28-30 Haziran 2000, Erzurum.

Sari, M. (2001). *Assessment of Stock Sizes of Lake Van Pearl Mullet and Determination of the Basics of Fishery Management*, CEKÜL Foundation, Scientific Series, 150 pp. ISBN-975-6825-01-04, Istanbul.

Sari, M. (2003). *The Pearl Mullet of Lake Van: information and law 2003,* CEKÜL and Game Wild Life Publications, Istanbul.

Sari, M. (2004). Achieving Sustainable Productivity in the Fishery Management of the Pearl Mullet Endemic to Lake Van Project ,Final Report, UNDP-GEF/SGP Final Report, Ankara.

Sari, M. (2006). Threatened Fishes of the World: *Chalcalburnus tarichi* (Pallas 1811) (Cyprinidae) living in the highly alkaline Lake Van, *Environmental Biology of Fishes, http://ebfi.edmgr.com/, (Online) DOI 10.1007/s10641-006-9025-4, 79:69-70,*

Smith, R. C. & K. S. Baker (1981). The bio-optical state of ocean waters and remote sensing, *Limnol. Oceanogr.,* 23, 247-259.

Simpson, J.J. (1994). Remote Sensing in Fisheries: A Tool for Better Management in the Utilization of a Renewable Resource, *Can. J. Fish. Aquat. Sci.,* 51:743-771.

13

The Application of Membrane Separation Processes as Environmental Friendly Methods in the Beet Sugar Production

Zita Šereš[1], Julianna Gyura[1], Mirjana Djurić[1], Gyula Vatai[2], Matild Eszterle[3]
[1]*University of Novi Sad, Faculty of Technology*
Serbia

[2]*Corvinus University, Department of Food Engineering*
Hungary

[3]*Research Institute of Hungarian Sugar Industry*
Hungary

1. Introduction

The sugar industry is one of the key segments of the food industry. This industry is also well known as one of the most energy-intensive in the field of food and chemical industry. In 1999, the total quantity of produced sugar was summarized. It was reported that the production reached 16 700 000 tones, while the total value of the produced sugar was 8924 million EUR. A significant quantity of thermal energy was consumed for the evaporation and beet pulp drying, as well as electrical energy needed for the pumps and for driving the centrifuges. According to CEFS, specific energy consumption was 31.49 kWh/100 kg beet in 1998 (IPPC, 2003). In a Greek study, a figure of 280 kWh/t is given for the electrical part of the energy consumption in sugar manufacturing (IPPC, 2003). While the overall water used is about 15 m^3/t sugar beet processed, the consumption of fresh water is 0.25 - 0.4 m^3/t sugar beet processed, or even less in modern sugar factories. Water consumption depends on the activities of each installation, e.g. more water is consumed in an installation that extracts and refines sugar beet compared to the one that does only one of these activities (IPPC, 2003). For example, the consumption of water in Austria was reported at a level of 1.5 m^3/t of sugar beet processed, which is equivalent to 9 m^3/t of produced sugar (IPPC, 2003). The transport water has high organic contamination due to the soil and sugar from damaged beets. Its COD is 5000 - 20000 mg/L. Waste water with high BOD levels is produced in large volumes (IPPC, 2003).

Despite the fact that the sugar industry is one of the causes of the environmental pollution, not enough has been done on its improvement. The technology applied in almost all European sugar factories is based on the traditional principles and methods. The major steps in the traditional sugar beet processing are (Poel Van der et al., 1998):

i) <u>Pre treatment</u> - Washing and slicing of the sugar-beets into cosettes are the initial operations;

ii) Extraction - The sugar is counter-currently extracted from beet cosettes to obtain raw juice and beet pulp. The raw juice is thermally unstable at temperatures above 85 °C. The beet pulp can be used as cattle feed or can be modified to obtain fibers for human feed;
iii) Beet juice purification - Milk of lime and karbonation gas CO_2, both produced in a separate facility, are applied. Coke and limestone are used for the production of CaO and CO_2. The lime usage of the conventional process is about 2%/beet. Classical juice purification consists of liming, carbonation, sludge separation and sulphitation. However, this process removes only a part of non sugars from the sugar juice (proteins, pectins, inorganic salts and colouring substances);
iv) Beet juice concentration - By multi-effect evaporation the thin juice with a dry substance content of 14-16% is concentrated to thick juice with 60 – 75% of dry matter;
v) Crystallization - Further evaporation of water leads to crystallization and growth of crystals. Sugar crystals are separated by centrifuge from the syrup. The separation occurs in three stages. The molasses is the by-product from which the crystallization is not possible.

Governments of the developed countries have tried to increase the pressure on the largest waste producers in order to reduce the undesired environmental pollution. For example, the Commission of the European Communities introduced the Integral Pollution and Prevention Control Directive (Council Directive, 1996). The purpose of the directive is to achieve integrated prevention and the control of pollution arising from the particular activities listed in its Annex I. Among others, the directive defines the Best Available Techniques (BAT) as the most effective and advanced stage in the development of activities and their operation methods which indicate the practical suitability of particular techniques for providing in principle the basis for emission limit values designed to prevent and, where that is not practicable, generally to reduce emissions and the impact on the environment. The European IPPC Bureau published a relevant document (BREF) where the BAT for the Food, Drink and Milk Industry are presented (IPPC, 2003). To a larger extent, the general techniques commonly used in this industry are described. However, no BAT is described relating specially to sugar beet production. Detailed information can be found in chapter 4, under the title Techniques to Consider in the Determination of BAT. The chapter contains a list of various pollution prevention, waste minimization and energy efficiency techniques applied in industry that are described everywhere, e.g. in books, journals, leaflets, the internet, etc. Recent research in the sugar industry has been focused on the development and implementation of new technologies and/or the particular unit operations, which would replace the traditional ones. The separation operation deserves special attention because of its significant consumption of water end energy.
Up to now, very few large-scale membrane separation processes have been applied in the sugar industry worldwide despite the encouraging results of numerous investigations published in the literature (Kwok, 1996; Willet, 1997; Chou, 2002; Lipnizki et al., 2006). In principle, all the separations in sugar production from beet or cane juices could be performed with appropriate membrane separation processes. In practice, however, regarding the high capacities of the sugar factories and relatively low price of sugar, the investment in an imaginary factory, operating exclusively with membranes or with other new environmentally friendly separation methods, would still be to high. The other approach is to find those membrane separation processes which could be advantageously embedded into or combined with the traditional technology to increase the effectiveness of the sugar production as a whole. This latter concept is closer to the actual state of the industry.

2. Membrane Separation Processes

2.1 Introductory Comments

Membrane separation is a pressure driven filtration technique in which a solution is forced through a porous membrane. Some of the dissolved solids are held back because their molecular size is too large to allow them to pass through. The size range depends upon the type of membranes used. Fractionation of the feed stream occurs, with some molecules being concentrated on the upstream side of the membrane, which is known as the concentrate or retentate. The smaller molecules pass through the membrane into the permeate stream. The variety of membrane separation techniques can be characterized by their membrane pore size.

Cross-flow microfiltration (MF) membranes can be used to remove non-sucrose compounds, or to fractionate the retentate rich in colourants. Ultrafiltration (UF) membranes can be applied to concentrate the relevant juices in sugar industry and to remove non-sucrose compounds. This membrane pore size ranges from about 10 to 100 nm. Applications of UF can be extended to the removal of oil from waste water and the removal of turbidity and colour colloids from juices. Nanofiltration (NF) membranes have selective permeability for minerals and some small organic and inorganic molecules and NF is used predominantly for concentration and pre-demineralization (removal of salts) of juices and waste water in sugar industry; the pore size of the NF membrane ranges from about 1 to 10 nm. Reverse osmosis filtration (RO) membranes are permeable to water but not minerals and are therefore used for dewatering and for removing heavy metals and pesticides. They are also used for refinement of NF permeates or evaporator condensate as well as in water treatment, such as softening and salt removal; this membrane pore size ranges from about 0.1 to 1 nm (Cheryan, 1986). Membranes can also be classified according to the material used for their production. So, polymeric, inorganic - ceramic and metallic- membranes are well known.

Along with the membrane characteristics, operating conditions are also significant. Among possible regimes, cross flow filtration has proven the best. Also important are following independent variables: transmembrane pressure, flow rate of the liquid phase, its temperature and process duration. Their optimization is a difficult task which can be performed either empirically or by solving adequate mathematical models. Certain number of papers is dealing with the mathematical modelling of the process, based on theoretical background as well as on statistical processing the measured data. As for the required goal, the quality of permeate, expressed through its colour, purity, turbidity, etc., is given the priority.

Sugar has to satisfy rigorous quality demands; particularly important are demands related to the colour of its crystals. Ensuring colour quality parameters of white sugar (EU-1 and EU-2) used to be difficult especially when the quality of the processed beet is poor (Šereš et al., 2004; Gyura et al., 2004). The limits of the traditional purification operations are well known; by using purification with lime, only 35% of non-sucrose compounds can be removed. Also, assurance of colour quality is expensive due to the high energy consumption of the sugar house: heat and electricity, or even if enough energy could be provided, limited capacities of equipment in the sugar house might cause stresses in the operation of the factory. In such a situation it seems reasonable to look for an additional non-sucrose compound elimination method by which the required colour, to obtain EU-2 white sugar from the standard liquor (feed of the 1st evaporating crystallizer) can be provided.

In the membrane separation technique water is used periodically to clean the separation equipment. The frequency of cleaning and the volume of water used vary depending on the product and equipment. Waste water from washing as well as from the process itself (in the form of separated waste products) contains dissolved non-sucrose compounds. Membrane separation as a pressure driven process requires electrical energy. The achieved environmental benefits are as follows: reduction of levels of the suspended, colloidal and dissolved solids in water, reduction of phosphorus levels and collection of waste water streams, thus reducing their volume prior to further possible treatment (IPPC, 2003). Also, it is possible to recover the expensive ingredients for re-use as well as to recover water for re-use. Problems may arise from the fouling of membranes and gel polarization. Since the flux rates through membranes are relatively low, an extensive membrane area is required.

2.2 Short History of Investigations

Investigations of membrane separation processes have started through their combinations with the traditional processes. It was already mentioned, the traditional raw juice purification involves non-sugars precipitation with lime. Surplus $Ca(OH)_2$ is eliminated by the CO_2 gas. Sludge produced is then filtrated through filters and pressed in the filter presses. Not only waste is produced in this process but an additional lime production plant would be needed, where solid waste (stones, sand, etc.), wastewater and waste gasses are produced. Urbaniec (Urbaniec et al., 2000) proposed an alternative process instead. Raw juice purification by the lime is replaced by screening and micro-filtration followed by low temperature evaporation. It is also well known that the conventional concentration of thin juice by evaporation requires about 50% of the total energy. In order to decrease this amount, nanofiltration and reverse osmosis has been considered as pre-concentration step before evaporation (Cartier et al, 1997; Gosh et al., 2000; Gosh & Balakrishnan, 2003). An overall concept for treatment or raw juice with polymeric hollow fiber microfiltration is the A.B.C. process which combines a continuous screening with ultrafiltration followed by softening and alkaline adjustment before evaporation and adsorption (Wilett, 1997). Madsen, Nielsen and Attridge studied the raw beet juice purification by ultrafiltration and microfiltration (Madsen, 1973; Nielsen et al., 1982; Attridge et al., 2001) using polymeric membranes. Apart from using polymeric membranes, the use of inorganic - ceramic and metallic- membranes to achieve commercially interesting fluxes and permeate purities has been investigated. Hinkova et al. (Hinkova et al., 2002) observed that the filtration of diluted raw juice concentrate with ceramic membrane led, at a pilot scale, to a significant decrease in colloids and colour matters resulting in juice suitable for direct crystallization. Schrevel (Schrevel, 2002) compared the results obtained on different membranes in different module configurations. He concluded that supplementary treatments of the beet juice, such as pre-filtration and pre-liming with carbonation, aid the ultrafiltration and guarantee the required sugar quality. Tebble et al. (Tebble et al., 2002) have proposed the integration of membranes, in a side stream approach, to reduce the need for lime and to boost the overall capacity of the raw juice purification. Kochering et al. (Kochering et al., 2003) have combined microfiltration and ultrafiltration for lime free raw juice purification using an ion exclusion process. Here membrane filtration is used after a clarifier and followed by softening, evaporation and chromatography for the elimination of non-sucrose and colour. The result is high quality of white sugar. Koekoek et al. (Koekoek et al., 1998) tested nanofiltration spiral wound and tubular modules for the concentration of thin juice. In spite of the acceptable fluxes at the beginning of the process, the long term performance of the membranes was disappointing due to fouling. Cleaning could not restore the initial fluxes. Similar results were obtained by Hinkova et al. (Hinkova et al., 2002). Beside application of membrane separation processes in sugar juice purification, there where attempts to use reverse osmosis to recycle pulp press water (Bogliolo et al., 1996) or to recover pectin from sugar beet pulp (Hatziantoniou & Howell, 2002).

Our early investigations on membrane separation of sugar-beet juices has started with the experiments on polymeric membranes (Eszterle et al., 2000; Gyura et al., 2001; Gyura et al., 2002a; Gyura et al., 2002b; Eszterle et al., 2003) at temperatures whose limit was 60 °C, according to the physical characteristics of the applied polymer materials (polyethersulfone and polyamide). Further investigations were performed on ceramic membranes which allow ultrafiltration of sugar juices at 80 °C. In this way, a simulation of real process was possible. At the same time, a turbulence promoter was introduced and the efficiencies of processes, without and with the application of static mixer, were compared (Šereš et al., 2005, Šereš et al., 2006a; Šereš et al., 2006b). Analysis of working conditions as significantly important factors in ultrafiltration and nanofiltration processes were investigated at a laboratory level and the obtained results were used for suggesting optimal conditions (Gyura et al., 2004; Djurić et al., 2004; Gyura et al., 2005). The suggestions mostly rely on the analysis of experimental values, but some of them suggested optimal solutions by using adequate mathematical models. Djurić et al. (Djurić et al., 2004) proposed the flux models as functions of concentration factor, flow rate, temperature and transmembrane pressure as independent variables. The suggested mathematical model enables prediction of separation time if the permeate fluxes as well as the initial and final concentrations of undesired nonsucrose compounds are known.

Similar to the research related to the purification of raw beet juice, the use of microfiltration and ultrafiltration for the purification of cane juices was investigated (Hamachi et al., 2003, Martoyo et al., 2000). Experiences acquired through these investigations might be of broader importance and will be presented shortly. In 1994, the first New Applexion Process was installed in a cane sugar factory (Kwok, 1996). The ultrafiltration stage was designed to treat 380 L/min of pre-filtered limed clear juice after the clarifier. A modification of the above mentioned process was developed by Chou (Chou, 2002). Also the A.B.C. process has been adopted for cane juice purification (Willet, 1997). Martoyo (Martoyo et al., 2000) described the successful application of spiral wound ultrafiltration elements for the pre-filtered/screened raw juice with fluxes in economically interesting range (up to 120 L/m^{2}h). The applicability of spiral wound modules was approved by pilot tests but with lower fluxes (Gosh et al., 2000; Gosh & Balakrishnan, 2003). More recently, membrane distillation became one of the latest technologies introduced to cane industry. By this process two streams with different temperatures are separated through a non-wetted microporous membrane. The driving force of the mass transfer is the vapour pressure difference resulting from the temperature gradient across the membrane (Nene et al., 2002). Despite purification steps before the concentration, cane sugar juice still contains high-molecular mass components. Microfiltration and ultrafiltration have been proposed as decolourization and purification methods. Dornier et al. (Dornier et al., 1995) focused their investigations on the performance of tubular ceramic microfiltration membranes. Cartier (Cartier et al., 1996a and 1996b) studied the influence of flocculation which preceded the microfiltration and ultrafiltration with tubular membrane. The most promising was the 300 kDa membrane allowing fluxes of 65 L/m^{2}h. The decolourization rate was 50% while the removal of turbidity was 90%. Decloux et al. (Decloux et al., 2000) concluded that ceramic membrane with 15 kDa has the best decolourization rate at 60 °C. In 2003, Hamachi and co-authors highlighted the limitations of using ultrafiltration for decolourization (Hamachi et al., 2003). It was found that even with membranes of 1 kDa colour removal would not exceed 60% while the flux decreased below 30 L/m^{2}h. Key success factors for an efficient application are both pre-treatment and operating conditions (Lipnizki et al., 2006).

2.3 New Trends

The commercial application of ultrafiltration, for the separation of non-sucrose compounds, is limited because of the concentration polarization and progressive fouling of the membrane. It can be expected that the molar mass of colourants increases while the density of sugar juice increases due to the polycondensation reactions favored at low water contents. According to the gel chromatography measurements (Anyos, 1984; Godshall et al., 2002), approximately 35% of all colourants, present in the initial sugar juice, have the molar mass higher than 4000 g/mol, while 51% and 81% of colourants, with the mentioned molar mass, are present in solution of the 2nd and the 3rd sugar. It can be concluded that elimination of colourants by ultrafiltration would be more efficient in the case of the raw sugar solutions of the 2nd or 3rd crystallization then in the case of the thick juice. The second important reason for applying ultrafiltration on thin juices might be the smaller quantity of these juices compared to the thick juice. Namely, the required membrane surface is proportional to the quantity of treated solution. In the industrial application permeate is to be sent to the 1st stage of crystallization and the darker coloured retentate into the 3rd stage of crystallization.

The described phenomena cause decline in permeate flux during the ultrafiltration of sugar juice. To make membrane separation process practically applicable, permeate flux should be increased. Among all the hydrodynamic methods used for improving mass transfer in cross flow membrane filtration, an increase in cross flow velocity represents the simplest way to create turbulence and reduce membrane fouling. Such an increase in cross flow velocity and thus enhanced permeate fluxes has been obtained by applying static turbulence promoters with or without superimposing pulsations for creating unsteady flow (Krstić et al., 2006). Kenics static mixers are the most common static mixers applied in the industry. Some studies on the use of Kenics static mixers, as turbulence promoters in cross flow membrane process, have been reported in the literature (Šereš et al., 2006a). A significant decrease of a gel layer concentration at the membrane surface has been observed (Vatai & Tekić, 1995). On the other hand, increased power consumption for fluid flow was required because of the increase of pressure drop along the membrane module with inserted static mixer. Most of these techniques are applied to ceramic membranes.

Only a few investigations were reported on the problem of ultrafiltration of sugar juice in the presence of static mixer as turbulence promoter (Šereš et al., 2005; Šereš et al., 2006a; Šereš et al., 2006b). We will present some results (at laboratory level) related to the removing undesired molecules from the sugar juice by the use of ceramic membranes, with and without presence of static mixer, under various working conditions. The process efficiency is quantified through the achieved values of permeate flux and its colour, while the working factors were: fluid flow rate, temperature, trans-membrane pressure and process duration. As treated raw sugar syrup (ca. 60% dry substance) belongs to viscous fluids, static mixer was expected to improve the permeate flux as well as the separation process as a whole.

3. Material and Methods

Non-affinated sugar from the 2nd stage of crystallization, diluted to 60°Bx of dry matter, was used as material for the investigation of coloured matter separation by UF and for examining the influence of static mixer on the separation process. Its basic characteristics corresponded to regular technological quality (Reinefeld & Schneider, 1978); its colour was 1957 and its purity was 98.51%.

The laboratory UF equipment was set up at the Faculty of Food Industry, "Corvinus" University in Budapest. The flow diagram of the setup is shown in Fig. 1.

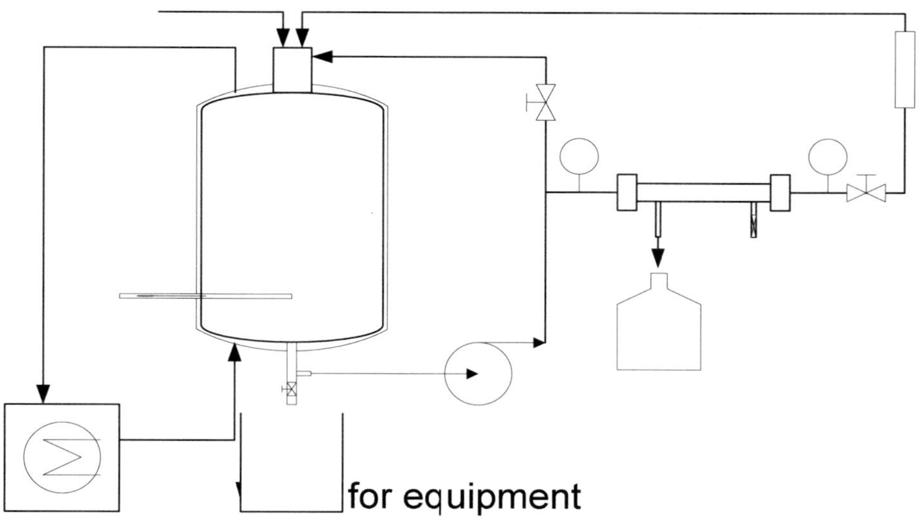

Fig. 1. Laboratory setup for ultrafiltration: 1 - feed tank, 2 - thermostat, 3 - pump (0.25 kW), 4 - module with membrane, 5 - vessel for permeate, 6 - vessel for retentate, 7, 8, 9 - valves, 10 - thermometer, 11, 12 - manometer, 13 - rotameter

The cross-flow filtration was realized on the membrane of 20 nm pore diameter (Membralox ceramic tubular membrane, SCT, Bazet, France), made of a zirconium oxide layer on an aluminum oxide support. The membrane was a single channel type 250 mm long, with 6.8 mm inner diameter. The useful membrane surface was 4.62×10^{-3} m^2. The effect of turbulence promotion on filtration performances was investigated by using Kenics static mixer (FMX8124-AC, Omega), consisting of 30 elements and having aspect ratio (length to diameter) of 1.

Experiments were performed in accordance with the plan presented in Table 1, where the lower and the upper boundaries of the independent variables are given. Experiments were performed without and with the application of static mixer, i.e. in non-static mixer mode (NSM- mode) and in static mixer mode (SM- mode).

Independent variables	NSM- mode		SM- mode	
	Lower level	Upper level	Lower level	Upper level
Q, L h^{-1}	150	400	50	400
t, ^{o}C	70	80	70	80
P, bar	6	10	3	10
τ, h	0	2	0	2

Table 1. Plan of experiments – boundaries of independent variables

A full factorial design was applied and flow rate (Q), temperature (t) and trans-membrane pressure (P) were kept at two levels, while time (τ) was continually measured together with the measurements of two dependent variables: permeate flux and colour change. The permeate colour change is expressed as a difference between the colours of the permeate and juice, divided by the colour of the initial juice. The colour is quantified by the absorbance, measured on a spectrophotometer at 420 nm. As for the reproducibility of the results only those measurements were repeated (more than two times) which gave significantly different values when twice repeated.

4. Results and Discussion

The results of the investigations are presented in Figs. 2-8. The experimental values of the presented variables were taken at the beginning and at the end of ultrafiltration process, after 0.5 h and after 2 h. In this way, the process could be analyzed in real time and more accurate conclusions could be drawn. From the Fig. 2, it can be seen that the greatest flux, through the ceramic membrane without static mixer, could be reached at temperature of 80°C, flow rate 400 L h^{-1} and pressure of 10 bar, after 0.5 h of ultrafiltration. It can also be seen that the best flux through the membrane supplied by static mixer could be reached under the same conditions. Also, when the pressure was held at 6 – 10 bar and when the flow rate was lower than 100 L h^{-1}, permeate flux (in SM-mode of operation) got high value. Comparison of fluxes typical of NSM- and SM- mode of operation, at the beginning of process, did not show significant difference. In Fig. 3, fluxes at the end of ultrafiltration process were presented. Trends similar to those at the beginning were noticed, but after 2 h lower absolute values of fluxes were achieved in both modes. However, it could be concluded that about 30% higher flux was reached in the presence of static mixer, at lower transmembrane pressure, which is important for energy saving. Flux values measured at 70°C, which were not graphically presented due to limited space, were higher in SM-mode of operation as follows: i) 50% after 0.5 h, ii) 60% after 1 h and iii) 65% after 2 h. As expected, static mixer has great effect on ultrafiltration of juice with higher viscosity.

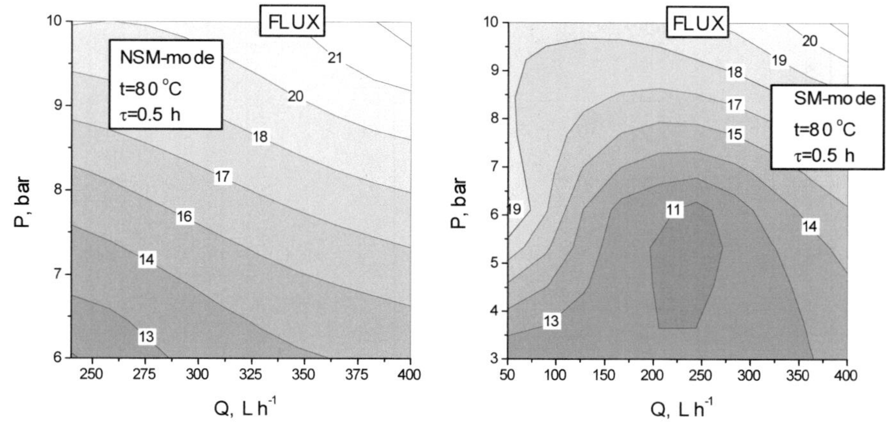

Fig. 2. Permeate flux as a function of flow rate and transmembrane pressure at 0.5 h of ultrafiltration

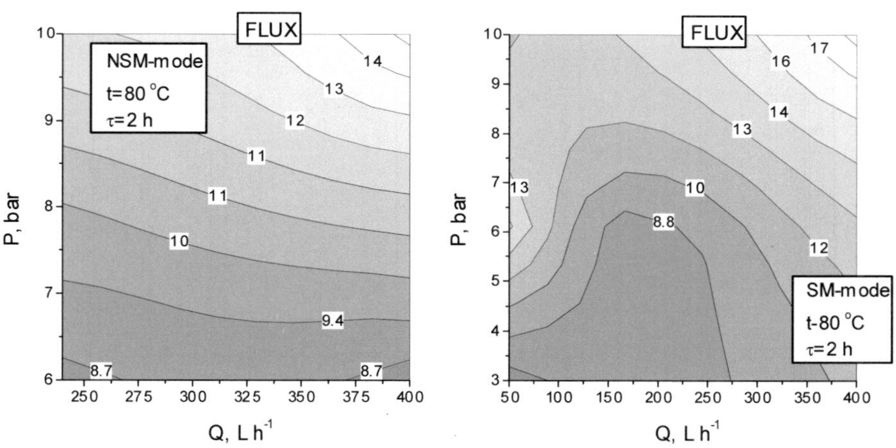

Fig. 3. Permeate flux as a function of flow rate and transmembrane pressure at 2 h of ultrafiltration

By ultrafiltration, in the presence of ceramic membrane with pore size of 20 nm without static mixer, permeate colour decreased approximately 35% - 40% (see Figs. 4 and 5). These values could be reached at NSM- mode, while pressure was held at 6 bar and flow rate has value lower than 250 L h^{-1}. However, in the presence of static mixer, following reduction of colour matter could be achieved: i) 45% after 0.5 h of ultrafiltration and ii) 60% after 2 h of ultrafiltration. Optimal working conditions are: flow rate between 100 - 200 L h^{-1} and pressure above 6 bar (see Figs. 4 and 5). Obviously, when a static mixer was used the decolourization of permeate was highly efficient. Comparison between NSM-mode and SM-mode of operation leads to a conclusion that SM-mode gives greater effects, 30% after 0.5 h and 55% after 2 h.

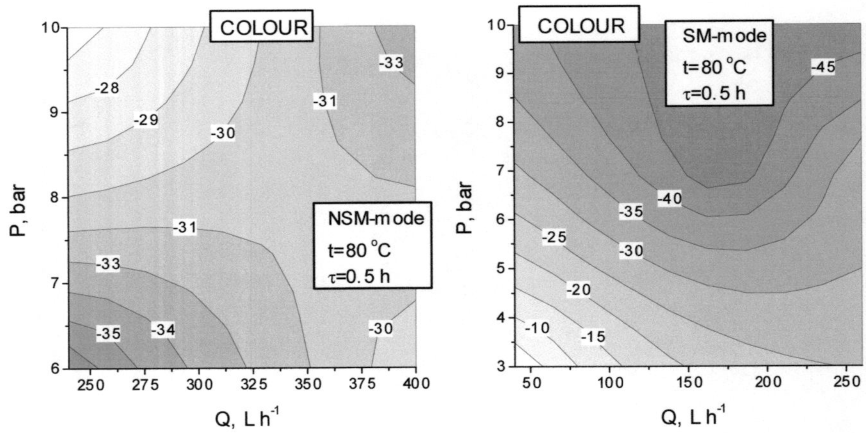

Fig. 4. Permeate colour as a function of flow rate and transmembrane pressure at 0.5 h of ultrafiltration

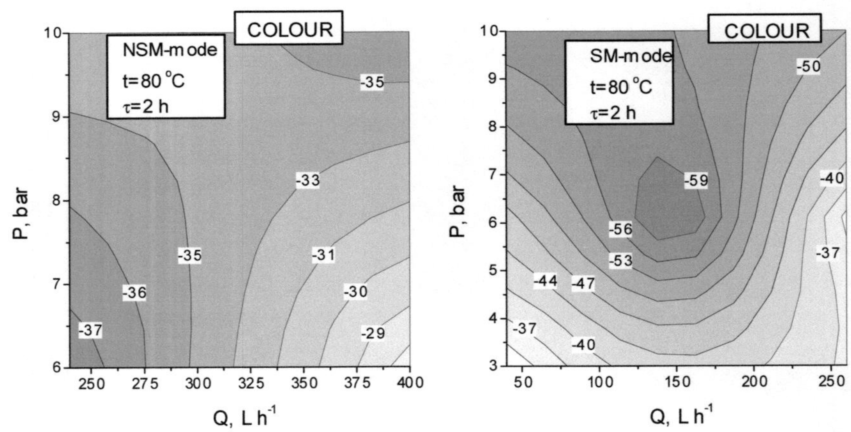

Fig. 5. Permeate colour as a function of flow rate and transmembrane pressure at 2 h of ultrafiltration

Analysis of the diagram in Fig. 6. showed that the juice colour decreased from 1080 IU to 750 IU. On the other hand, the retentate colour increased up to 1200 IU. At the same time, the permeate purity (defined as a ratio obtained by dividing the measured values of sucrose and quantity of dry matter content) increased from 98.9% to 99.7%, while the retentate purity decreased from 98.9 to 97%. Such a change of purity shows that membrane with 20 nm pore sizes does not reject sucrose molecules and there is no change in permeate dry matter content (because the main part of dry matter is sucrose).

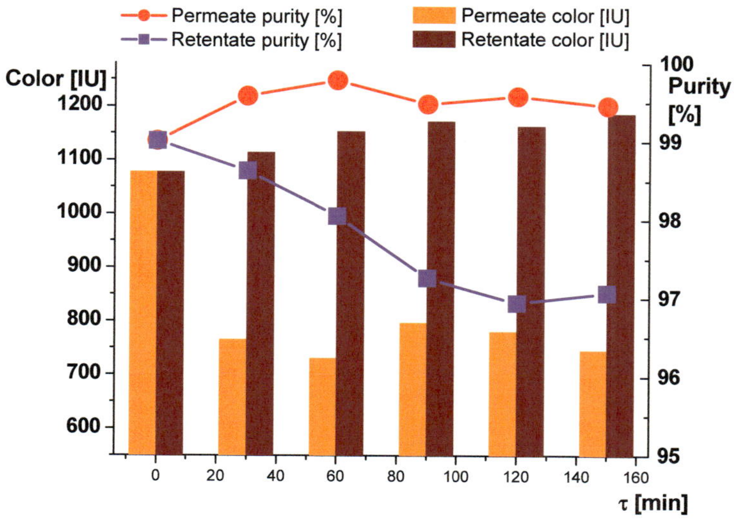

Fig. 6. Colour and purity of permeate and retentate in NSM- mode at flow rate 400 L/h, pressure at 6 bar and temperature at 80 °C

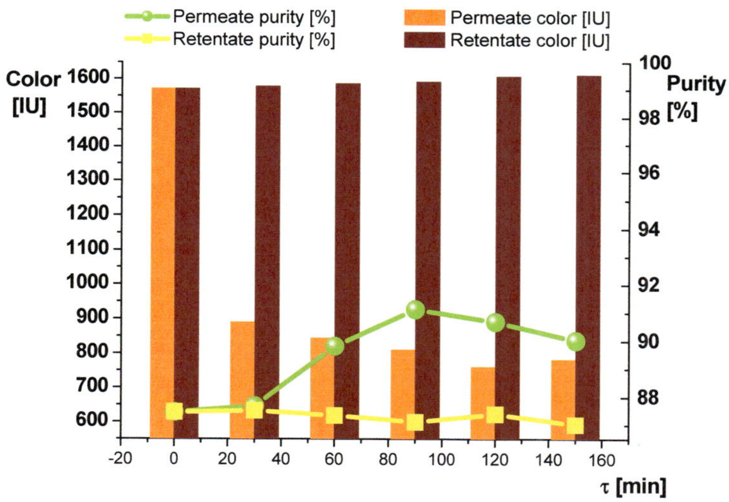

Fig. 7. Colour and purity of permeate and retentate in SM- mode at flow rate 400 L/h, pressure at 6 bar and temperature at 80 °C

When static mixer was used, the decolourization of permeate was more efficient (see Fig. 7.). The permeate colour decreased for about 50% while its purity increased for 2%.

The difference between pressures at opposite sides of the UF- module, as a function of average working pressure, is shown in the case of SM- mode of operation at 80°C in Fig. 8. It is interesting to observe that an increase of flow rate from 25 to 400 L h^{-1} causes an increase of the pressure difference from 0.1 to 3.7 bar. It could be concluded that flow rate has significant influence on the ultrafiltration when static mixer is used. High pressure drop along the length of the membrane (Fig. 8.) and therefore the increase in energy consumption per unit volume of permeate (E) is usually the main disadvantage of configurations with static turbulent promoters.

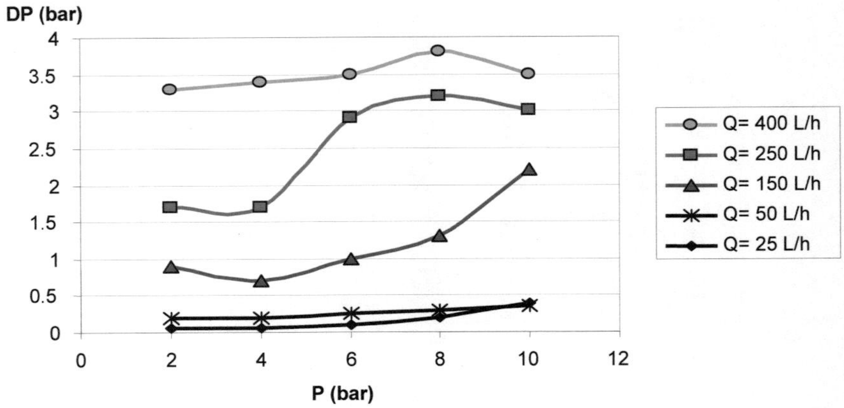

Fig. 8. Pressure drop along the length of the membrane (measured before and after UF module) as a function of the applied pressure in SM- mode at 80°C

The loss of hydraulic pressure (HP), due to the pressure drop along the length of the membrane in the SM-mode and NSM-mode of operation, is presented in Table 2. It is obtained by multiplying the pressure drop along the length of the membrane to the liquid velocity. As it is obvious, the loss of hydraulic pressure is much greater in the ultrafiltration process performed without static mixer compared to the process with the mixer at the same flux values. Namely, when static mixer is used higher fluxes can be achieved at lower velocities. At lower velocity, the pressure drop has also lower value. When the permeate flux increases the loss of hydraulic pressure in NSM- mode and SM- mode become similar. The results of an analysis of energy consumption per unit volume of permeate (E), in both modes of operation, are presented in Table 2 as well. Finally, the energy reduction (ER) for the sake of using static mixer is given. As it is obvious from Table 2, energy reduction is proportional to the permeate flux values (the greater the flux the higher the energy reduction). Under the investigated conditions, the energy reduction in SM- mode of operation reached approximately 85% comparing to NSM- mode.

J [L/m^{2}h]	HP [W]		E [kWh/m^3]		ER [%]
	SM	NSM	SM	NSM	
8	0,21	0,56	5,68	15,15	62,51
16	0,28	1,25	3,79	16,91	77,58
20	0,42	2,70	4,55	29,22	84,43

Table 2. Loss of hydraulic pressure due to the pressure drop along the length of the membrane (HP), energy consumption per unit volume of permeate (E) and energy reduction due to the presence of static mixer (ER)

5. Conclusion

The data collected on the impact of the traditional sugar beet processing on the environment pollution have shown that the greatest disadvantage is related to very high amount of required energy (mostly for the evaporation). The second important disadvantage is associated with the high level of water consumption (mostly for the extraction). The third disadvantage is connected to the purification of sugar juice and removal of non-sucrose compounds, undesired from the point of view of sugar quality. This chapter is related to the third disadvantage, which can be avoided or at least decreased by introducing environmentally friendly operation- membrane separation, instead of traditional operation- purification by chemically induced precipitation of undesired macromolecules.
Membrane separation is applied in food and chemical industry for several decades but its application in purification of sugar juices is still under investigations. Aware of the fact that membrane separations have great potential, many scientists are dealing with their adjustment to the requirements of sugar-industry. This chapter offers a review of results on the purification of sugar- beet juices and sugar- cane juices by membrane separations, mostly at a laboratory level. Despite the encouraging character of these results, very few large-scale membrane separation processes have been applied in the sugar industry worldwide. The reason lies in the fact that some weaknesses of the membrane separation processes have to be reduced before the separation can be applied on a large-scale. One of the greatest weaknesses is a gradual decrease of permeate flux through membrane caused by the precipitation of rejected macromolecules on the membrane surface. Among all the hydrodynamic methods used for improving mass transfer in membrane filtration, an increase in flow velocity represents the simplest way to create turbulence and reduce membrane fouling. Such an increase can be obtained by applying static turbulence promoters- static mixers. Some of our results on static mixer application together with the results on the same system without mixer application are presented comparatively. It is concluded that this kind of improvement makes membrane separation process closer to its industrial application which is highly desired from the point of view of environment protection.

6. Acknowledgment

These investigations are part of the project No. 142045, supported by the Ministry of Science, Serbia.

7. References

Anyos, E. (1984). Clarification of sugar-house products by liming and carbonatation, Thesis, Budapest

Attrige, L.; Erings, A. & Jaferey, I. (2001). Ultrafiltration of beet diffusion juice using spiral and tubular polymeric membranes, ASSBT 2001, *Proceedings of 3rd General meeting*, pp. Vancouver, 2001

Bogliolo, M.; Bottino, A.; Capanelli, G.; De Petro, M.; Servida, A.; Peyyi, G. & Vallini, G. (1996). Clean water recicle in sugar extraction process: performance analysis of reverse osmosis in treatment of press water. *Desalination*, Vol. 108, 261-271, 0011-9164

Cartier, S.; Theoleyre, M.A. & Decloux, M. (1996a). Combination of flocculation and crossflow filtration for removing colored impurities in raw cane sugar solutions. *Proceedings of the 1996 S:P:R:I. workshop on separation processes in the sugar industry,* New Orleans, USA, April 14-17, Sugar Processing Research Institute inc. (spri), www.spriinc.org

Cartier, S.; Theoleyre, M.A.; Lancrenon, X. & Decloux, M. (1996b). Membranes Technology in the Sugar Industry. *Proceedings of the 1996 S:P:R:I. workshop on separation processes in the sugar industry,* New Orleans, USA, April 14-17, Sugar Processing Research Institute inc., (spri), pp. 55, www.spriinc.org

Cartier, S.; Tatoud, L.; Theoleyre, M.A. & Decloux, M. (1997). Sugar refining process by coupling flocculation and crossflow filtration. *Journal of Food Engineering,* Vol. 32, 155-166, 0260-8774

Cheryan, M. (1986). *Ultrafiltration Handbook,* Technomic Publishing Co. Inc., 87762-456-9, Lancaster, Basel

Chou, C.C. (2002). White and refind sugar production from cane sugar factories. *Proceedings of First Biennial World Conference on Recent Developments in Sugar Technologies,* pp. 5, Delray Beach, Florida USA, May 2002, Dr. Chou Technologies, Inc. New York, USA and South China University of Technology, Guangzhou, China, www.esugartech.com

Council Directive 96/61/EC. (1996). Concerning Integrated Pollution Prevention and Control, *Official Journal of the European Communities,* No. L 257/26, 10.10.1996.

Decloux, M. & Tatoud, L. (2000). Importance of the control mode in ultrafiltration: case of raw sugar remelt. *Journal of Food Engineering,* Vol. 44, 119-126, 0260-8774

Djurić, M.; Gyura, J.; Zavargo, Z.; Šereš, Z. & Tekić, M. (2004). Modelling of ultrafiltration of non-sucrose compounds in sugar beet processing. *Journal of Food Engineering,* Vol. 65, 73-82, 0260-8774

Dornier, M.; Petermann, R. & Decloux M. (1995). Influence of start-up procedure on crossflow microfiltration of raw cane sugar. *Journal of Food Engineering,* Vol. 24, 213-224, 0260-8774

Eszterle, M., Šereš, Z. & Vatai, Gy. (2003). Elimination of colouring substances from sugar factory syrups by ultra- and nano-filtration. *Proceedings of International Conference of Visegrad Countries, Permea,* pp. 69, Tatranska Matliare, Slovakia, September 2003, Slovak Society of Chemical Engineering, Slovak University of Technology, CD-ROM

Eszterle, M.; Šereš, Z.; Vatai, Gy. & Gyura, J. (2000). Elimination of colouring substances from sugar industrial solution by ultrafiltration. *Proceedings of the XI International Conference of Membrane Techniques,* pp. 16-17, ISBN 963 8203 97 8, Tata, Hungary, October 2001, Műszaki és Természettudományi Egyesületek Szövetsége, Tatabánya

Godshall, M.A.; Vercelotti, J. & Triche, R. (2002). Effect of Macromolecules on Sugar processing: Comparition of Cane and Beet Macromolecules. *Proceedings of 9th Symposium AVH Association,* pp. 23-30 , Reims, France

Gosh, A.M.; Balakrishnan, M.; Dua, M. & Bhagat, J.J. (2000). Ultrafiltration of sugarcane juice with spiral wound modules: on-site pilot trials. *Journal of Membrane Science,* Vol. 74, 205-216, 0376-7388

Gosh, A. M. & Balakrishnan M. (2003). Pilot demonstration of sugarcane juice ultrafiltration in an Indian sugar factory. *Journal of Food Engineering*. Vol. 58, 143-150, 0260-8774

Gyura, J.; Šereš, Z. & Eszterle, M. (2001). Effects of ultra- and nano-filtration on separation of non-sucrose compounds in sugar industry syrup. *Proceeding of 17th International Congress of Nutrition*, pp. 370, ISBN 3-8055 7294-8, Vienna, Austria, August 2001, Karger Medical and Scientific Publishers, Basel

Gyura, J.; Šereš, Z.; Vatai, Gy. & Békási-Molnár, E. (2002a). Separation of non-sucrose compounds from the syrup of sugar beet processing by ultra- and nanofiltration using polymer membranes, *Desalination*, Vol. 148, 49-56, 0011-9164

Gyura, J.; Šereš, Z. & Békási-Molnár, E. (2002b). Separation of nonsucrose compounds from the syrup of sugar-beet by ultrafiltration and nanofiltration. *Proceeding of International Congress on Membranes and Membrane Processes*, pp. 432, Toulouse, France, July 2002, Europian Comission, Paris

Gyura, J.; Šereš, Z. & Eszterle, M. (2004). Influence of operating parameters on the separation of colorants from sugar house juices by ultra- and nanofiltration. *Zuckerindustrie* , Vol. 129, 174-180, 0344-8657

Gyura, J.; Šereš, Z. & Eszterle, M. (2005). Influence of operating parameters on separation of green syrup colored matter from sugar beet by ultra- and nanofiltration. *Journal of Food Engineering*, Vol. 66, 89-96, 0260-8774

Hamachi, M.; Gupta, B.B. & Ben Aim, R. (2003). Ultrafiltration: a means for decolorization of cane sugar solution. *Separation and Purification Technology*, Vol. 30, 229-239, 1383-5866

Hatziantoniou, D. & Howell, J.A. (2002). Influence of the properties and characteristics of sugar-beet pulp extract on its fouling and rejection behaviour during membrane filtration. *Desalination*, Vol. 148, 67-72, 0011-9164

Hinkova, A.; Bubnik, Z.; Kadlec, P. & Pridal, J. (2002). Potentials of separation membranes in the sugar industry. *Separation and Purification Technology*, Vol. 26, 101-110, 1383-5866

Integrated Pollution Prevention And Control (2003). *Draft Reference on Best Available Techniques in the Food, Drink and Milk Industry*, European Commission, Directorate-General JRC, Institute for Prospective Technological Studies, European IPPC Bureau, Seville, http://eippcb.jrc.es

Kochering, V.; Kearney, M.; Jacob, W. & Pryor, T. (2003). Chromatografic purification of raw beet juice: A lime free technology. *Zuckerindustrie*, Vol. 128, 667-672, 0344-8657

Koekoek, P.J.W.; Vannispen, J. & Vermuelen, D.P. (1998). Nanofiltration of thin juice for improvement of juice purification. *Zuckerindustrie*, Vol. 123, 122-126, 0344-8657

Krstić, D.; Koris, A. & Tekić, M. (2006). Do static turbulence promoters have potential in cross-flow membrane filtration applications?. *Desalination*, Vol. 191, No. 1-3, 371-375, 0011-9164

Kwok, R.J. (1996). Production of super VLC raw sugar in Hawaii. Experience with the new NAP ultrafiltration /softening process. *International Sugar Journal*, Vol. 98, 490-496, 0020-8841

Lipnizki, F.; Carter, M. & Tragardh, G. (2006). Applications processes in the beet and cane sugar production. *Zuckerindustrie* , Vol. 131, No. 1, 29-38, 0344-8657

Madsen, R. F. (1973). Application of UF and RO to cane juice. *International Sugar Journal*, Vol. 75, 163-167, 0020-8841

Martoyo, T.; Hino, M.; Nagase, H. & Bachtiar, A. (2000). Pilot test on ultrafiltration of cane raw juice at the Indonesian plantation white Kedawoeng sugar factory. *Zuckerindustrie,* Vol. 125, 787-792, 0344-8657

Nene, S.; Kaur, S.; Sumod, K.; Joshi, B. & Ragharavarao, K.S.M.S. (2002). Membrane distillation for the concentration of raw cane sugar syrup and membrane clarified sugarcane juice. *Desalination,* Vol. 147, 157-160, 0011-9164

Nielsen, W.K.; Kristensen, S. & Madsen, R.F. (1982). Prospects and possibilities in application of membrane filtration systems within beet and cane sugar industry. *Sugar Technology Reviews,* Vol. 9, 59-117, 0081-9204

Reinefeld, E. & Schneider, F. (1978). *Analitische Betriebskontrolle der Zuckerindustrie (Teil B),* Verlag Dr. Albert Bartens, Berlin, 978-3-87040-019-4

Schrevel, G. (2002). Separation of macromolecules in beet sugar processing. *Zuckerindustrie ,* Vol. 127, 768-769, 0344-8657

Šereš, Z.; Gyura, J.; Eszterle, M. & Vatai, Gy. (2004). Coloured matters removal from sugar beet industry syrup by ultra- and nano-filtration. *Acta alimentaria,* Vol. 33, No. 2, 119-127, 0139-3006

Šereš, Z.; Gyura, J.; Eszterle, M. & Vatai, Gy. (2005). Separation of non-sucrose compounds from syrup as a part of the sugar-beet production process by ultrafiltration with ceramic membrane. Membrane Science and Technology Conference of Visegrad Group - Permea 2005, Polanica Zdroj, Poland, CD

Šereš, Z.; Gyura, J.; Eszterle, M. & Djuric, M. (2006a). Separation of non-sucrose compounds from syrup as a part of the sugar-beet production process by ultrafiltration with ceramic membranes. *European Journal of Food Engineering,* Vol. 223, 829-636, 0260-8774

Šereš, Z.; Gyura, J.; Tekić, M. & Jokić, A. (2006b). The influence of the kenics static mixer on raw sugar solution ultrafiltration, Book of Abstracts of International Workshop on Quality of Food and Beverages in the Frame of EU Standards, p. 36, Bucharest, Romania, February 2006, Balkan Environmental Association (B.EN.A.) and Romanian Academy of Agricultural and Forestry Sciences (AAFS), Bucharest

Tebble, I.; Nelson, R. & Sargent D. (2002). The use of membrane filtration to reduce lime usage in purification. *Zuckerindustrie,* Vol. 127, No. 8, 600-603, 0344-8657

Urbaniec, K.; Zalewski, P. & Zhu X. X. (2000). A Decomposition Approach for Retrofit Design of Energy Systems in the Sugar Industry. *Applied Thermal Engineering,* Vol. 20, No. 15-16, 1431-1442, 1359-4311

Poel, Van der, P.; Schiweck, H. & Schwartz, T. (1998). *Sugar Technology – Beet and Cane Sugar Manufacture,* Verlag Dr. Albert Bartens, 978-3-87040-065-1, KG – Berlin

Vatai, Gy. & Tekic, M. (1995). Convection promotion and gel formation in an ultrafiltration process. *Chem. Eng. Comm.,* Vol. 132, 141-149

Willet, C.C. (1997). The A.B.C. process for the production of refined sugar from beets. *International Sugar Journal,* Vol. 99, 48-50, 0020-8841

14

Assessment of Air Quality in an Urban Area of Belgrade, Serbia

Mirjana Tasić[a], Slavica Rajšić[a], Milica Tomašević[a], Zoran Mijić[a], Mira Aničić[a], Velibor Novaković[a], Dragan M. Marković[a], Dragan A. Marković[b], Lazar Lazić[c], Mirjana Radenković[d] and Jasminka Joksić[d]

[a]*Institute of Physics, Belgrade, Serbia*
[b]*Faculty of Applied Ecology, Singidunum University, Belgrade, Serbia*
[c]*Institute of Meteorology, Faculty of Physics, Belgrade, Serbia*
[d]*Institute of Nuclear Science Vinča, Belgrade, Serbia*

1. Introduction

Clean air is considered to be a basic requirement for human health and well being. Various chemicals are emitted into the air from both, natural and anthropogenic sources. In spite of the introduction of cleaner technologies in industry, energy production and transport, air pollution remains a major health risk and tighter emission controls are being enforced by many governments. In humans, the pulmonary deposition and adsorption of inhaled chemicals from the air can have direct consequences on health. Public health can also be indirectly affected by deposition of air pollutants in environmental media and uptake by plants and animals, resulting in chemicals entering the food chain or being present in drinking water, and thereby constituting additional sources of human exposure. Furthermore, the direct effects of air pollutants on plants, animals and soil can influence the structure and function of ecosystems, including their self-regulation abilities, thereby and thus affecting the quality of life.

According to the most recent update of the WHO (World Health Organization) air quality guidelines, many studies were published that had investigated the effects of air pollution on human health. Particulate matter pollution is nowadays one of the problems of the most concern in great cities, not only because of the adverse health effects, but also for the reducing atmospheric visibility and affect to the state of conservation of various cultural heritages (Van Grieken & Delalieux, 2004). On a global scale, particulate matter (PM) also influences directly and/or indirectly the Earth's radiation energy balance, and can subsequently impact on global climate change (IPCC, 2001).

The measurement of the levels of atmospheric particulate matter is a key parameter in air quality monitoring throughout the world regarding the cause/effect relationship between exposure PM levels and health impacts (WHO, 2002; WHO, 2003). A number of epidemiological studies (Dockery & Pope, 1994, 2006; Schwartz et al., 1996, 2001) have demonstrated that acute and chronic health effects are related to the inhalable PM_{10} (aerodynamic diameter less than 10 μm) exposure in the urban environment, and some data also seem to indicate possible seasonal effects of the particulate matter on human health. This is especially important for urban aerosols, whose variety of size and composition make complete characterization a difficult task.

As a result of health and environmental impacts of PM, more rigorous regulations are in force in the USA and European countries. PM standards, issued by European Commission (EC), have included PM_{10} monitoring and limit values in the Air Quality Directive in 1999 (EC, 1999). Directive established in the first stage, annual limit value of 40 $\mu g\ m^{-3}$ and 24 h limit value of 50 $\mu g\ m^{-3}$ (not to be exceeded more than 35 times in a calendar year) to be met by 2005, and in the second stage annual limit value of 20 $\mu g\ m^{-3}$ and 24h limit value of 50 $\mu g\ m^{-3}$ (not to be exceded more than 7 times a calendar year) to be met by 2010.

Although the current focus on health-related sampling of particulate matter is on PM_{10}, recent research pointed out more serious health effect of fine particles, $PM_{2.5}$ (aerodynamic diameter less than 2.5 μm), and even $PM_{1.0}$ (aerodynamic diameter less than 1.0 μm) (Vallius et al., 2005; Ariola et al., 2006) and signed that the health effects associated with PM are related mostly to anthropogenic emission sources. EC has also established average $PM_{2.5}$ annual limit of 20 $\mu g\ m^{-3}$ (EN 14907, 2005) and current WHO Air Quality Guidelines set 10 $\mu g\ m^{-3}$ as annual average and 25 $\mu g\ m^{-3}$ as 24 h average (WHO, 2006).

Within the European Program for Monitoring and Evaluation of the Long-Range Transmission of Air Pollutants (EMEP), measurements of PM_{10} and trace metals, as highly toxic species, have been introduced. Spatial and temporal variation of atmospheric aerosol particles also gained in significance and resulted in an increased interest in the use of analytical techniques capable of measuring the size, morphology, and chemical composition of individual aerosol particles. Such data are essential for an understanding of particle formation, transport, transformation and deposition mechanisms, as well as, the impact of particles inhaled by a respiratory system.

The studies of the transport and mobilization of trace metals up to now have attracted attention of many researchers (Nriagy & Pacyna, 1988; Pacyna et al., 1989; Alcamo et al., 1992). Trace metals are persistent and widely dispersed in the environment and interacting with different natural components results in toxic effects on the biosphere.

Trace elements are released into the atmosphere by human activities, such as combustion of fossil fuels and wood, high temperature industrial activities and waste incinerations. The combustion of fossil fuels constitutes the principal anthropogenic source for Be, V, Co, Ni, Se, Mo, Sn, Sb, and Hg. It also contributes to anthropogenic release of Cr, Mn, Cu, Zn, and As. High percentages of Ni, Cu, Zn, As, and Cd are emitted from industrial metallurgical processes. Exhaust emissions from gasoline may contain variable quantities of Ni, Cu, Zn, Cd, and Pb (Samara et al., 2003). Several trace metals are emitted through the abrasion of tires (Cu, Zn, Cd) and brake pads (Sb, Cu), corrosion (V, Fe, Ni, Cu, Zn, Cd) lubricating oils (V, Cu, Zn, Mo, Cd) or fuel additives (V, Zn, Cd, Pb) (Pacyna & Pacyna, 2001; Ward, 1990; Sutherlan & Tolosa, 2000). The platinum group of elements, Rh, Pd and Pt, represent a relatively new category of traffic related trace metals in the environment, specially urban one, due to their application in automobile catalytic converters since the beginning of the 1980s (Haus et al., 2007).

Most of the trace metals are emitted in particulate form (Molodovan et al., 2002) and are present in almost all aerosol size fractions, but mainly accumulated in the smaller particles (Espinosa et al., 2001). This has a great effect on the toxicity of metals since the degree of respiratory penetration depends on particle size (Dockery and Pope, 1994, 2006; Espinosa et al., 2001). Urban anthropogenic particles are mainly in the $PM_{2.5}$ range and their sampling diminishes the interference of natural sources and reduces the loss of potentially volatile components such as ammonia and chloride. They could remain in the air with relatively

long residence time and could efficiently penetrate human lungs, and cause greater response in epithelial cells of human respiratory tract (Li et al., 2002, 2003). In addition to the PM mass limit values, also based on health impact criteria, recent European Union (EU) standards set target (Ni, As, Cd) and limit (Pb) values for metals and polycyclic aromatic hydrocarbons (PAHs) (EC, 1999; Directive 2004/107). Environmental technologies may have to be adopted in specific industrial spots to reach the target values. For aimed reduction of PM_{10} or $PM_{2.5}$ levels detailed knowledge of sources and their respective contribution to the PM levels, is required.
Most trace elements in terrestrial ecosystems originate from atmospheric wet and dry deposition. From a biogeochemical perspective, the characterization of total atmospheric deposition is relevant in order to identify the variability and sources of the atmospheric pollutants (Azimi et al., 2005). Direct collection of atmospheric deposition using bulk sampling devices offers a practical approach to monitor atmospheric heavy metal deposition providing valuable information on the influences of atmospheric inputs of heavy metals on the surface environment (Morselli et al., 2003; Azimi et al., 2003). Bulk sampling has been extensively used, since the samplers are easier to operate than wet-and-dry ones. The limitation of this sampling method is mainly possible under-estimation of fluxes; advantages are integration of samples over 1 month and the possibility of large-scale application with low-cost equipment.
Studies on atmospheric contamination have frequently been limited by high cost of instrumental monitoring methods and difficulties in carrying out extensive sampling in time and space. For these reasons, there is an increasing interest in using indirect monitoring methods such as the use of organisms that may act as bioaccumulators. Biomonitoring of trace elements from atmospheric deposition can be currently evaluated by environmental biomonitors such as mosses, lichens and higher plants (Rühling & Tyler, 1971; Steinnes et al., 1992; Markert, 1993; Bargagli, 1998; Bargagli et al., 2002; Adamo et al., 2003). Native mosses and lichens have often been used in passive biomonitoring, and have several advantages as compared to higher plants. They lack a developed root system, so they rely on atmospheric wet and dry deposition for their mineral nutrition; they have a high surface/volume ratio and cation exchange capacity; unlike many other plants, they lack variability of morphology through the growing season and they have no cuticle (Tyler, 1990; Bargagli, 1998).
The heavy metals in mosses survey, first introduced in Scandinavia (Rühling & Tyler, 1968), has been repeated since 1980, at five-years intervals, with an increasing number of participating countries (http://icpvegetation.ceh.ac.uk/intro.htm; UNECE ICP Vegetation, 2003). The survey has provided data on heavy metals concentration in naturally growing mosses throughout Europe, and there is substantial database for assessment of pollution level and identification of pollution sources. The standardized procedures for obtaining moss samples included collecting the preferred moss species: *Hylocomium splendens, Pleurozium schreberi, Hypnum cupressiforme*. As previously recommended species are not widespread in arid areas of southern European countries, it is necessary to find corresponding alternate moss species for monitoring studies.
In highly polluted areas of industrial or urban environment, terrestrial plants can act as appropriate bioindicators and biomonitors. Although biomonitoring of air quality using plants has been practiced for many years, in many European countries, it has still not been applied at a satisfactory level, due to different, and even opposite results, depending, first of all, on plant species. Therefore, efforts towards setting up the large European projects to

biomonitor air quality have been made recently (Klumpp et al., 2002). Trees are very efficient at trapping atmospheric particles, and they have a special role in reducing the level of fine, "high risk" $PM_{2.5}$, with the potential to cause serious human health problems. Thus, the use of plant leaves, primarily, as accumulative biomonitors of trace metal pollution has attained great ecological importance (Markert, 1993; Bargagli, 1998; WHO, 2000; Mignorance & Rossini, 2006).

Leaves of various tree species, both evergreen and deciduous, have been tested for this purpose in urban areas (Alfani et al., 1996; Monaci et al., 2000), including a search for sensitive tree species and approval of the validity of using such leaves as biomonitors. It is well known that metal pollution leads to physiological disturbances in plants and affects the biogeochemical balance and stability of their habitats. Metal uptake in higher vascular plants takes place through their root system, additionally through the leaves and, therefore, it is difficult to distinguish whether the accumulated elements originate from the soil or from the air (Harrison & Johnston, 1987; Verma & Singh, 2006). The research of heavy metals contamination of vegetation requires the use of standard methodological procedures (Markert, 1993; Bargagli, 1998). One of the most important is the representative sampling of plant material.

Ground level ozone (O_3) and other photochemical oxidants have become pollutants of concern because elevated concentrations are known to cause detrimental effects and threaten human health (WHO, 2003; Mulholland et al., 1998), vegetation and objects. Moreover, in recent years there have been numerous reports of an association between increases in particle air pollution (PM_{10}) and ozone concentration (Meng et al., 1997; Mulholland et al., 1998; Ying & Kleeman, 2003). In order to protect human health and ecosystem, EU has set limits for ozone concentrations (Directive 2002/03/EC). The information threshold is the same as in the previous Directive being 180 $\mu g\ m^{-3}$ (the hourly average concentration of 240 $\mu g\ m^{-3}$, measured over three consecutive hours is set as an alert threshold). High ozone levels are mainly observed during periods with warm and sunny weather in combination with stagnant air masses and the build-up of precursor substances, such as nitrogen oxides (NO_x), carbon monoxide (CO) and volatile organic compounds (VOCs).

Although PM, above all PM_{10} or $PM_{2.5}$, is of great concern for public health, no systematic studies have been performed in Belgrade until recently. The studies on the quality of air in urban atmosphere related to suspended particles PM_{10} and $PM_{2.5}$, and the first measurements of their mass concentrations have been initiated in Serbia in 2002, and are still in progress. The results of preliminary investigations revealed the need for the continuous and long-term systematical sampling, measurements and analysis of interaction of the specific pollutants - PM_{10}, $PM_{2.5}$ and trace metals in the ground level (Tasić et al., 2001; Rajšić et al., 2004a, b).

In 2002, the national project "Air Quality Studies in Belgrade Urban Area: Suspended Particles, Heavy Metals and Radionuclides", financed by The Ministry for Science and Environmental Protection of the Republic of Serbia, has started. At present, the project "Emission and Transmission of Pollutants in an Urban Atmosphere", is running, and includes measurements of trace and other elements (Al, V, Cr, Mn, Fe, Ni, Cu, Zn, As, Cd, Pb, etc.) concentration in particulate matter, PM_{10} and $PM_{2.5}$, bulk atmospheric deposition, soil, plant leaves, mosses, and natural and man made radionuclides (Be-7, Cs-137, Pb-210), and ground level ozone. The objective of the project is to assess air quality and to identify the main sources by multivariate receptor modeling (PCA, CA), enrichment factor (EF) calculation, meteorological conditions and air back trajectories analysis.

In this review, we report some of the results of the integral monitoring of air quality in Belgrade urban area in order to evaluate the impact of airborne trace metals on the pollution load for the period from 2002 to 2006. Some of the results concerning suspended particle mass and trace metal concentrations in ambient air of the Belgrade will be presented, including physical and chemical characterization, with the aim to examine elemental associations and to indicate the main sources of trace and other metals in the city. The results of this long-term project of the pollution monitoring could be used as the baseline data for analysis of health risks due to inhalation of suspended aerosols, and to provide scientific evidence for setting up an air pollution control strategy. This information is crucial in environmental quality assessment, and can lead to the determination of a possible exceedance of the critical loads.

Also, the aim of our research was to set up a reliable methodological approach in sampling and analytical procedures for investigation of moss and deciduous leaves biomonitoring, and to evaluate the reliability of investigated species for biomonitoring of trace metal pollution in Belgrade urban areas. The higher plants leaves, horse chestnut (*Aesculus hippocastanum* L.) and Turkish hazel (*Corylus colurna* L.) widely spreaded in the city have been sampled and their trace metal accumulation abilities analysed. Some physical and chemical characteristics of particles deposited on leaf surfaces have been studied as well. Mosses *Brachythecium sp.* and *Eurhynchium sp.*, used in this study, more common in Serbia, were investigated for their accumulation capability.

The first data set for ozone and correlation to relevant meteorological parameters obtained in Belgrade during several sampling periods in Belgrade will be presented, and consequently, important information about the local air quality.

Meteorological conditions favorable for the build up of ozone are frequent, in Belgrade and the surrounding area, from early spring to early autumn. During this period, photochemical smog events often show different features and are difficult to assign to a particular pattern. Although ozone levels were measured in Belgrade sporadically during the early 80-ties, and each season starting from 1991, few of these data have been published and the information is scant and difficult to find (Vukmirović et al., 1987). Therefore, it would be useful to present the more complete results of our measurements, stressing the main characteristics of the photochemical episodes recorded in the Belgrade area. As there is no available data on ozone measurements from the area of the former Yugoslavia (except Slovenia and FYROM) in European Environmental Agency reports, it is important to increase the geographical coverage of the current state of knowledge with respect to ozone seasonal cycles in the troposphere over Europe.

2. Experimental Methods and Procedures

2.1 Studied Sites and Sampling

Belgrade, (H_s = 117 m, $\varphi = 44^0 49'14''$N and $\lambda = 20^0 27' 44''$E) the capital of Serbia, with about 2 million inhabitants, is situated at the confluence of the Sava and Danube rivers (Fig. 1). In the winter, severe air pollution in the form of aerosol smog occurs frequently in the urban area of Belgrade, particularly under the meteorologically calm (wind speed < 2 m s^{-1}) and stable conditions. The total number of vehicles in the year 2002 was more than 350,000, including 22,000 heavy-duty vehicles and over 1,000 city buses using diesel. The average

age of passenger cars is more than 15 years, which means that leaded gasoline (0.4 g l^{-1} Pb) is still in wide use in the country. There are many old buses and trucks on the streets and it could be the significant major source of ambient PM_{10}. There are 18 bigger heating plants with a total capacity of 2018 MW, run with natural gas or crude oil and 59 smaller plants run only with crude oil (approximately 193 MW). Fuel used for domestic heating consists mainly of coal or crude oil and natural gas in last few years.

The climate of Belgrade is moderate continental with fair cold winters and warm summers. The prevailing wind is N-NW, but characteristic wind "Košava" (SE-ESE) blows with an annual frequency of 26% and an average speed of 4 m s^{-1} (Unkašević, 1999). Fortunately, "Košava" comes from relatively unpolluted area. This wind effectively improves the horizontal dispersing and dilution of pollutants in the ground-level atmosphere of Belgrade city.

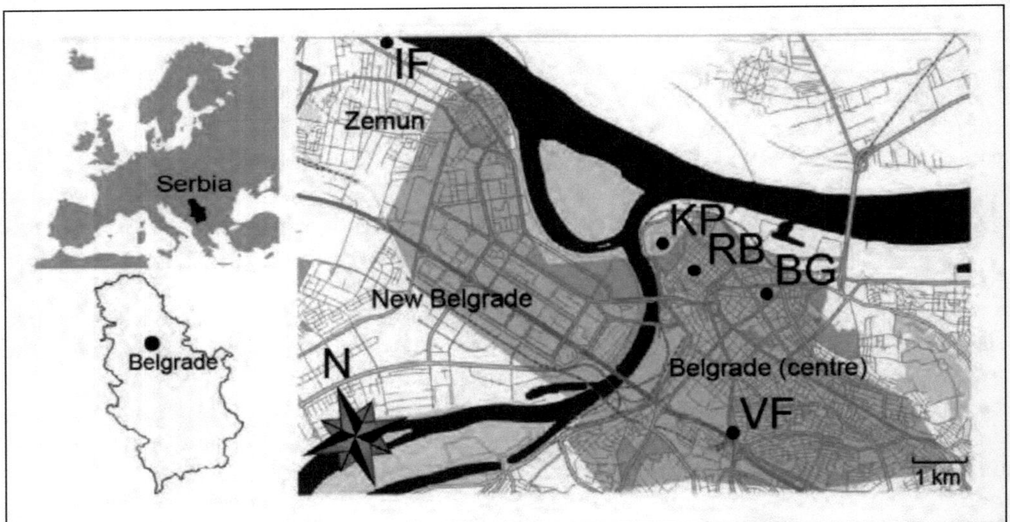

Fig. 1. Location of the sampling sites in Belgrade urban area: Rector's Office building (RB); Botanic Garden (BG); Autokomanda (VF); Institute of Physics- Zemun (IF); Kalemegdan Park (KP)

2.1.1 Particulate Matter

Sampling of particulate matter PM_{10} and $PM_{2.5}$ started on three sites in the very urban area of Belgrade in June 2002 and has continued afterwards. The first sampling point was on the roof of the Rector's Office building of Belgrade University on Student Square (RB), at a height of about 20 m, near a small city-park. The square has high traffic density and a bus terminal. As this sampling point is in the very city center, on the rooftop where the airflow is not blocked by any direction, it can be considered as representative for urban-background concentrations. The second sampling location was at about 6 m height in the Botanic Garden (BG) about 50 m far from heavy-traffic streets. The third sampling site was the platform above the entrance steps to the Faculty of Veterinary Medicine (VF) at a height of about 4 m from the ground, 5 m away from a street with heavy traffic and close to the big Autokomanda junction with the main state highway. The traffic is controlled by street lights. This point can be considered as traffic-exposed. From time to time, samples were taken at a control suburban site in Zemun, on the right bank of the Danube River, near the Institute of Physics (IF). During the sampling, meteorological parameters including temperature, relative humidity, rainfall, wind direction and speed were provided by the Meteorological Station of the Hydro-Meteorological Institute of the Republic of Serbia (H_s = 132 m, φ = 44°48′ N and λ = 20° 28′ E), located inside the central urban area, very close (≈200 m) to the Autokomanda (VF) sampling site.

Suspended particles were collected on preconditioned (48 h at 20°C and constant relative humidity around 50%) and pre-weighed Pure Teflon filters (Whatman, 47 mm diameter, 2 μm pore size) and Teflon-coated Quartz filters (Whatman, 47 mm diameter) using two MiniVol air samplers (Airmetrics Co. Inc., 5 l min^{-1} flow rate) provided with PM_{10} and $PM_{2.5}$ cutoff inlets and positioned at 2 m height. The sampling time was 24 h, yielding a sample volume of 7.2 m^3. Routine maintenance of the samplers and calibration of the flow meters were often conducted in order to ensure the sampling quality. After particle collection, the filters were sealed in plastic bags and kept in portable refrigerators, in a horizontal position during transportation back to the laboratory where they were reconditioned for another 48 h. The sampling methodology used in this study was described in detail by Rajšić et al. (2004a).

2.1.2 Total Atmospheric Deposition

The total atmospheric deposition (TD) collection was performed using an open polyethylene cylinder (29 cm inner diameter and 40 cm height) fixed in a basket on a pole 2 m above the ground to avoid the collection of re-suspended dust from the surface. The devices collected both dry deposition and precipitation continuously for 4 - week periods from June 2002 to December 2006 Rector's Office building, Botanic Garden (BG) and Autokomanda (VF) (Fig. 1). The collection bottles were filled before each sampling period with 20 ml of 10% HNO_3 (Suprapure, Merck).

2.1.3 Biomonitoring

Deciduous leaves for trace metal deposition and accumulation analyses were sampled from horse chestnut (*Aesculus hippocastanum* L.) and Turkish hazel (*Corylus colurna* L.) trees in the Belgrade urban area at three locations, BG, RB and VF (Fig.1). Leaf samples for the metal accumulation study were collected at the beginning and the end of the seasonal vegetation cycles. Ten leaves growing at 2 m height were cut off with Teflon coated stainless steel scissors. Measurements were performed at the single leaf level. Each leaf was placed horizontally in a polycarbonate Petri dish and transferred to the laboratory. Sampling and handling of all plant material were carried out using polyethylene gloves and bags.

Collection of native moss, *Brachythecium sp.* (*B. rutabulum* and *B. salebrosum*) and *Eurhynchium sp.* (*E. hians* and *E. striatum*), for passive biomonitoring of atmospheric trace and other elements pollution was performed according to standardized procedure (UNECE ICP Vegetation, 2003). Mosses were collected at two parks in Belgrade, IF and KP (Fig. 1), within a 30 x 30 m area, at least 100 m away from main roads, and 50 m from smaller roads and houses. The samples were taken at least 5 m from the base of any tree so as not to be directly exposed to throughfall precipitation. In laboratory, the samples were carefully cleaned from all dead material and attached litter, then only green and green-brown moss upper parts up to two/three-years old were analyzed. The samples were dried for 48 h at 35°C to constant weight prior to analysis of elements content.

2.1.4 Ozone

The tropospheric ozone concentrations were measured using UV photometric O_3 analyzer Model 108-AH Dasibi Environmental Corporation, at the same points in Belgrade urban ares as suspended particles, total atmospheric deposition and plant leaves in 2002. In June, September and October, the measurements were conducted at 20 m above the ground, on the roof of Belgrade University Rector's Office Building (RB), Student Square, Belgrade. In July, the measurements were conducted at 3 m above the ground in Botanic Garden (BG). In July, November and December, the measurements were performed at the height of 3 m on the platform above the entrance stairs to the Faculty of Veterinary Medicine (VF).

2.2 Analytical Procedures

2.2.1 Mass Concentrations

Daily PM samples were handled and processed in a Class 100 clean laboratory, at the Institute of Physics, Belgrade. Particulate matter mass concentration was determined by weighting of the filters using a semi-micro balance (Sartorius, R 160P), with a minimum resolution of 0.01 mg. Loaded and unloaded filters (stored in Petri dishes) were weighed after 48 hours conditioning in a desiccator, in the clean room at a relative humidity of 45-55% and a temperature of 20 ± 2 °C. Quality assurance was provided by simultaneous measurements of a set of three ''weigh blank'' filters that were interspersed within the pre- and post- weighing sessions of each set of sample filters and the mean change in "weigh blank" filter mass between weighing sessions was used to correct the sample filter mass changes.

2.2.2 Trace Metal Analysis

Atomic absorption spectroscopy (AAS)

The elemental composition (Al, V, Cr, Mn, Fe, Ni, Cu, Zn, As, Cd, and Pb) of the aerosol samples and total atmospheric deposition, was measured by the atomic absorption spectroscopy method (AAS). After completion of gravimetric analysis, PM samples were digested in 0.1 N HNO_3 on an ultrasonic bath. An extraction procedure with dilute acid was used for the evaluation of elements which can become labile depending on the acidity of the environment. This procedure gives valid information on the extractability of elements, since the soluble components in an aerosol are normally dissolved by contact with water or acidic solution in the actual environment (Kyotani & Iwatsuki, 2002). Depending on concentration levels, samples were analyzed for a set of elements by flame (FAAS) (Perkin Elmer AA 200) and graphite furnace atomic absorption spectrometry (GFAAS) using the transversely-heated graphite atomizer (THGA; Perkin Elmer AA 600) with Zeeman-effect background correction. The THGA provided a uniform temperature distribution over the entire tube length, rapid heating and an integrated L'vov platform, which gave an improved signal/interference ratio and high analytical sensitivity. Analyte injection (20 μl) and the atomization were done in five steps controlled by the appropriate software and auto-sampler.

Total atmospheric deposition samples were evaporated to dryness, digested with 50 ml 0.1 N HNO_3 on ultrasonic bath, the digested solution was filtered through 0.45 μm porosity Sartorius membranes and analyzed using a flame and graphite furnace atomic absorption spectrometer. Laboratory blanks were analyzed in the same manner as field samples and the heavy metal concentration values were below the detection limit values for all analyzed

metals. Data treatment included the calculation of Al, V, Cr, Mn, Fe, Ni, Cu, Zn, As, Cd, and Pb monthly deposition fluxes.
For calibration, standard solutions containing all metals of interest were prepared using Merck certified atomic absorption stock standard solutions containing 1000 mg l^{-1} metal in 0.5 N HNO_3 and Milli-Q quality deionized water, with no matrix modifier addition. Detection limits for the trace elements were found to be: 0.04 ng ml^{-1} for Cd, 0.1 ng ml^{-1} for Cr, 0.2 ng ml^{-1} for Cu, 0.5 ng ml^{-1} for Pb, 2 ng ml^{-1} for Zn, 0.4 ng ml^{-1} for Ni, 0.2 ng ml^{-1} for Mn, 5 ng ml^{-1} for V, 0.5 ng ml^{-1} for Fe, 2 ng ml^{-1} for Al, and 0.5 ng ml^{-1} for As. Standard practices for the handling of trace metal samples were implemented. For quality assurance, NIST 2783 standard reference material was used.

Differential pulse anodic stripping voltammetry (DPASV)
Leaf samples were transferred to a Class 100 clean room under the specific conditions required for analysis of low concentrations of trace metals (Vukmirović et al., 1997). Leaves were rinsed in double distilled water, and the samples were then dried at 105 °C to constant weight. Ashing was carried out for 6 hours at 450 - 500 °C, with a gradual temperature increase of 50 °C per hour to eliminate organic matter without losing some constituents from the samples by volatilization. The ash was dissolved in 0.1 N HNO_3 prior to analysis. All chemicals and standard solutions employed were of ultra pure quality. An electrochemical method, differential pulse anodic stripping voltammetry with a hanging mercury drop electrode (DPASV) was used for determination of Cu, Zn, Cd and Pb contents in a single leaf. Measurements were performed with an EDT, ECP 140 Polarograph and the analytical technique was described in detail previously (Vukmirović et al., 1997; Tomašević et al., 2004). The detection limits (ng ml^{-1}) were 0.5, 1.0, 0.1 and 1.0 for Cu, Zn, Cd and Pb, respectively.

Instrumental neutron activation analysis (INAA)
Heavy metal and other element concentrations in the native moss samples were determined by instrumental neutron activation analysis (INAA). INAA was performed at the Frank Laboratory of Neutron Physics, Joint Institute for Nuclear Research (FLNP JINR), Dubna, Russian Federation (Frontasyeva & Pavlov, 2000). The moss was analyzed on 36 elements. Approximately 0.3 g of well homogenized moss was taken for analyzing by INAA and most element concentrations were determined with detection limits within the range of 0.01 - 10 µg g^{-1}. The short-term irradiation (2 min) was used for short-lived radionuclides (Mg, Al, Cl, K, Ca, Ti, V, Mn, I, and Dy). The long irradiation (100 h) was used to determine elements associated with long-lived radionuclides (Na, Sc, Cr, Fe, Co, Ni, Zn, As, Se, Br, Rb, Sr, Mo, Sb, Cs, Ba, La, Ce, Sm, Tb, Hf, Ta, W, Hg, Th, and U). Gamma-ray spectra were measured four times using a high-purity Ge detector after decay periods of 5 and 10 min following the short irradiation and after three and 20 days following the long irradiation, respectively. The low temperature during irradiation of samples (60 – 70 °C) provides determination of elements present in the samples in volatile form.
To provide quality control, content of elements yielding short- and long-lived isotopes in moss samples was determined using certified reference materials issued by the International Atomic Energy Agency (IAEA): lichen (IAEA-336), cabbage (IAEA-359) and standard reference material SRM-1575 (pine needles) from the National Institute of Standards and Technology (US NIST). For the short irradiation, the three reference materials were irradiated together with 10 experimental samples. In the case of long irradiation, the three reference materials were packed and irradiated together with 7 – 9 samples in each transport container.

2.2.3 Scanning Electron Microscopy

Scanning electron microscopy (SEM) coupled with Energy-Dispersive X-ray analysis (EDX) was used for the characterization (size, size distribution, morphology and chemistry of particles) and source apportionment of suspended atmospheric particulate matter and particles deposited on leaves.

One PM sample per sampling episode was analyzed with the SEM/EDX (JOEL JSM-5300 SEM) according to the US-EPA Guidelines (2002). Prior to analyses three small sections of the filters (5 mm x 5 mm) were mounted on the SEM stubs and then coated with 10 nm layer of high purity gold using vacuum evaporator (Balzers/Union FL-9496). The SEM observations were carried out at magnifications up to 15,000X; the electron beam energy was 30 keV, and probe current of the order of 100 μA. Ten photomicrographs were arbitrarily taken under low resolution conditions and about 300 particles per PM sample were assessed for their morphology and about 50 particles for the X-ray spectral analysis. The elemental composition of selected particles in the secondary electron images was deduced from an energy dispersive X-ray spectrum in the energy range up to 20 keV, collected from the selected particles for a spectrum acquisition time of 100 s. The elements observed were: C, N, Na, Mg, Al, Si, P, S, Cl, K, Ca, Ti, V, Cr, Mn, Fe, Ni, Cu, Zn, As, Cd, and Pb, with detection limit of 1 wt % (Tasić et al., 2006).

An SEM Philips XL30 apparatus equipped with a thin-window EDAX DX4 system for energy dispersive X-ray microanalysis was used to analyze the particles deposited on the leaf samples. Leaf samples were dried in air in the clean room. To minimize charge build-up on the samples from exposure to the SEM electron beam the samples were coated with (10 – 15) nm layer of high purity carbon using vacuum evaporator (Balzers/Union FL-9496) prior to analyses. The SEM observations were carried out at magnifications up to 2000X while the electron beam energy was fixed at 20 KeV, and the working distance in most cases was about 10 mm and probe current was 100 pA. Particles were observed by backscattered electron images. Three different leaf discs of the adaxial and abaxial surfaces for both tree species were examined in the same way. Ten photomicrographs were randomly taken of each 0.03 mm^2 area at 624X magnification and about 1800 particles per species were assessed to their morphology and about 900 for X-ray spectra analysis. For each tree species about 0.025% of the original leaf surface was examined.

An energy dispersive X-ray spectrum (EDS) was collected from the selected particles in the range up to 15 keV for a preset time (live time) of 10 s to 20 s. The total X-ray count rate was between 1000 and 2000 counts s^{-1}. The relative elemental composition of the particles, were computed directly with EDAX software, using the "ZAF" (atomic number, absorption, fluorescence) correction. As the particles deposited on leaves have complex shapes, quite different from an ideal flat sample, there may be over- or underestimation of the actual atomic concentration, but this does not prevent identification of the most important particle types. Periodical checks of the X-ray by peak identification were conducted. EDX Spectrometer gain calibration was accomplished using a gold/copper standard since X-ray lines from these two elements span almost the entire spectral range of the detector.

2.3.4 Multivariate Receptor Modelling

Principal component analysis (PCA) and cluster analysis (CA) were used to identify the possible emission sources of trace elements and correlations among them in suspended particulate matter and total atmospheric deposits. The extracted principal components were interpreted as source categories contributing to PM concentrations at the sampling site and total deposition as well. The identification of source categories was done by examination of the profiles of the principal components, i.e. loadings of the elements and other variables on the Varimax (orthogonally) rotated principal components. Factor loadings > 0.71 are typically regarded as excellent and < 0.32 as very poor (Nowak, 1998). In this study, all principal factors extracted from the variables with eigenvalues > 1.0 were retained, according to the Kaiser criterion (Kaiser, 1960). When PCA with Varimax normalized rotation was performed, each PC score contained information on the metal elements, while the loadings indicated the relative contribution each element made to that score.

Cluster analysis (CA), as a complementary analysis to PCA, was performed to classify elements of different sources further on the basis of their chemical properties. CA was applied to the concentration data using Ward's method, with Euclidean distances as the criterion for forming clusters of elements and also to determine when two clusters were sufficiently similar to be linked. In general, this form of CA is regarded as very efficient, although it tends to create small clusters. As the variables had large differences in scaling, standardization was performed before computing.

2.3.5 Enrichment Factor

Enrichment factor (EF) analysis was used to differentiate between the elements originating from human activities and those of natural origin and to assess the degree of anthropogenic influence. By convention, the average elemental concentration of the natural crust is used instead of the continental crust composition of the specific area, as detailed data for different areas are not easily available. There is no rule for the reference element choice and Si, Al, and Fe have been used as the most common elements for this purpose (Manoli et al., 2002; Gao et al., 2002). In this study, Al was used as the reference element with upper continental crustal composition given by Mason (1966). EF represents the ratio of the fraction of the element E with respect to reference element R in the samples (aerosols, atmospheric deposition, moss...) $(E/R)_{sample}$ to the fraction of E with respect to the same R in the crust $(E/R)_{crust}$:

$$EF = \frac{(E/R)_{sample}}{(E/R)_{crust}} \tag{1}$$

According to the degree of enrichment the elements may be grouped as follows: highly enriched ($EF > 100$); intermediately enriched ($10 < EF < 100$) and less enriched ($EF < 10$) (Berg et al., 1994; Wang et al., 2005). If the EF approaches unity, the crustal material is likely the predominant source for element; if $EF > 1$, the element has a significant fraction contributed by non-crustal sources.

2.3.6 Air Back Trajectories

The analysis of air back trajectories for high PM concentrations episodes, in Belgrade, has been performed. Theoretical and experimental evidence was based on numerical weather prediction model and trajectory model so-called Eta model. The model used for simulation and air back trajectory calculation in this study is a regional weather prediction primitive equation model for synoptic and meso-scale processes (Mesinger et al., 1984, 1988, Janjić et al., 1990, 1994). In this study model with 3.2 km horizontal resolution and 32 layers in the vertical was used. The boundary conditions were updated every 6 hours obtained from European Centre for Medium-Range Weather Forecast (ECMWF). Construction of three-dimensional atmospheric trajectories provides a valuable diagnostic tool for illustrating and studying three-dimensional flow fields and associated transports. Trajectories are calculated from simulated wind fields, with both horizontal and vertical wind components derived from the Eta model.

Trajectories can be calculated forwards and backwards in time. Air back trajectories are calculated by specifying final parcel locations and time, and then tracing the parcels with decreasing time to ascertain their origins. The model has been used for research in entire Serbia region and boundary regions of the other countries in the neighborhood.

3. Results and Discussion

3.1 Particulate Matter

A first assessment of PM_{10} and $PM_{2.5}$ particulate level in the ambient air of Belgrade

Daily mass concentrations of 96 PM samples (PM_{10} and $PM_{2.5}$) were determined by gravimetric analysis of filters that were exposed to urban air in Belgrade during the year 2002. The PM_{10} mean 24-hours mass concentration value, over whole measuring period was 77 μg m^{-3}, almost twice as much as the annual limit in European Union (40 μg m^{-3}) and 62% of days had mean daily concentrations above limit value of 50 μg m^{-3}. Average $PM_{2.5}$ mass concentration exceeded the EC annual limit of 20 μg m^{-3} (EN 14907, 2005) by a factor of 3 (Rajšić et al., 2004; Tasić et al., 2005).

PM_{10} and $PM_{2.5}$ mass and trace metal concentrations

During the next sampling episode, between June 2003 and July 2005, daily mass (μg m^{-3}) and trace and other element (ng m^{-3}) concentrations were calculated in PM_{10} and $PM_{2.5}$ and already presented in detail (Rajšić et al., 2007; Todorovic et al., 2007). A total of 273 (209 PM_{10} and 64 $PM_{2.5}$) valid samples were taken during the 2-year period. The high mean and maximum levels of PM_{10} and $PM_{2.5}$ were observed; the PM_{10} mean mass concentration during the 2-year period (68.4 μg m^{-3}) exceeded the proposed EC annual limit of 40 μg m^{-3} (EC, 1999). Of more concern was the average $PM_{2.5}$ concentration of 61.4 μg m^{-3} for the 2-year period, which was three times higher than the EC annual limit of 20 μg m^{-3} (ES 14907, 2005)

The results for the total mean concentrations of individual metals indicate iron as the most abundant metallic element (1462.9 ng m^{-3}) in the PM_{10}. Zinc and Al concentrations in this fraction were very high, amounting to 1389.2 ng m^{-3} and 873.8 ng m^{-3}, respectively. The highest mean concentration in $PM_{2.5}$ was for Zn (1998.0 ng m^{-3}), followed by Al (1180.3 ng m^{-3}) and Fe (1081.2 ng m^{-3}). Zinc is reliable tracer of unleaded fuel and diesel oil powered motor vehicle emissions (Monaci et al., 2000) and besides, it could be released in large amounts from tired friction or various industrial activities. Concerning Cu, a heavy metal characterized by its toxicity, relatively high mean values of 71.3 ng m^{-3} in PM_{10} and 20.8 ng m^{-3} in $PM_{2.5}$ were obtained. This trace element is associated with industrial activities, but in urban areas, road traffic (diesel engines and wearing of brakes) could be the most important source. Aluminum concentration was higher in $PM_{2.5}$ than in PM_{10}. Although Al and Fe are typically crustal elements, if coupled with other elements, they can indicate the presence of anthropogenic sources, such as the steel production industry. The mean Ni concentration of 28.4 ng m^{-3} in the $PM_{2.5}$ fraction was above the target value of 20 ng m^{-3} for PM_{10} (Directive 2004/107/EC). Mean concentrations of V (36.6 ng m^{-3}), Mn (20.8 ng m^{-3}), Cd (1.4 ng m^{-3}), and Pb (46.5 ng m^{-3}), did not exceed the current air quality guideline values (WHO, 2002). The seasonal variations of the trace metals in PM_{10} and $PM_{2.5}$ were also analyzed. In winter, when domestic heating becomes a significant source of particles in the area, the amounts of all elements were elevated. The sources for the elements exhibiting winter enrichment are mostly connected with fossil fuel combustion in heating units. The exceptions were Cr, Cu, and Cd, which suggests that some local industrial source of these elements is more influential during the summer.

3.2 Total Atmospheric Deposition

A total of 141 atmospheric deposits was collected monthly from June 2002 to Decembar 2006 in three sites in the urban area of Belgrade - RB, BG and VF (Fig. 1) and trace and other metal (Al, V, Cr, Mn, Fe, Ni, Cu, Zn, As, Cd, and Pb) monthly fluxes calculated. Table 1 presents average daily atmospheric deposition of heavy metals (μg m^{-2} day^{-1}) in the Belgrade urban area for all sampling sites for the period of 2003 to 2006.

	Al	V	Cr	Mn	Fe	Ni	Cu	Zn	As	Cd	Pb
RB	807.13	40.20	1.91	64.08	1723.19	23.92	113.03	117.65	1.59	0.58	55.67
BG	976.60	49.66	2.32	78.95	2062.74	29.07	61.84	138.24	9.78	0.58	62.35
VF	1071.59	58.89	2.77	88.84	2549.86	38.64	97.22	148.19	40.43	0.70	74.18

Table 1. Average daily atmospheric deposition of heavy metals (μg m^{-2} day^{-1}) in the Belgrade urban area for 2003 - 2006

Besides Fe and Al, the most abundant trace metal in bulk deposition was Zn, followed in order of abundance by V, Mn, Ni, Cu, Pb, As, Cr and Cd. Cu and Zn have multiple anthropogenic sources, including high temperature combustion processes and the most probable source of Cu and Zn enrichment are vehicle-related particles. The impact of human activities could be seen on several metals like Zn, Cu, Pb, Cd, Mn, Al, Cr, and Fe. This influnce was more obvious for Cu, Cd and Pb which have only anthropogenic sources, while other four metals originate from many natural sources as well (Pacyna & Pacyna, 2001).

As expected, the TD values were highest at Autokomanda site (VF) for almost all metals except Cu. High Zn and Cu fluxes in total deposition samples could also be related to roof covering of Zn metal sheets nearby and could point to the presence of point sources, which are clearly site specific. High deposition of Cu in samples at Rector's Office building (RB) comparing to other locations indicates the possible local source influence.

Seasonal variation for element concentrations in bulk deposition have been analysed. Maximum concentrations of V (Fig. 2) and Ni were in winter periods, while seasonal variations of the other elements were not pronounced.

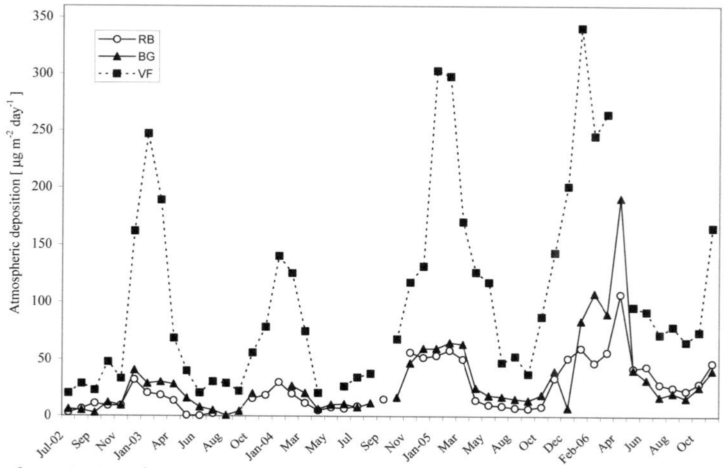

Fig. 2. Seasonal variation for V concentrations in total atmospheric deposits in Belgrade urban area for the period July 2002 - November 2006

3.3 Biomonitoring

3.3.1 Higher plants

The level of trace metals, particularly Pb, accumulated in the leaves of *A. hippocastanum* showed a high correspondence to their increased atmospheric concentrations, indicating this plant species as suitable biomonitor for trace element atmospheric pollution (Tomašević et al., 2004)

The results, presented on Fig. 3, illustrate this correspondence in two successive experimental years with different traffic and industrial emissions. Trace metal concentrations of Cu, Zn, Cd and Pb in the leaves from *A. hippocastanum* and *C. colurna* were analyzed at the beginning and in the end of the vegetation periods. Concentrations of Cd (0.02 to 0.06 μg g^{-1}) were below the detection limit in most of the samples. The increased atmospheric trace metal concentrations in the Belgrade down-town area affected their amounts found in the investigated deciduous tree leaves. While the average accumulations of trace metals in *C. colurna* were very similar for both experimental years, the levels found in *A. hippocastanum* leaves were considerably higher in September 1997 as compared to September 1996.

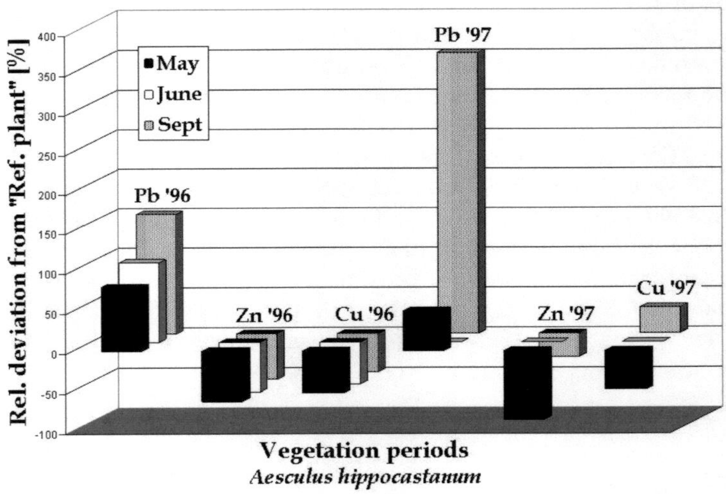

Fig. 3. Chemical fingerprints of *Aesculus hippocastanum* after normalization against the "Reference plant" system for comparison (Markert, 1992)

Similarly, over the same time interval, higher concentrations of the investigated trace metals were also found in the water-soluble fraction of deposits on leaves of *A. hippocastanum* in comparison to *C. colurna* (Tomašević et al., 2005).

Fig. 3 shows "the chemical fingerprints" of *A. hippocastanum* obtained for the second experimental year as normalized against the "reference plant" system for comparison (Markert, 1992). In both years, Pb concentrations were much higher than the "reference plant" value, and markedly increased in the second experimental year. As the chemical fingerprint may be assumed to represent the background concentrations, it offers some advantages for a quick assessment of the pollution level and allows comparison between different species and vegetation types, such as mosses, herbaceous plants and trees (Markert, 1992; Djingova et al., 1994).

Metal uptake in higher plants takes place through the roots and somewhat via the leaves, which makes it difficult to distinguish whether the accumulated elements in leaves originate from the soil or from the air (Markert, 1993; Bargagli, 1998; WHO, 2000). Pb in leaves is considered to originate mainly from atmospheric deposition (Tjell, 1979; Harrison& Johnston, 1987), while the soil contributes to Cu and Zn in leaves (Kim&Fergusson, 1994; Palmieri et al., 2005). As there were hardly any other emission sources during the investigated vegetation seasons, the results obtained here may suggest that the Pb found in the leaves mainly originated from traffic emissions. Unfortunately, leaded gasoline is still the prevailing traffic fuel in Belgrade, and many diesel engine vehicles are old and not maintained well. In the second experimental year Zn concentrations increased in *A. hippocastanum* leaves, while no increase occurred in the leaves of *C. colurna*. This result also implies a species-specific element accumulation. Moreover, the highest Pb concentrations in leaves also occurred in *A. hippocastanum* among a group of urban deciduous tree species in Istanbul (Baycu et al., 2006).

3.3.2 Moss

The element concentrations in both investigated moss genera, *Brachythecium sp.* and *Eurhynchium sp.*, were at a similar level (Table 2). For a majority of the 36 determined elements, the difference between the moss contents in the two genera was within the range of specific experimental error, except for Co, Ta, Ce, Sm, Tb, Th, and U which concentrations were close to the detection limits of used INAA, and hence it would not be reasonable to compare such data (Aničić et al., 2007). The previous result suggests that both moss species, found in sufficient quantity for sampling in the urban area of Belgrade, could be combined for biomonitoring purposes.

Especially interesting for evaluation, as carcinogenic and toxic elements (EEA, 2005), have been the obtained moss concentrations of V, Cr, Ni and As. The median values of these elements in native moss samples from this study were presented in relation to some other corresponding data (UNECE ICP Vegetation, 2003) as shown in Fig. 4.

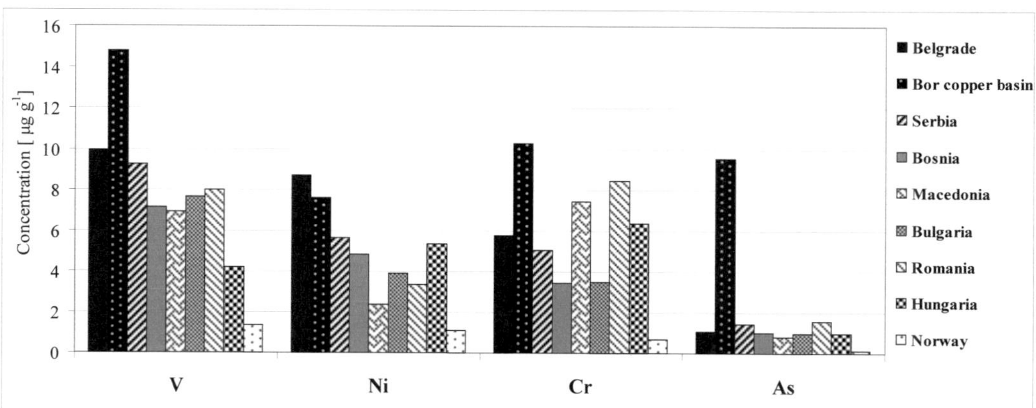

Fig. 4. Median concentrations (µg g^{-1}) of V, Cr, Ni and As in moss from Belgrade area and some European countries

Such comparison gave an insight into a level of heavy metal and other element air pollution in Belgrade urban area in relation to the highest polluted industrial area in Serbia (Bor copper basin), base-level moss content (data from Norway) and the concentration levels in adjacent countries. In general, the concentrations of V, Cr, Ni, and As obtained for the Belgrade urban area correlated to the results from the neighbouring countries where fossil fuel is still a major energy source resulting in higher heavy metal and other element air pollution. However, the moss content for the above-mentioned elements was up to 10 fold higher than background levels (Norway).

	Zemun		Kalemegdan Park	
Element	***Brach.sp.***	***Eurhin.sp.***	***Brach.sp.***	***Eurhin.sp.***
Na	545	757	313	307
Mg	16950	22760	10173	8290
Al	5000	6685	2147	1940
Cl	642	565	817	332
K	10705	10637	9760	7264
Ca	12590	14937	18647	12340
Sc	0.90	1.0	0.27	0.33
Ti	329	539	137	160
V	10.0	14	9.9	9.5
Cr	7.0	9.2	4.5	4.5
Mn	90	114	56	57
Fe	3000	3504	1057	1295
Co	0.80	1.3	0.42	0.68
Ni	8.2	13	6.8	9.2
Zn	34	37	41	40
As	0.95	1.6	1.0	1.1
Se	0.06	0.09	0.12	0.14
Br	1.8	2.5	2.9	3.7
Rb	8.1	11	6.7	7.7
Sr	29	36	50	42
Mo	0.52	0.59	1.2	0.75
Sb	0.18	0.22	0.29	0.35
I	0.19	0.26	0.30	0.31
Cs	0.39	0.59	0.22	0.22
Ba	44	64	33	28
La	3.0	3.7	1.1	1.5
Ce	3.9	7.0	1.7	3.1
Sm	0.40	0.77	0.20	0.25
Tb	0.05	0.09	0.02	0.03
Dy	0.81	1.2	0.42	0.38
Hf	0.42	0.78	0.22	0.24
Ta	0.05	0.10	0.02	0.03
W	0.32	0.29	0.16	0.19
Hg	0.48	0.43	0.18	0.38
Th	0.62	1.1	0.27	0.34
U	0.07	0.14	0.23	0.10

Table 2. The element concentrations (μg g^{-1}) in *Brachythecium sp.* and *Eurhynchium sp.* at Zemun and Kalemegdan Park

3.4 Factor Analysis

Principal Component Analysis (PCA) with Varimax rotation on the dataset of selected metals and particle mass concentrations in PM_{10} and $PM_{2.5}$ was performed for the source identification. Table 3 presents four rotated factor loadings with eigenvalues >1, embodying and explaining more than 73% of total variance for the case of PM_{10}. The first factor, explaining most of the variance (26%), has high loadings for Mn, Zn, Fe, Al, and Ni, can be attributed to road dust. Its bulk matrix is soil, while correlation with other metals indicates some other sources, such as tire tread, brake-drum abrasion, yellow paint, etc. Therefore, this factor is interpreted as representing road dust resuspension, which includes soil dust mixed with traffic related particles. Zn could be released from wear and tear of vulcanized vehicle tires and corrosion of galvanized automobile parts (Li et al., 2002, 2003; D`Alessandro, 2003). Adriano (2001) also reported that corrosion of galvanized steel is a major source of Zn emission in the surface environment. This is probably a significant source, as numerous old tracks, buses, cars and tires are present on the Belgrade streets. The second factor, with 17% of the total variance, shows high loading for fuel oil markers V and Ni. The third factor, including Cu, Cd and Pb, also accounted for 17%, while Factor 4 accounted for 13% with Cr and Pb as the main components. Factor 3 may be associated with road traffic emission. Cu and Cd are associated with diesel engines and wearing of brakes. Pb probably comes from exhaust emission, since road vehicles use leaded gasoline or diesel fuel. Factor 4 with the Pb component is most likely due to traffic exhausts.

	PM_{10}				$PM_{2.5}$				Total atmospheric deposition		
	Fac 1	Fac 2	Fac 3	Fac 4	Fac 1	Fac 2	Fac 3	Fac 4	Fac 1	Fac 2	Fac 3
Pb	-0.07	0.04	0.41	**0.72**	0.27	**0.85**	-0.05	0.06	0.51	0.31	**0.56**
Cu	0.01	0.13	**0.86**	0.02	-0.1	0.14	0.46	**0.78**	-0.08	0.1	**-0.74**
Zn	**0.78**	0.09	-0.12	0.01	**0.85**	0.17	0.25	-0.24	**0.96**	-0.08	-0.02
Mn	**0.84**	0.16	0.08	0.27	**0.8**	0.22	0.27	0.03	**0.76**	0.27	0.32
Fe	**0.77**	-0.03	0.04	-0.13	**0.78**	0.16	-0.09	0.16	**0.75**	0.42	0.11
Cd	-0.08	-0.11	**0.79**	0.03	0.1	-0.25	-0.36	**0.79**	**0.64**	0.48	0.02
Ni	0.32	**0.85**	0.19	-0.01	0.06	0.1	**0.88**	0.08	0.25	**0.84**	-0.11
V	0.02	**0.94**	-0.13	-0.05	0.3	-0.15	**0.71**	-0.11	0.26	**0.86**	0.22
Al	**0.74**	0.23	-0.14	0.22	**0.84**	0.09	0.08	-0.01	**0.71**	0.49	0.13
Cr	0.24	-0.11	-0.2	**0.81**	0.16	**0.91**	0.05	-0.15	-0.01	0.71	0.57
As	-	-	-	-	-	-	-	-	0.42	0.66	-0.33
%Variance	26.1	17.4	16.8	13.2	29	17.8	17.7	13.5	32.3	29.3	13.4

PCA loadings > 0.5 are marked in bold

Table 3. Principal component analysis after Varimax rotation for the trace elements analyzed in PM_{10}, $PM_{2.5}$ and total atmospheric deposition

PCA analysis and the following Varimax rotation were conducted on element concentrations dataset in total deposits. Three factor loadings explaining 75% of total variance are presented in Table 3. The first factor has high loadings for most of the elements and represents resuspended road dust, which includes soil dust mixed with traffic related particles. Factor 2 has high loading for Ni, V, As and Cr emitted from fossil fuel combustion processes and Factor 3 has high loadings for Pb and Cu. Cu shows the most independent behavior as it is almost on the third factor with negative loading indicating its specific source.

The results of CA for the variables, trace elements in PM, were obtained as dendograms displaying four main clusters. In the dendogram for PM_{10} (Fig. 5a) the first group containing the variables Pb and Cu with Cd, is associated with traffic emissions; the second group includes Zn and Fe mostly originating from abrasion of mechanical parts of road vehicles. The third cluster containing Ni and V is mainly associated with oil burning and emission from heavy fuel oil, while the fourth group includes Mn, Al and Cr with a soil origin. All groups are connected at some distance, suggesting that the main sources of metals in PM_{10} are fossil fuel combustion (traffic or stationary units) and resuspended dust, which is a mixture of soil and road dust.

The dendogram for trace elements in $PM_{2.5}$ (Fig. 5b) points to several groups connected to each other at some distance. The first group contains only Pb and Cr, probably from leaded gasoline vehicular exhaust and may be also from the oil refinery in Pančevo, 25 km northeast of Belgrade. The second group containing Zn and Mn is closely connected with Al and Fe and, all together, they represent road dust. The third group includes Cu and Cd and is connected with the fourth group (V and Ni) at a higher level suggesting perhaps a common source. These results imply that the main source of trace elements in urban $PM_{2.5}$ is traffic, with a considerable portion of resuspended road dust, and products of other fossil fuel combustion processes.

The results of cluster analysis on trace metals in PM_{10}, $PM_{2.5}$ and total atmospheric deposition (Fig. 5c) are in good agreement with PCA and correlation study, showing that metals with common sources have a strong inter-relationship; emission of metals most associated with traffic (Zn, Cu, Fe, Mn, Pb, Al) is probably more related to suspension or resuspension of road dust, which includes soil dust mixed with traffic related particles, than to direct exhaust emission (Vukmirović et al., 1997; Rajšić et al., 2006).

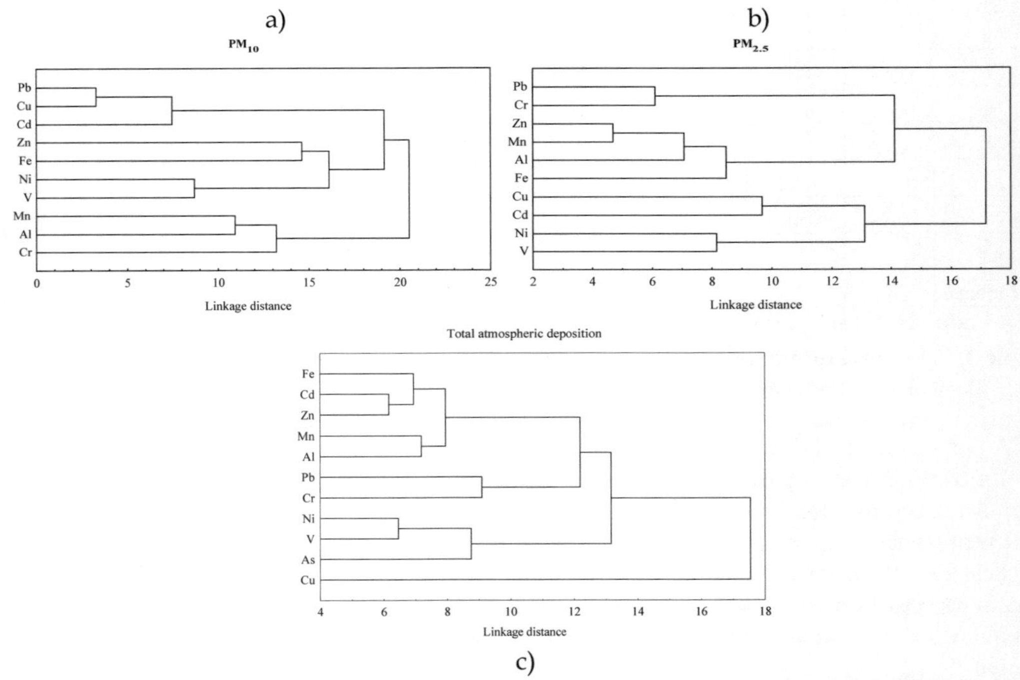

Fig. 5. Dendogram resulting from Ward's method of hierarchical cluster analysis of the trace elements in PM_{10} (a); $PM_{2.5}$ (b) and total atmospheric deposition (c)

3.5 Enrichment Factor

Enrichment factors (EFs) for the mean concentration of trace elements in suspended particles, PM_{10} and $PM_{2.5}$, total atmospheric deposition and moss samples were calculated according to the earth's crustal mean abundances of the elements given by Mason (1966) and using Al as a reference element and presented in Fig. 6.

The EF sequence related to particulate matter in the Belgrade urban area for the sampling period was: Zn > Cd > Pb > V > N > Cu > Cr > Mn > F e> Al. The highly enriched elements, primarily emitted from high-temperature processes (e.g. fossil fuel combustion and smelting), are usually associated with small and medium sized aerosol particles, and can be transported to remote areas.According to the degree of enrichment, the elements are grouped as follows (Wang et al, 2005): Zn, Cd and Pb, elements with a toxic character, were highly enriched (EF > 100) confirming that anthropogenic sources prevail over natural inputs for these elements; Cu was intermediately enriched (EF between 10 and 100); EFs for V and Ni were higher than 10 during the heating season and in the $PM_{2.5}$ fraction; Cr was less enriched (EF less than 10) probably attributed to both natural and anthropogenic sources. EF values higher than 1 were found for Mn and Fe, suggesting a mainly crustal origin, although an earlier analysis implicated an important influence of anthropogenic sources on the amounts of these metals. Because dilute acid was used for the extraction, the concentrations would be slightly underestimated for some common crustal elements, especially for the coarse particle size range (Pakkanen et al., 1993, 2003).

For total atmospheric deposition, the calculated enrichment factors for trace elements show the same pattern as in the particulate matter. Additionally, As (484) and Cu (155) were highly enriched.

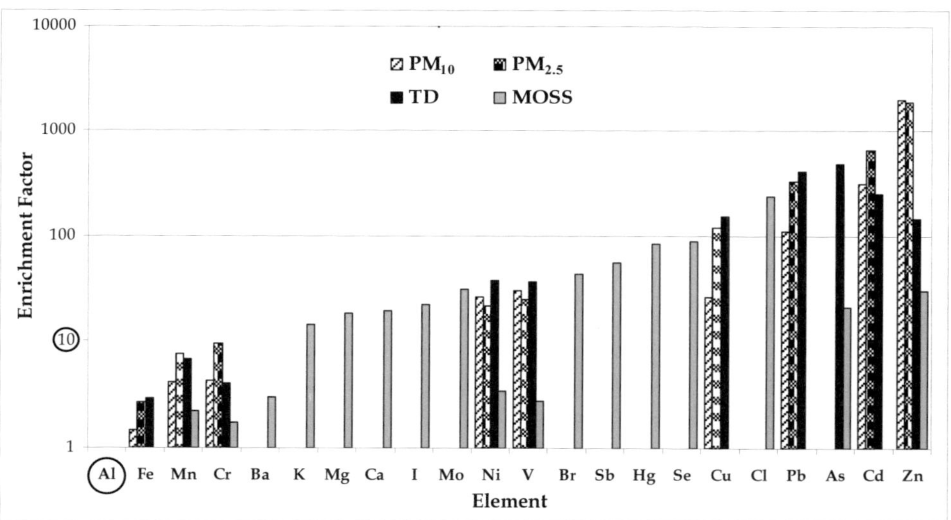

Fig. 6. Enrichment factor (EF) of elements in PM_{10}, $PM_{2.5}$, total atmospheric deposition and moss

High EFs for K, Mg, Ca, As, I, Zn, Mo, Br, Sb, Se, Hg and Cl (from 14 to 238) in moss samples were calculated. The highest EFs in moss samples were observed for Cl, Hg and Se which are important tracers for coal combustion (Watson et al., 2001). The fuel used in the complex of coal-fired power plants, 20 km SW from Belgrade, as well as for local domestic heating in Belgrade, is mainly lignite-brown coal (high in As) or crude oil. These fuels are

significant sources of the enriched elements at studied sites (Rajšić et. al., 2004). Elements such as Br, Sb, As, Mo and Zn are considered as indicators of emission from fossil fuel combustion processes, including vehicle exhausts (Arditsoglou & Samara, 2005; Pacyna & Pacyna, 2001). Kalemegdan Park is surrounded with heavy traffic roads, with trucks very frequently present. Leaded gasoline and diesel fuel (still widely used in Belgrade) contain a large amount of Br and Mo (Pacyna & Pacyna, 2001).
Tire and brake lining wear as well as other metallic parts of vehicles might be a significant source of Sb (Arditsoglou & Samara, 2005). Concentrations of the enriched elements from this study were in accordance with some previous investigations of air quality of the urban Belgrade in the vicinity of Kalemegdan Park (Tomašević et al., 2004; 2005; Vukmirović,et al. , 1997).

3.6 Correlation Factors

In order to investigate the extent to which metal concentrations are related to road traffic, the relevant data set for carbon monoxide (CO), nitrogen oxides (NO_x) and sulphur dioxide (SO_2) for the whole PM sampling period were provided by the Institute of Public Health of Belgrade and correlated with PM data. Pearson's correlation coefficients between meteorological parameters, combustion-related gases, PM mass and trace element concentrations in PM_{10} and $PM_{2.5}$ were calculated, and presented in Table 4. As products of fossil fuel combustion, CO, NO, NO_2, and SO_2 were the most closely correlated (r = 0.70 to 0.95). The highest correlations among the trace elements in PM_{10} were between V and Ni (r = 0.69) (elements associated with oil combustion), Al and Mn, elements of mostly crustal origin, as well as Zn - Mn and Fe - Mn (emission from traffic and possibly steel production). Copper was most correlated to Cd, which could have originated from brake linings and coal combustion in stationary sources, or some local industry. The mass concentration of PM_{10} was mainly correlated to the concentrations of gases related to combustion processes and, among the trace elements, to V as a product of fossil fuel combustion.
Higher correlation was found during the winter, with the most significant correlation coefficient between V and Ni (r = 0.72), while for Pb - Cu the correlation coefficient was 0.56. Pb was also significantly correlated with gases, Cd and Cr. Copper was more closely associated with Pb, than with Cd and gases, which are good indicators of combustion-related sources. These correlations support the recent finding that Cu is one of the metals most closely related to vehicle circulation in urban areas. Zinc was more correlated with Cr (r = 0.65) than with Mn, Al, Fe and Ni.
Generally, mass concentrations of $PM_{2.5}$ were positively correlated with the pollution gases, CO (r = 0.80), NO (r = 0.74), NO_2 (r = 0.70) and SO_2 (r = 0.65) as well as V (r = 0.40). The correlation of NO_2 with particle mass was more prominent than for SO_2, showing that particles from traffic emissions predominated. Lead was most closely connected to Cr, Zn to Mn and Al.
Regarding correlations between metal concentrations and wind speed, only Cu and Cd in $PM_{2.5}$ were significantly negatively correlated with wind speed, suggesting that Cu and Cd are mostly from local sources. Vanadium and Ni were strong negatively correlated with temperature. This result supports the presumption that those elements originate from combustion of fossil fuels in heating units. The absence of good correlations between Pb and most meteorological parameters (temperature, relative humidity and wind speed) suggests that those factors may not be influential enough to control the Pb levels.
The analysis of the correlation coefficients between all metal fluxes (Table 5) in total deposition (TD) have shown the similar connections as in the case of PM (Fe - Mn=0.61, V -

Ni=0.68, As - Ni=0.65; Zn - Fe=0.57, Ni - Cd=0.58). As in the case of PM_{10}, high correlation between V and Ni was obtained indicating the origin in oil combustion processes.

	PM_{10}	P_a	T_{max}	T	RH	W_{av}	Pb	Cu	Zn	Mn	Fe	Cd	Ni	V	Al	Cr	NO	NO_2	SO_2	CO
PM_{10}	1.00	**0.27**	**-0.28**	**-0.30**	0.00	**-0.26**	**0.15**	**0.19**	0.11	**0.35**	**0.25**	-0.02	**0.36**	**0.44**	**0.15**	-0.11	**0.58**	**0.39**	**0.56**	**0.61**
P		1.00	**-0.30**	**-0.31**	0.00	**-0.34**	-0.02	-0.03	0.05	0.12	**0.16**	0.01	0.05	0.09	0.00	**-0.20**	**0.18**	0.07	**0.25**	0.12
T_{max}			1.00	**0.99**	**-0.50**	-0.12	0.05	0.09	**-0.25**	-0.02	0.01	0.01	**-0.45**	**-0.54**	**-0.21**	**0.19**	-0.07	**0.32**	**-0.30**	0.02
T				1.00	**-0.45**	-0.14	0.03	0.10	**-0.27**	-0.05	0.00	0.00	**-0.46**	**-0.55**	**-0.23**	**0.19**	-0.09	**0.29**	**-0.32**	0.00
RH					1.00	**-0.17**	-0.04	0.08	-0.10	**-0.21**	**-0.21**	0.10	**0.25**	**0.34**	-0.03	-0.07	0.10	**-0.34**	0.03	0.07
W_s						1.00	-0.07	-0.14	**0.18**	-0.12	-0.09	-0.05	-0.01	-0.11	0.02	0.02	**-0.24**	**-0.24**	-0.06	**-0.25**
Pb							1.00	**0.28**	-0.02	**0.16**	-0.01	**0.25**	0.04	-0.03	0.00	**0.25**	**0.17**	**0.28**	0.12	0.12
Cu								1.00	-0.13	0.12	0.02	**0.45**	**0.29**	-0.04	-0.07	-0.09	0.13	**0.14**	0.07	0.11
Zn									1.00	**0.56**	**0.41**	-0.09	**0.31**	0.11	**0.54**	**0.16**	**-0.22**	**-0.19**	-0.10	**-0.15**
Mn										1.00	**0.55**	-0.05	**0.37**	**0.15**	**0.66**	**0.35**	-0.02	0.02	0.01	0.12
Fe											1.00	-0.12	**0.24**	0.05	**0.36**	0.10	0.01	0.01	0.01	0.03
Cd												1.00	0.02	**-0.14**	-0.10	-0.08	-0.02	-0.07	0.01	-0.05
Ni													1.00	**0.69**	**0.35**	0.01	0.07	-0.07	**0.16**	0.08
V														1.00	**0.22**	-0.12	**0.19**	-0.03	**0.30**	**0.18**
Al															1.00	**0.30**	**-0.15**	**-0.18**	-0.08	-0.03
Cr																1.00	-0.14	0.10	-0.07	0.01
NO																	1.00	**0.73**	**0.83**	**0.94**
NO_2																		1.00	**0.70**	**0.67**
SO_2																			1.00	**0.71**
CO																				1.00

a)

	$PM_{2.5}$	P_a	T_{max}	T_{av}	RH	W_{av}	Pb	Cu	Zn	Mn	Fe	Cd	Ni	V	Al	Cr	NO	NO_2	SO_2	CO
$PM_{2.5}$	1.00	0.01	-0.29	-0.30	0.06	-0.17	0.19	0.20	-0.17	-0.03	-0.16	-0.14	**0.26**	0.40	-0.16	-0.17	**0.74**	**0.70**	**0.65**	**0.80**
P		1.00	-0.16	-0.15	-0.04	-0.20	-0.25	-0.06	**-0.35**	**-0.28**	-0.17	**0.34**	-0.13	-0.15	**-0.37**	**-0.33**	0.14	0.14	0.22	0.04
T_{max}			1.00	**0.99**	**-0.68**	0.08	-0.12	-0.13	-0.07	-0.19	0.07	**-0.28**	**-0.29**	**-0.55**	-0.20	0.14	-0.02	0.08	-0.23	-0.02
T_{av}				1.00	**-0.65**	0.05	-0.14	-0.11	-0.09	-0.21	0.08	-0.25	**-0.28**	**-0.56**	-0.20	0.13	-0.04	0.06	-0.25	-0.04
RH					1.00	-0.26	0.08	0.18	-0.11	-0.05	0.00	**0.27**	**0.27**	**0.37**	0.21	-0.18	0.01	-0.20	0.02	-0.02
W_{av}						1.00	0.14	**-0.36**	0.03	-0.01	-0.11	**-0.34**	-0.01	-0.19	0.02	**0.28**	-0.19	-0.15	-0.07	-0.17
Pb							1.00	0.04	0.15	**0.25**	0.08	-0.12	0.04	0.10	0.20	**0.56**	0.12	0.14	0.13	0.16
Cu								1.00	0.04	0.08	0.04	0.20	**0.49**	0.16	0.01	0.08	0.13	0.06	0.18	0.14
Zn									1.00	**0.82**	**0.48**	0.12	0.08	0.00	**0.66**	**0.37**	**-0.31**	-0.18	-0.24	-0.19
Mn										1.00	**0.37**	-0.05	0.17	0.23	**0.50**	**0.34**	**-0.25**	-0.06	-0.14	-0.15
Fe											1.00	**0.29**	-0.13	-0.17	**0.43**	0.22	-0.17	-0.11	-0.19	-0.08
Cd												1.00	-0.24	-0.21	0.14	-0.01	-0.22	**-0.36**	-0.08	-0.15
Ni													1.00	**0.50**	0.02	-0.03	0.01	0.07	-0.01	0.02
V														1.00	0.07	**-0.26**	0.04	0.05	0.07	0.08
Al															1.00	0.17	-0.24	-0.20	-0.22	-0.14
Cr																1.00	-0.15	-0.05	-0.07	-0.14
NO																	1.00	**0.90**	**0.85**	**0.95**
NO_2																		1.00	**0.86**	**0.86**
SO_2																			1.00	**0.80**
CO																				1.00

b)

P_a - atmospheric pressure; T_{max} - maximal temperature; T - temperature; RH-relative humidity; W_s – daily wind speed

Correlation coefficients significant at 0.05 level are marked in bold

Table 4. Pearson's correlation coefficients between meteorological parameters, combustion-related gases, mass and trace element concentrations in PM_{10} (a) and $PM_{2.5}$ (b)

	Fe	Cd	Pb	Cu	Ni	Zn	Cr	Mn	Al	V	As
Fe	1.00	**0.49**	**0.44**	**-0.18**	**0.57**	**0.57**	**0.45**	**0.61**	**0.78**	**0.63**	**0.58**
Cd		1.00	**0.39**	0.15	**0.58**	**0.33**	**0.30**	**0.41**	**0.51**	**0.57**	**0.48**
Pb			1.00	-0.05	**0.29**	0.08	**0.20**	**0.34**	**0.46**	**0.41**	**0.21**
Cu				1.00	0.01	-0.15	**-0.20**	**-0.19**	-0.12	-0.07	0.11
Ni					1.00	0.15	**0.28**	**0.32**	**0.65**	**0.68**	**0.65**
Zn						1.00	**0.26**	**0.59**	**0.54**	**0.17**	**0.35**
Cr							1.00	**0.58**	**0.45**	**0.55**	**0.25**
Mn								1.00	**0.65**	**0.40**	**0.40**
Al									1.00	**0.61**	**0.50**
V										1.00	**0.57**
As											1.00

Correlation coefficients significant at 0.05 level are marked in bold

Table 5. Pearson's correlation coefficients of trace element concentrations in total atmospheric deposition

3.7 Air Back Trajectories

During the period of June 2002 to December 2004 the PM_{10} and $PM_{2.5}$ mass concentrations in Belgrade urban area had high average values (83 and 75 μg m^{-3}), and 72% of PM_{10} samples exceeded the level of 50 μg m^{-3}. To make an identification of possible pollution sources and assess the influence of meteorological parameters on PM mass concentrations, air back trajectories for high PM concentrations episodes were analyzed (Mijić et al., 2006). Three selected, sixty hours backward trajectories, starting at 12 00 UTC, are shown on Fig. 7.

Fig. 7a shows backward trajectories calculated for 21/12/2002., when the highest mass concentration occurred (362 μg m^{-3}). This case was characterized by anticyclone (1030 mb) at the surface and very slow flow caused by small gradient of the pressure field. Wind varied from north at higher levels, over western and southwestern at low levels (contribution from the complex of coal power plant in Obrenovac).

The episode for 2/4/2005 (Fig. 7b) was characterized with trajectories which at all levels show very weak flow changing from north-eastern at higher levels to eastern at lower levels (Pančevo). Synoptic situation was characterized by ridge of high pressure (1020 - 1025 mb) from the north, with weak pressure gradients caused by eastern flow in our region.

On Fig. 7c, trajectories corresponding to 27/11/2005 are placed from south-western direction at highest levels to south-eastern direction at lowest levels (Smederevo). They are presented for the last 12-20 h only, because the flow is strong at all levels, and particles quickly leave domain. Synoptic situation: A cyclone is placed on the west of the domain, caused southern magnified flow in the region of interest. Calm wind conditions were found to have an increment effect on particle concentrations and were also associated with the appearance of persistent episodic events. All cases analysed in this study were characterized by very slow flow field caused by small gradient of the pressure field. At the surface layer high pressure was placed over observed region.

The main sources of particulate matter were of local origin: traffic emission, and individual heating emissions. When the air masses were coming from the SW direction, the contribution from the power plants in Obrenovac was also evident, from NE-E contribution of Pančevo-rafinery and chemical industry, and from SE direction influence of Smederevo-steel industry. Episodes of high mass PM concentrations were observed in Belgrade throughout the year, although, they were more prominent during winter period. The prevalence of stagnant or week flow regimes favors the suspension and accumulation of particles produced locally, resulting at the elevation of PM levels.

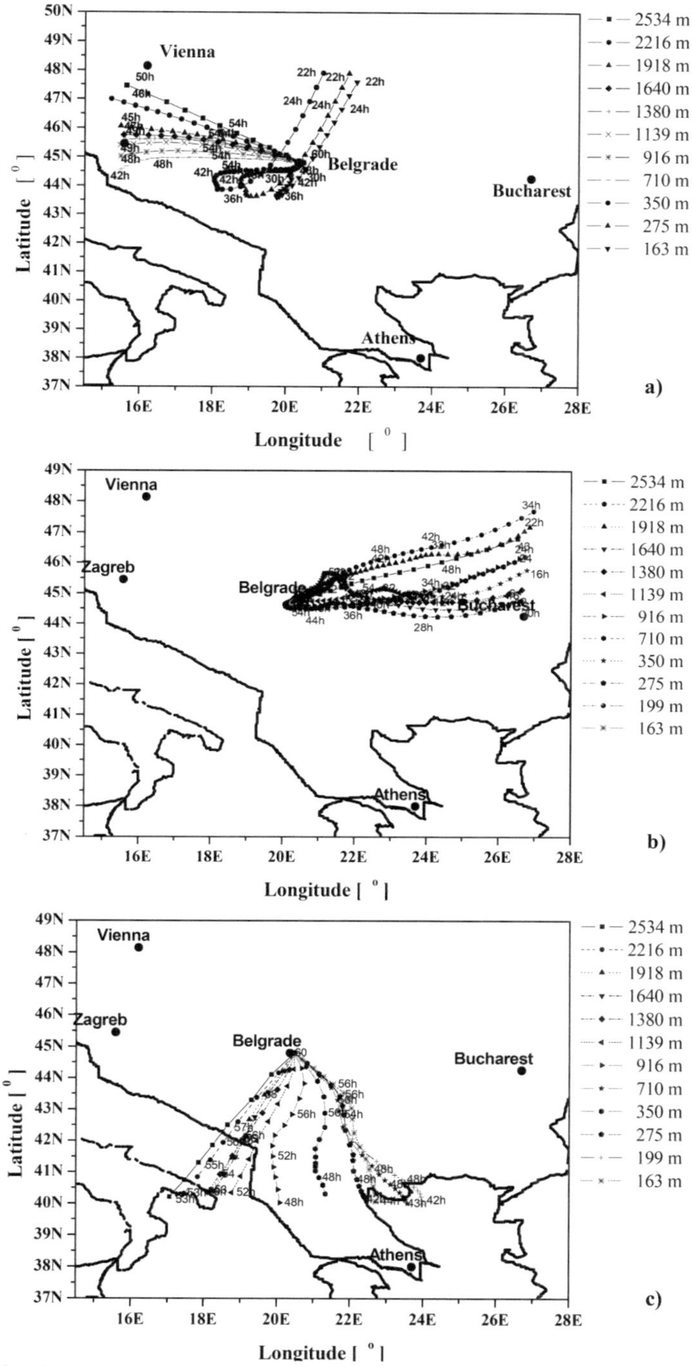

Fig 7. Air mass backward trajectories for the high PM concentrations episodes on: a) 21/12/2002; b) 02/04/2005 and c) 27/11/2005

3.8 Scanning Electron Microscopy

Atmospheric PM_{10} and $PM_{2.5}$ sampled at three representative sites in the urban area of Belgrade were analyzed with scanning electron microscopy coupled with energy-dispersive X-ray analysis with the aim to identify their origin. Classification of the present particles was based on the morphology (McCrone & Delly, 1973) and chemical composition of particles, typically expressed in terms of EDX peak-to-background values for the elements of interest, as well as to the particle classification rules described in US-EPA (2002). According to their morphology, two main particle categories were observed. Particles of natural sources include materials of organic origin (pollen, bacteria, fungal spores etc.). This category also includes suspended soil dust (mostly minerals) such as the angular-shaped material. Particles from anthropogenic sources, mostly emitted from high temperature combustion processes are characterized by their spherical shapes and smooth surfaces. This type of particles occur as individual particles but also in an aggregate form, as agglomerates of similar-sized particles and individual large particles carrying several smaller attached particles.

Related to the chemical composition and morphology, the analyzed particles were classified into the most abundant groups such as soot, Si-rich particles, sulfates, metal-rich and biological particles. The SEM photomicrographs of some characteristic particles and their X-ray spectra are presented in Fig. 8 - 9. *Soot* is present as agglomerates of many fine spherical primary particles. This kind of aggregate has an irregular morphology of various shapes. The X-ray microanalyses show traces of and sometimes of Na and K. The surface of carbonaceous particles acts as a catalyst for SO_2 photochemical oxidation producing ammonium and alkaline metal sulfates. C-rich particles are mainly resulting from the vehicular traffic and, during winter, from the heating systems. The most of *silica particles* (probably Si oxides) and aluminosilicates (containing Al, Si, K, Fe, and Ca) present in the coarse fractions have irregular forms and come from soil. Spherical aluminosilicates that dominate in the size fraction below 1 μm are anthropogenic fly ash (e.g. coal combustion) (Chen et al., 2006; Conner et al., 2001). In Belgrade urban area, this type of particles originates mostly from individual heating units.

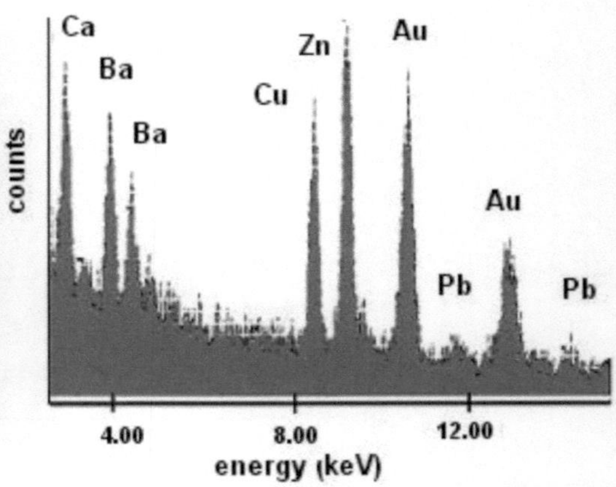

Fig. 8. High resolution SEM image and X-ray microanalysis spectrum of fly ash particles agglomerate from fossil fuel combustion process emissions and coal-fired power plant ("Nikola Tesla" A, B, Obrenovac)

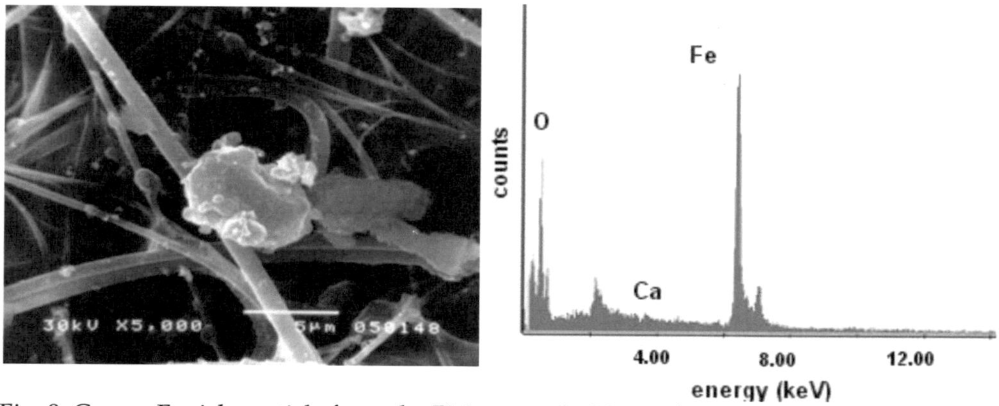

Fig. 9. Coarse Fe rich particle from the $PM_{2.5}$ sample, November 2003 and X-ray microanalysis spectrum

Sulfates are characterized by a strong S line in the X-ray spectrum and mostly by the presence of Ca, or Fe, Pb and K. These particles are formed as a result of the reaction in the atmosphere between sulfur compounds and other substances. They predominate in the fine fraction and have round forms. Sulfate clusters, often with sharp edges are mainly composed of Ca sulfates; they arise from the reactions between Ca carbonate materials and sulfurous compounds and have been found in the coarse particle range.

Oxides of Al, Zn, Cu, Ni, Pb, Ti, with spherical morphology, abundant in the sub-micrometer range, are identified as anthropogenic fly ash. Many particles, which could not be classified into one of these groups, found in the coarse particle range, were mixed aggregates, irregularly shaped, consisting of soil and road dust: Si, Al with minor constituents such as C, Fe, Mg, Ca, Ba, Pb, K, S, Zn, Ni, Cu, Ti. Coarse metal-rich particles of irregular form, liberated in industrial processes, were also found (Fig. 9). Biological particles (pollen, bacteria, fungi, spores, etc.) are recognized (by their characteristic morphology) as important constituents of the coarse particle fraction in the PM_{10} samples, especially during summer period.

Scanning electron photomicrographs of particles deposited on leaf surfaces showed a general appearance of adaxial and abaxial surfaces (Fig.10) of hypostomatic leaves of *A. hippocastanum* and *C. colurna* sampled in a dry period, at the end of the vegetation seasons (Tomašević et al., 2005). Approximately, 10 - 15% the leaf surface was covered with deposited particles. A greater density of particles was observed on *A. hippocastanum* than on *C. colurna*. This could be explained by different epidermal characteristic such as surface roughness, which was more prominent in *A. hippocastanum* than for *C. colurna*. Particles were present at a higher density on the adaxial leaf surfaces of both tree species. They were not homogeneously distributed and mean particle density approximately varied from 5 000 to 20 000 mm^{-2}.

Particles were present in a wide range of diameters up to 50 µm, but the analyses of the particle size distribution for both species showed that 50 - 60% of the analyzed particles were of a diameter less than 2 µm (fine particles) mainly originating from anthropogenic acitivities.

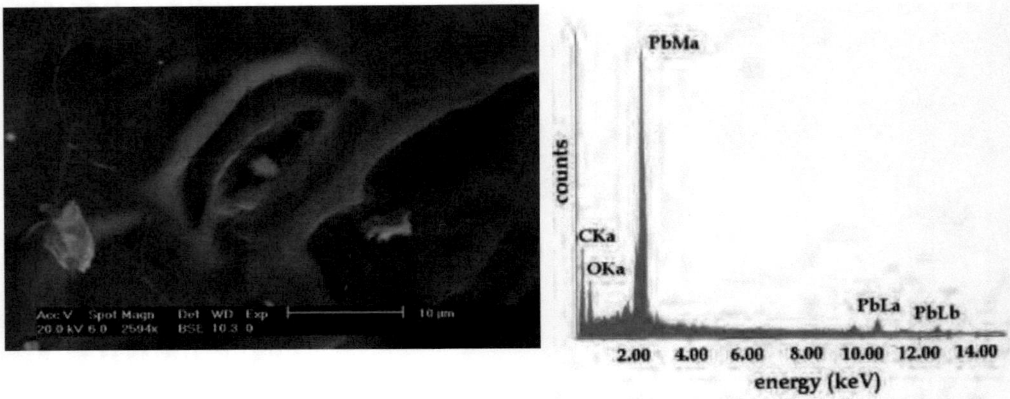

Fig. 10. Scanning electron micrograph of a stomatal area of leaf surface of *A. hippocastanum* and EDX-spectrum of a lead-rich fine particle in the stomatal opening

The chemical composition of the particles deposited on both plant species suggested that the most abundant particles were: soot (C) and soil dust with characteristic matrix elements (Si, Al, Fe, Mg, N, S, Ca, K, Cl); fuel oil particles rich in Al, Si, Ca, Ni, Fe, V and Pb; coal ash particles containing C, Al, Si, K, Ca; and industrial particles such as Fe, Zn, Ni, Cu or Pb-rich. Among the particles containing trace metals, the most abundant were particles in aggregates form, where Pb is the major element associated with lower concentrations of S, Fe, Cd, Cu, As and Zn.

3.9 Ozone

Twenty four–hour fluctuations of O_3 concentrations are recorded at all measuring sites. The obtained results mainly indicate usual afternoon maxima and night and early morning minima. The differences between maximal and minimal concentrations of O_3 are more marked during warmer periods when main daily temperatures are increased, which are correlated with maximal O_3 concentrations. At all measuring sites, ozone concentrations were for at least 1 h equal or higher than 90 ppbv what exceeds the EU threshold (EU Directive 92/72/EEC on air pollution by ozone) requiring public anouncement.

In the Fig. 11, daily changes of ozone concentrations recorded at the location BG during July 2002 are presented indicating cyclic dependence of afternoon maxima and night and early morning minima.

In Fig. 12 maximal daily O_3 concentrations recorded during measurements at the measuring point BG are presented together with mean daily air temperatures showing a tendency towards increasing ozone as the temperature increases.

Our data indicated that temperature did not play an important role in the ozone variation when high values of ozone concentration were reached (> 100 ppbv). This may be easily seen from the lack of correlation between average or maximum temperature and maximal ozone concentration (Fig. 11).

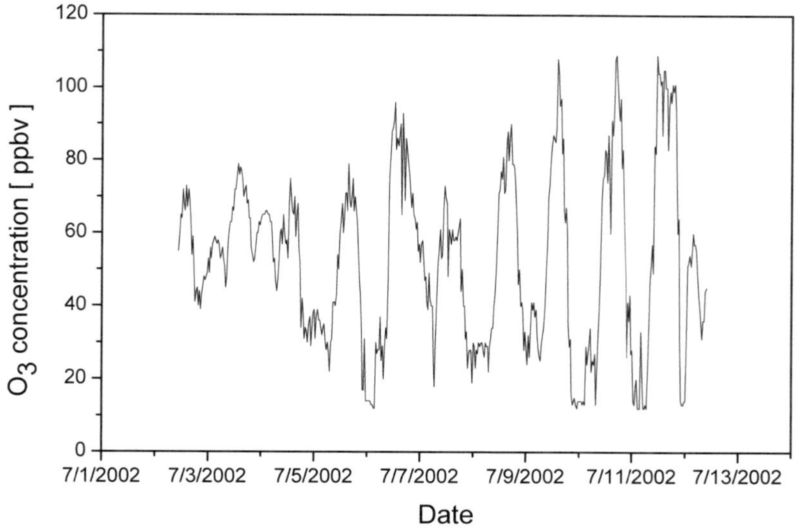

Fig. 11. Daily cycles of ozone; measuring point BG, July 2002

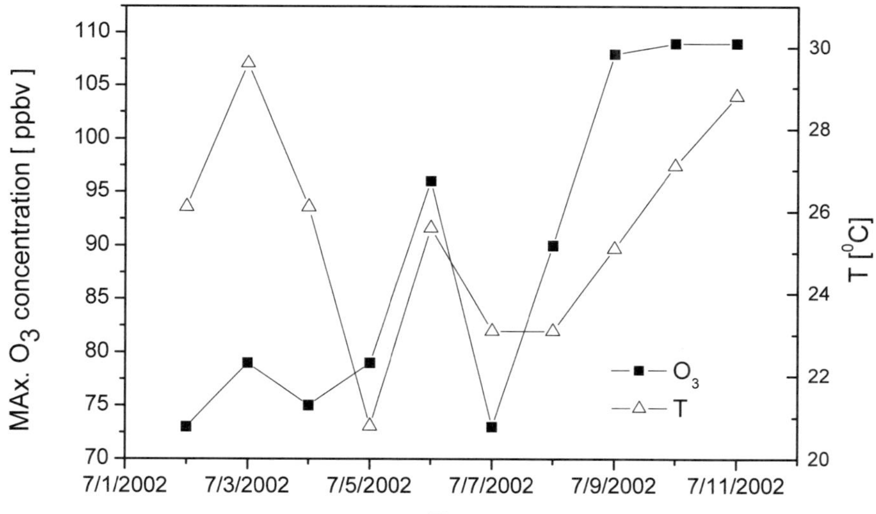

Fig. 12. Maximal ozone concentrations and mean daily temperatures; measuring point BG, July 2002

During the days when the daily maximum ozone concentration was greater than, or equal to 100 ppbv, the surface wind speed was less than, or equal to 1 m s^{-1}. This could imply that the surface wind speed plays an important role in the processes that control variation in ozone concentration and indicates that the highest ozone concentrations are reached when the wind speed decreases.

The mean daily relative humidity during O_3 measurements at the measuring point BG did not change much and it amounted approximately to 60%. It is important to mention that in the first three days of O_3 measurements, highest values of mean daily wind speed of 4 ms^{-1} were registered, while in the subsequent four days they reached 1.5 m s^{-1} on average. During the period when maximal O_3 concentrations were recorded, average wind speed was lowest and it amounted to 1.0 m s^{-1}, mainly from the north direction. The results of O_3 concentration measurements, obtained at the points RB and VF are to a high degree in accordance with the results obtained at the point BG.

Lower values of O_3 concentrations were recorded in September and October 2002, at the measuring point RB. Mean values of maximal O_3 concentrations varied between 59 ppbv and 36 ppbv.

4. Conclusion

Air quality monitoring is used to determine the air pollution levels in urban environments and concentration measurements can be compared with the national and international guideline values. This review presents an attempt to assess air quality in Belgrade urban area and identify the possible pollution sources. An integrated monitoring approach has been applied related to the urban Belgrade locations with the aim to estimate the level of specific pollutants: particulate matter, trace metals and ozone. The measurements included the PM_{10} and $PM_{2.5}$ mass and trace metals concentrations, total atmospheric deposition fluxes, as well as, biomonitoring of deposition and accumulation of trace elements.

The obtained results gave us the first impression of concentration level, characterization and sources of inhalable particulates with aerodynamic diameter less than 10 μm in Belgrade ambient air. The PM_{10} and $PM_{2.5}$ values had high average values, compared to EU standard and results for other European cities. The highest mean metal concentration in $PM_{2.5}$ was for Zn followed by Al and Fe. The mean Ni concentration of in the $PM_{2.5}$ fraction was above the target value for PM_{10} issued by EU. Mean concentrations of Pb, Cd, V, and Mn did not exceed the current WHO air quality guideline values.

Strong seasonal modulation existed and lower summer average decreased the annual average concentrations. The explanation for the seasonal pattern could be possibly attributed to the higher traffic density and combustion of fossil fuels for heating during winter, as well as, to the prevailed meteorological conditions e.g. temperature inversion layers, low temperature and stagnation of air masses. During the exceptional high pollution days, both in the summer and winter, high production of secondary aerosols occurred, as can be seen from increase of $PM_{2.5}$ in respect to PM_{10}. This was a result of high relevance for health aspects, as the fine particles were those of highest negative impact on health and had the longest residence time in the atmosphere.

Episodes of high mass PM concentrations were observed in Belgrade throughout the year, although, they were more prominent during winter period. Air back trajectories analysis showed that the prevalence of stagnant or week flow regimes (calm conditions) favors the suspension and accumulation of particles produced locally, resulting at the elevation of suspended particles levels. The appearance of stronger W, SE and NE flows was responsible for the particles transported from the thermoelectric plants, refineries, chemical and steel industry.

Most of traditional studies on atmospheric pollution have been limited by problem of high cost and the difficulty of carrying out extensive sampling, in terms of both time and space. There is, thus, an increasing interest in using indirect monitoring methods such as analysis of bioaccumulators.

In this study, higher plants and mosses were as biomonitors of atmospheric pollution in urban area. Variability in trace metal concentrations in deposits was estimated in *Aesculus hippocastanum* and *Corylus colurna* leaves. The method of analysis at the level of single leaves was chosen in order to assess more precisely the contribution of atmospheric deposition to trace metal accumulation. It has been shown that the effect of elevated air pollution can be reflected in the amount of deposits on leaves, and in trace metal concentrations in deposits, especially on *A. hippocastanum* leaves. Analyses of some trace metal contents (Cu, Zn, Cd, and Pb) in single leaves pointed to a high correspondence with the level of atmospheric pollution measured by instrumental monitoring.

Physical and chemical characterisation of deposits, both on filters and leaf surfaces, also indicated anthropogenic origin of particles.

The comparison of accumulation abilities of two moss genera, *Brachythecium sp.* and *Eurhynchium sp.*, showed that they could alternately be used for biomonitoring purposes in urban areas due to very close levels of the measured trace elements. The obtained data for moss collected in urban area of Belgrade are in accordance to some instrumental monitoring data for the same area.

The moss concentrations of some cancerogenic and toxic elements such as V, Cr, Ni, and As, obtained in this study, were up to 10 fold higher than moss-background levels (Norway) and correlated with results from the adjacent countries (Bosnia, Hungaria, Romania, Bulgaria, etc.) where coal is still widely used as an energy source.

High values of enrichment factors for the trace elements (Mo, Ni, V, Br, Sb, Hg, Se, Cu, Cl, Pb, As, Cd, Zn) characteristic for fossil fuel combustion were calculated for both the total atmospheric deposition and moss samples.

Ozone concentration measurements showed that in the area of Belgrade during summer of 2002, moderate to high ozone concentrations were measured and lasted long enough to represent a risk for human health and vegetation. This was accompanied with manifestations of all usual characteristics of O_3 photochemical cycle with marked influence of transport on high level of O_3 concentrations during the night. The results of ozone measurements confirmed the dependence of O_3 concentration on meteorological parameters. Ozone concentrations equal to or higher than 90 ppbv were registered at low wind speeds of about 1.5 m s^{-1} mostly from the N which indicates that the O_3 was produced at the very site of measurement.

In this work we demonstrated how to identify the sources of atmospheric trace metals by trace metal analysis, multivariate receptor modelling, air back trajectories, meteorological analysis, and biomonitoring. The complete analysis showed that the main sources of air pollution in the city of Belgrade are combustion processes, e.g. emissions from mobile and stationary units, with the major contribution from traffic and resuspended road dust. Monitoring ambient air quality also provides important input data for epidemiological studies, which are crucial in establishing associations between health outcomes and concentrations of ambient air pollution. Due to measured high PM mass concentrations, it is obvious, that it would be very difficult to meet the EU standards, by 2010. There are some necessary measures to start with: switch to unleaded gasoline, replace old cars and substitute coal by gas, light heating oil or district heating.

Urban areas are often considered as "anthropogenic deserts" in sense of plant species recommended for biomonitoring purposes. Therefore, active biomonitoring with transplanted moss (or lichen) can be applied to assess urban air pollution. "Moss bags" technique has been developed having a number of advantages: well-defined exposure time; known background concentrations of elements in the moss; flexibility in site selection and number of sampling sites to be chosen; uniformity of entrapment surface, etc.

With the aim of more comprehensive insight in air quality in urban area at Belgrade, the further investigations will include indoor measuremens, $PM_{1.0}$, volatile organic compounds (VOCs), active moss biomonitoring, etc.

5. References

Adamo, P.; Giordano, S.; Vingiani, S.; Castaldo, C.R. & Violante, P. (2003). Trace element accumulation by moss and lichen exposed in bags in the city of Naples (Italy). *Environmental Pollution*, 122, 91–103

Adriano, D. C. (2001). Trace Elements in Terrestrial Environments-Biogeochemistry, Bioavailability, and Risks of Metals. 2nd Edition. Springer–Verlag, New York

Alcamo,J.; Bartnicki, J.; Olenrzynski, K.& Pacyna, J. (1992). *Atmospheric Environment*, 27A, 3355

Alfani, A.; Maisto, G.; Iovieno, P.; Rutigliano, F. A. & Bartoli, G. (1996). Leaf contamination by atmospheric pollutants assessed by elemental analysis of leaf tissue, leaf surface deposit and soil. *Journal of Plant Physiology*, 148, 243

Aničić, M.; Frontasyeva, M.V.; Tomašević, M. & Popović, A. (2007). Assessment of Atmospheric Deposition of Heavy Metals and Other Elements in Belgrade Using the Moss Biomonitoring Technique and Neutron Activation Analysis. *Environmental Monitoring and Assessment*, 129, 207–219

Arditsoglou, A. & Samara, C. (2005). Level of total suspended particulate matter and major trace elements in Kosovo: A source identification and apportionment study. *Chemosphere*, 59 (5), 669–678

Ariola, V.; Alessandro, A. D.; Lucarelli, F.; Marcazzan, G.; Mazzei, F.; Nava, S.; Garcia-Orellana, I.; Prati, P.; Valli, G.; Vecchi, R.&Zucchiatti, A. (2006). Elemental characterization of PM_{10}, $PM_{2.5}$ and PM_1 in the town of Genoa (Italy). *Chemosphere*, 62(2), 226 - 232

Azimi, S.; Rocher, V.; Muller, M.; Moilleron, R. & Thevenot, S. (2005). Sources, distribution and variability of hydrocarbons and metals in atmospheric deposition in an urban area (Paris, France). *Science of the Total Environment*, 337, 223-239

Azimi, S.; Ludwig, A.; Thevenot, D.R. & Colin, J-L. (2003). Trace metal determination in total atmospheric deposition in rural and urban areas. Science of the Total Environment, 308, 247-254

Bargagli, R. (1998). *Trace Elements in Terrestrial Plants*: An Ecophysiological Approach to Biomonitoring and Biorecovery. Springer-Verlag Berlin, Heidelberg, N. Y.

Bargagli, R.; Monaci, F.; Borghini, F.; Bravi, F. & Agnorelli, C. (2002). Mosses and lichens as biomonitors of trace metals. A comparison study on *Hypnum cupressiforme* and *Parmelia caperata* in a former mining district in Italy. *Environmental Pollution*, 116, 279-287

Baycu, G.; Tolunay, D.; Ozden, H. & Sureyya, G. (2006). Ecophysiological and seasonal variations in Cd, Pb, Zn and Ni concentrations in the leaves of urban deciduous trees in Istanbul. *Environmental Pollution*, 143, 545–554

Berg, T.; Røyset, O. & Steinnes, E. (1994). Trace elements in atmospheric precipitation at Norwegian background stations (1989–1990) measured by ICP-MS. *Atmospheric Environment*, 28(21), 3519–3536

Chen, Y.; Shah, N.; Huggins, F. E. & Huffman, G. P. (2006). Microanalysis of ambient particles from Lexington, KY, by electron microscopy. *Atmospheric Environment*, 40, 651 - 663

Conner , T. L.; Norris, G. A.; Landis, M. S. & Williams, R. W. (2001). Individual particle analysis of indoor, outdoor, and community samples from the 1998 Baltimore particulate matter study. *Atmospheric Environment*, 35, 3935

D`Alessandro, A.; Lucarelli, F.; Mando, P.A.; Marcazzan, G.; Nava, S.; Prati, P.; Valli, G.; Vecchi, R.&Zucchiatti, A. (2003). Hourly elemental composition and sources identification of fine and coarse PM_{10} particulate matter in four Italian towns. *Journal of Aerosol Science*, 34, 243-259

Directive 2002/3/EC of the European Parliament and of the Council of 12 February 2002 relating to ozone in ambient air. OJ L 67, 3 March 2002

Directive 2004/107/EC of the European Parlament and of the Council of 15 december 2004 relating to arsenic, cadmium, mercury, nickel and polycyclic aromatic hydrocarbons in ambient air. Off. J. of the European Union L 23/3

Djingova, R.; Kuleff, I. & Markert, B. (2004). Chemical fingerprinting of plants. *Ecological Research*, 19, 3-11

Dockery, D. W.& Pope, III C.A. (1994). Acute respiratory effects of particulate air pollution. *Annu Rev Public Health*, 15, 107-132

Dockery, D.W.& Pope, III C, A. (2006). Critical Review: Health Effects of Fine Particulate Air Pollution: Lines that Connect. *J Air&Waste Manage Associarion*, 56, 709-742

EEA (2005). Environment and Health. *Environmental assessment report*, no. 10.

EC 1999 Air Quality Directive 1999/30 EC of the European Parliament and of the Council of 22 April 1999 relating to limit values for SO_2, NO_2 and NOx, particulate matter and lead in ambient air. *Off J Eur Communities* L163, Brussels

European Standard EN 14907. (2005). Ambient Air Quality - Standard Gravimetric Measurement Method for the Determination of the $PM_{2.5}$ Mass Fraction of Suspended Particulate Matter. European Committee for Standardization, Brussels

Espinosa, A.J.F.; Rodriguez, M. T.; Barragan de la Rosa, F. J.& Sanchez, C. J. (2001). Size distribution of metals in urban aerosols in Seville (Spain). *Atmospheric Environment*, 35, 2595-2601

Frontasyeva, M. V. & Pavlov, S. S. (2000). Analytical investigations at the IBR-2 reactor in Dubna. *JINR Preprint*, E14-2000-177, Dubna

Gao, Y.; Nelson, E. D.; Field, M. P.; Ding, Q.; Li, H.l; Sherrell, R. M.; Gigliotti, C. L.; Van Ry, D. A.; Glenn, T. R.& Eisenreich, S. J. (2002). Characterization of atmospheric trace elements on $PM_{2.5}$ particulate matter over the New York-New Jersey harbor estuary. *Atmospheric Environment*, 36, 1077-1086

Harrison, R.M. & Johnston, W.R. (1987). Experimental investigation of the relative contribution of atmosphere and soils to the lead content of crops. In: *Pollution*

Transport and Fade in Ecosystems. Coughtrey P.J., Martin M.H., Unsworth, M.H. (eds). Oxford: Blackwell Science Publicatipon, 277-289

Haus, N.; Zimmermann, S.; Wiegand, J.&Sures, B. (2007). Occurrence of platinum and additional traffic related heavy metals in sediments and biota. *Chemosphere,* 66, 619 - 629

IPCC 2001 Intergovernmental Panel on Climate Change, Third Assessment Report. Cambridge University Press. Cambridge UK

Janjić, Z.I. (1990). The step-mountain coordinate: physical package. *Monthly Weather Review,* 118, 1429-1443

Janjić, Z.I. (1994). The step-mountain eta coordinate model: Further developments of the convection, viscous sublayer and turbulence closure schemes. *Monthly Weather Review,* 122, 927-945

Kaiser, H.F. (1960). The application of electronic computers to factor analysis. *Educ Psychol Measure,* 20,141-151

Kim, N. D. & Fergusson, J. E. (1994). Seasonal variations in the concentrations of cadmium, copper, lead and zinc in leaves of the Horse Chestnut (*Aesculus hippocastanum* L.). *Environmental Pollution,* 86(1), 89–97.

Klumpp, A.; Ansel, W.; Klumpp, G.; Belluzzo, N.; Calatayud, V.; Chaplin, N.; Garrec, J.P.; Gutsche, H.-J.; Hayes, M.; Hentze, H.-W.; Kamberzidis, H.; Laurent, O.; Penuelas, J.; Rasmussen, S.; Ribas, A.; Ro-Poulsen, H.; Rossi, S.; Sanz, M. J.; Shang, H.; Sifakis, N. & Vergne, P. (2002). Eurobionet: A Pan-European Biomonitoring Network for Urban Air Quality Assessment. *Environmental Science and Pollution Research,* 9 (3), 199-203.

Kyotani, T.& Iwatsuki, M. (2002). Characterization of soluble and insoluble components in $PM_{2.5}$ and PM_{10} fractions of airborne particulate matter in Kofy city, Japan. *Atmospheric Environment,* 36(4), 639-649

Li, N.; Kim, S.; Wang, M.; Froines, J.; Sioutas, C. & Nel, A. E. (2002). Use of a stratified oxidative stress model to study the biological effects of ambient concentrated and diesel exhaust particulate matter. *Inhal. Toxic.,*14 459-486

Li, N.; Sioutas, C.; Cho, A.; Schmitz, D.; Misra, C.; Sempf, J.; Froines, J. R.& Nel, A. (2003). Ultrafine particulate pollutants induse oxidative stress and mitochondrial damage. *Environ Health Perspect,* 111, 455-460

Manoli, E.; Voutsa, D.& Samara, C. (2002). Chemical characterization and source identification/apportionment of fine and coarse air particles in Thessaloniki, Greece. *Atmospheric Environment,* 36(6), 949-961

Markert, B. (1992). Establishing a 'reference plant' for inorganic characterization of different plant species by chemical fingerprinting. *Water, Air and Soil Pollution,* 64, 533–538

Markert, B. (1993). Instrumental analysis of plants. In: *Plants as Biomonitors. Indicators for Heavy Metals in Terrestrial Environment.* Weinheim VCH, Markert, B. (ed.), pp. 65-103

Mason, B. (1966). *Principles of Geochemistry*. Wiley. New York

McCrone, W. J.& Delly, G. (Eds), (1973). *The Particle Atlas,* Ann Arbor Science Publishers, Michigan

Meng, Z.; Dabdub, D. & Seinfeld, J.H. (1997). Chemical Coupling between Atmospheric Ozone and Particulate Matter. *Science,* 277, 116-119

Mesinger, F. (1984). A blocking technique for the representation of mountains in atmospheric models. *Meteor. Aeronautica*, 44, 195-202

Mesinger, F.; Janjić, I. Z.; Ničković, S.; Gavrilov, D.& Deaven, G.D.(1988). The step-mountain coordinate: model description and performance for cases of Alpine lee cyclogenesis and for a case of an Appalachian redevelopment. *Monthly Weather Review*, 16, 1493–1518

Mignorance, M. D. & Rossini, O. S. (2006). Heavy metals content in Nerium oleander leaves as urban pollution assessment. *Environmental Monitoring Assessment*, 119, 57–68

Mijić, Z.; Lazić, L.; Rajšić, S.; Tasić, M. & Novaković, V. (2006). Air back trajectories analysis for high PM concentration episodes. In: *Proc., of the First Accent Symposium, The Changing Chemical Climate of the Atmosphere*, 12-16. Sep. 2005. Urbino, Italy eds Sandro Fuzzi and Michela Maione (ARACHE editrice S.r.l.) CD

Moldovan, M.; Palacios, M. A.; Gomez, M.M.; Morrison, G.; Rauch, S.; McLeod, C.; Ma, R.; Caroli, S.; Alimonti, A.; Petrucci, F.; Bocca, B.; Schramel, P.; Zischka, M.; Pettersson, C.; Wass, U.; Luna, M.; Saenz, J. C. & Santamaris, J. (2002). Environmental risk of particulate and soluble platinum group elements released from gasoline and diesel engine catalytic converters. *Science of the Total Environment*, 296, 199 - 208

Monaci, F.; Moni, F.; Lanciotti, E.; Greichi, D. & Bargagli, R. (2000). Biomonitoring of airborne metals in urban environments: new tracers of vehicle emission, in place of lead. *Environmental Pollution*, 107(3), 321-327

Morselli, L; Olivieri, P., Brusori, B & Passarini, F. (2003). Soluble and soluble fractions of heavy metals in wet and dry atmospheric depositions in Bologna, Italy. *Environmental Pollution*, 124 (3), 457-469

Mulholland, J.A.; Butler, A.J.; Wilkinson, J.G.&Russell, A.G. (1998). Temporal and Spatial Distributions of Ozone in Atlanta: Regulatory and Epidemiologic Implications. *J.Air & Waste Management Association*, 48, 418-426

Nowak, B. (1998). Contents and relationship of elements in human hair for a non-industrialized population in Poland. *Science of the Total Environment*, 209, 59-68

Nriagy, J.O & Pacyna, J.M. (1988). *Nature*, 333, 134

Pacyna, J. M.& Pacyna, E. G. (2001). An assessment of global and regional emissions of trace metals to the atmosphere from anthropogenic sources worldwide. *Environmental Review* , 9, 269-298

Pacyna, J.M.; Martonova, A.; Cornille, P. & Maenhaut, W. (1989). *Atmospheric Environment*, 19, 2109

Pakkanen, T. A.; Hillamo, R. E. & Maenhaut, W. (1993). Simple nitric acid dissolution method for electrothermal atomic absorption spectrometric analysis of atmospheric aerosol samples collected by a Berner-type low-pressure impactor. *J. Anal. At. Spectro.* 8, 79

Pakkanen, A. T.; Kerminen, V.; Loukkola, K.; Hillamo, E. R.; Aarnio, P.; Koskentalo, T. & Maenhaut, W. (2003). Size distributions of mass and chemical components in the street-level and rooftop PM_1 particles in Helsinki. *Atmospheric Environment*, 37(12), 1673-1690

Palmieri, R. M.; La Pera, L.; Di Bella, G. & Dugo, G. (2005). Simultaneous determination of Cd (II), Cu(II), Pb(II) and Zn(II) by derivative stripping chronopotentiometry in Pittosporum tobira leaves: A measurement of local atmospheric pollution in Messina (Sicily, Italy). Chemosphere, 59(8), 1161–1168

Rajšić, F. S.; Tasić, D. M.; Novaković, T.V. & Tomašević, N. M. (2004a) First Assessment of the PM_{10} and $PM_{2.5}$ Particulate Level in the Ambient Air of Begrade City. *Environmental Science and Pollution Research*,11, 158-164

Rajšić, S.; Tasić, M.; Marković, D.; Novaković, V.; Marković, D. & Tomašević, M. (2004b). Measurements of atmospheric particulate matter and ozone in Belgrade urban area. *Cent. Eur J Occup. Environ. Med.*, 10(1) 25-29

Rajšić; S.; Mijić, Z.; Tasić, M.; Djordjević, D.; Radenković, M. & Joksić, J. (2006). Heavy metal content of PM_{10} and $PM_{2.5}$ within the urban area of Belgrade. Proc 8th International Conference on Fundamental and Applied Aspects of Physical Chemistry 2006. Sept. 26-29., Belgrade, Serbia, eds: Society of Physical Chemists of Serbia, 624

Rajšić, S.; Mijić, Z.; Tasić, M.; Radenković, M.&Joksić, J. (2007). Evaluation of the Levels and Sources of Trace Elements in Urban Particulate Matter. *Environmental Chemistry Letters*. DOI: 10.1007/s10311-007-0115-0

Rühling, A. & Tyler, G. (1968). An ecological approach to the lead problem. *Bot Notiser*, 121, 321-342

Rühling, Å. & Tyler, G. (1971). Regional differences in the heavy metal deposition over Scandinavia. *Journal of Applied Ecology*, 8, 497–507

Samara, C.; Kouimtzis, Th.; Tsitouridou, R.; Kanias, G.& Simeonov, V. (2003). Chemical mass balance source apportionment of PM_{10} in an industrialized urban area of northern Greece. *Atmospheric Environment*, 37(1), 41-54

Schwartz, J.; Dochery, D.W.& Neas, L.M, (1996). Is daily mortality associated specifically with fine particles. *J Air Waste Manage Assoc*, 46, 927-939

Schwartz, J.; Ballester, F.; Saez, M.; Perez-Hoyos, S.; Bellido, J.; Cambra, K.; Arribas, F.; Canada, A.; Perez-Boillos, M. J.& Sunyer, J. (2001). The concentration-response relation between air pollution and daily deaths. *Environmetal Health Perspective*, 109 1001

Steinnes, E.; Rambaek, J. P. & Hansenn, J. E. (1992). Large-scale multi-element survey of atmospheric deposition using naturally growing moss as biomonitor. *Chemosphere*, 25(5), 735–752.

Sutherland, R. A. & Tolosa, C. A. (2000). Multi-element analysis of road-deposited sediment in an urban drainage basin, Honolulu, Hawaii. *Environmental Pollution*, 110(3), 483-495

UNECE ICP Vegetation (2003). *Heavy metals in European mosses: 2000/2001 survey*. Centre for Ecology & Hydrology, University of Wales Bangor, United Kingdom, March, p. 45

Tasić, D. M.; Rajšić, F. S.; Novaković, T.V. & Tomašević, N.M. (2001) Contribution to the methodology of dry deposition measurements. *Fresenius Environmental Bulletin*, 10, 305

Tasić, D.M.; Rajšić, F.S.; Novaković, T.V.; Mijić, R.Z. & Tomašević, N.M. (2005). PM_{10} and $PM_{2.5}$ Mass Concentration Measurements in Belgrade Urban Area. *Physica Scripta*, 2005, T118, 29-30

Tasić, M., Djurić-Stanojevic, B., Rajšić, S., Mijić, Z. & Novaković, V. (2006). Physico-Chemical Characterization of PM_{10} and $PM_{2.5}$ in the Belgrade Urban Area. *Acta Chimica Slovenica*, 53 401-405

Tjell, J.C., Hovmand, M.F. & Mosbaek. H. (1979). Atmospheric lead pollution of grass grown in a background area in Denmark. *Nature*, 280, 425-426

Todorovic, D.; Popovic, D.; Rajsic, S. & Tasic,M. (2007). Radionuclides and Particulate Matter in Belgrade Air. In: *Environmental Research Trends*. Editor Marvin A. Cato. 2007 Nova Science Publishers, ISBN 978-1-60021-556-8

Tomašević, M.; Rajšić, S.; Đorđević, D.; Tasić, M.; Krstić, J. & Novaković, V. (2004). Heavy Metals Accumulation in Tree Leaves from Urban Areas. *Environmental Chemistry Letters*, 61, 753-760

Tomašević, M.; Vukmirović, Z.; Rajšić, S.; Tasić, M. & Stevanović, B. (2005). Characterization of trace metal particles deposited on some deciduous tree leaves in an urban area. *Chemosphere*, 61(6), 753-760

Tyler, G. (1990). Bryophytes and heavy metals: a literature review. Botanical journal of the Linnean Society 104, 231-253

Unkašević, M.; Mališić, J. & Tošić, I. (1999). Some aspects of the wind "Koshava" in the lower troposphere over Belgrade, *Meteorological Application*, 6, 69-80

US-EPA (2002). *Guidelines for the Application of SEM/EDX Analytical Techniques to Particulate Matter Samples*. EPA-600/R-02-070

Vallius, M.; Janssen, NAH.; Heinrich, J.; Hoek, G.; Ruuskanen, J.; Cyrys, J.; Van Grieken, R.; de Hartog, J. J.; Kreyling, W.G. & Pekkanen, J. (2005). Sources and elemental composition of ambient $PM_{2.5}$ in three European cities. *Science of the Total Environment*, 337, 147

Van Grieken, R.&Delalieux, F. (2004). X-ray spectrometry for air pollution and cultural heritage research. In: "Invited Lectures of the 5th Gen. Conf. Balkan Phys. Union, BPU-5", Eds., Serbian Physical Society, Belgrade, 234-246

Verma, A.& Singh, S. N. (2006). Biochemical and ultrastructural changes in plant foliage exposed to auto-pollution. *Environmental Monitoring and Assessment*, 120(1–3), 585–602

Vukmirović, Z. (1990). Expected levels of air pollution in pseudo-lognormal distribution. *Idojaras*, 94, 249

Vukmirović, Z.; Spasova, D.; Marković, D.; Veselinović, D.; Vukmirović, D.; Stanojević, Č.; Popović, M.& Hadžipavlović, A. (1987). Some characteristics of oxidant occurrence in the atmosphere of Belgrade. *Atmospheric Environment*, 21, 1637-1641.

Vukmirović, Z.; Marendić-Miljković, J.; Rajšić, S.; Tasić, M. & Novaković, V. 22 (1997). Resuspension of trace metals in Belgrade under drastically reduced emission levels. *Water, Air & Soil Pollution*, 93, 137-156

Wang, X.; Sato, S.; Xing, B.; Tamamura, S.& Tao, S. (2005). Source identification, size distribution and indicator screening of airborne trace metals in Kanayawa, Japan. *Journal of Aerosol Science*, 36, 197-210

Ward, N. I. (1990). Multielement contamination of British motorway environments. *Science of the Total Environment*, 93, 393

Watson, G. J.; Chow, C. J. & Houck, J. E. (2001). PM2.5 chemical source profiles for vehicle exhaust, vegetative burning, geological material and coal burning in northwestern Colorado during 1995. *Chemosphere*, 43(8), 1141–1151

World Health Organization WHO (2000). *Air hygiene report no.10. Biomonitoring of Air quality Using Plants*. Geneva.

World Health Organization WHO (2002). *Air quality Guidelines for Europe*

World Health Organization (WHO) (2003). Health aspects of air pollution with particulate matter, ozone and nitrogen dioxide. *Report on a WHO Working Group*, Regional Office for Europe; Bonn, Germany, 2003. EUR/03/5042688.

WHO (World Health Organization). 2006. WHO Air quality guidelines for particulate matter, ozone, nitrogen dioxide and sulphur dioxide, Global update 2005. (accesses April 6th, 2007)

Ying, Q.&Kleeman, M.J. (2003). Effect of aerosol UV extinction on the formation of ozone and secondary particulate matter. *Atmospheric Environment*, 37, 5047-5068

15

Environmental, Medical, Technogenic and Computer Technology: Modeling, Risk Assessment and Cost/Benefit Analysis of the Accidents

Yanenko V.M., Rykhtovsky V.O. and Yanenko N.V.
Glushkov Institute of Cybernetics of the National Academy of Science of Ukraine
Ukraine

1. Introduction

The modern environmental technology or "green technology" is developed to have a possibility for the united computer assessment of the condition indexes, risks of the irreversible changes of the technogenic, ecological, bio- and medical systems and to evaluate its reserve possibilities. That technology allows us to estimate the condition and risks of the personal and population illness at the handling with consequences of the accident, for instance, with the radioactive wastes of the Chernobyl Alienation Zone, to estimate a cost\benefit ratio and to assess risks of the dangerous situations, to restore damaged data etc.

1.1 Mathematical Model of the Region

The method of restoring damaged data is based on decision of a task of the identification for the model of the region. The model of the region has potentially dangerous objects with the technologies of production of the useful products, utilization and elimination of the contamination to research the ecological and economical situation in the region suffered by the technogenic accident (Lyashenko I., Yanenko N., 1999), (Yanenko V. et al., 2006). The mathematical model is an open system with intake means and resources and outside - products and contaminations. The model consists of the following system levels: V_1 - money equivalent of the funds in the region, V_2 - money equivalent of the funds produced in the region (integrated products), Z – level of the environmental contamination, M - power of binding of the pollution, V^+ - flow of money resources to the region, Q – integral index for quality of life of the personal-liquidators of the technogenic accident.
The model is based on the following assumptions: decreasing level M is directed to transiting a part of Z to the free condition and to increasing the common level of contamination. V^+ includes money resources directed to restoring the resources (U_M), to liquidating the pollution (U_Z), to reproducing the base capital (U_V), to preventing migration of the pollution for bounds of the region (U_T), to decreasing risk of the illness of the liquidators (U_Q), to progressing the production technology of the useful products and

technology of the wastes utilization and elimination of the contamination (U_C). I_Z is a flow out of the region such as fire, floods, wind, ground water etc.
The model is:

$$Z_M = Z_M(M_0/M), \quad \frac{dV_1}{dt} = -\frac{V_1}{T_V} + \frac{U_V V^+}{C_V(U_C)},$$

$$V_2 = \begin{cases} f_1(V_1), t < t_1 \\ f_2(V_1), t > t_1 \end{cases}, \quad \frac{dZ}{dt} = Z_V V_2 + Z_N Z_M - \frac{Z}{T_Z} - \frac{U_Z V^+}{C_Z(U_C)} - I_Z, \tag{1}$$

$$I_Z = I_N Z - \frac{U_T V^+}{C_T(U_C)}, \quad \frac{dM}{dt} = -M_N + \frac{U_M V^+}{C_M(U_C)}, \quad Q = Q_Z(\frac{Z_0}{Z}) Q_M(\frac{U_0}{U_Q}),$$

where V^+ - a function determined on the base of empirical data analysis; $C_M = C_M(U_C)$ - cost of the unit of the power renewal of binding of the pollution; $C_Z = C_Z(U_C)$ - cost of the unit of the pollution elimination; $C_V = C_V(U_C)$ - cost of the unit of recovery funds; $C_T = C_T(U_C)$ - cost of the migration prevention of the pollution unit for bounds of the region; T_V - a period of the depreciation of the funds; T_Z - a period of the disintegration of the unit of the technogenic pollution is an integral parameter that includes a time of the natural disintegration and constants of the speed of the pollution transformation in the eco-systems; Z_V - speed of the pollution generation at the unit of the produced funds; Z_N - a normal speed of the pollution in depending of the resources condition generation; Z_M - a table function determined in a result of solution of the identification task; I_N - a normal speed of the flow generation of the pollution out of the region; M_N - a speed of the depreciation of the resources; t_1 - a time moment corresponding to the accident conditioned the technogenic catastrophe; $Q_Z(\frac{Z_0}{Z})$ - a table function that determinates a level of the illness risk of the personal that liquidate the consequences of the technogenic catastrophe; $Q_M(\frac{U_0}{U_q})$ - a table function that determinates a decreasing of the illness risk at the increasing of the part of means U_q .

Cost/benefit analysis is researched by the modeling of the ecological and economics situation in the region. The following functions are introduced:

1) The function of the cost caused by migration of the technogenic pollution out of the region, binding pollution inside of the region and increased risk of the illness of the liquidators and suffered by the accident:

$$L = l_Z Z + l_I I_Z + l_Q Q_Z \quad ,$$

where Z is a free pollution; I_Z - a flow of the pollution out of the region; Q_Z - a function that determinates a level of the liquidators illness of the technogenic consequences; l_Z , l_Z , l_Q are coefficients that normalize.

2) The function of benefit

$$P = L - L_{opt} + V^{+}(1 - U_M - U_Z - U_V - U_T - U_Q - U_C),$$

where L - a determent is a result of solution of the dynamic modeling task; L_{opt} - a determent is a result of solution of the task of the optimal control;

$$V^{+}(1 - U_M - U_Z - U_V - U_T - U_Q - U_C) \text{ - unused means.}$$

At the numerical decision of the system of the fundamental differential equations we use the method of Runge-Kutt of the 4th array.

Target setting. The mathematic model (1) is given. The given sample consists of the values that correspond to the variables of the model within the time $[t_1, t_k]$:

$$\left\{X_i^j = X_i(t_j)\right\}, i = \overline{1..N}(N = 7), j = \overline{1..M}(t_j \in [t_1, t_k]),$$

where X_1^j corresponds to the value V_1 (money equivalent of the funds in the region) at the time t_j; X_2^j corresponds to the value Z (pollution of the technogenic nature); X_3^j corresponds to the value M (power of binding of the pollution); X_4^j corresponds to the value V^{+} (flow of money resources); X_5^j corresponds to the value V_2 (money equivalent of the funds produced in the region); X_6^j corresponds to the value I_Z (pollution flow out of the region); X_7^j corresponds to the value Q (index for illness of the liquidators of the technogenic accident).

The particularity of the given sample is that some parameters of the model are not determined at the time moments. Let us point as A is a lot of ordered pair of the indexes $\langle i,j \rangle$ for that the values X_i^j are determined.

The parameters of the model are given

$$D_i = D_i(t), i = \overline{1..K}(K = 8),$$

where D_1 is a value of the coefficient U_M (money resources directed to restoring of the resources); D_2 is a value of the coefficient U_Z (a part of the costs that is directed to changing a free pollution to the connected condition); D_3 is a value of the coefficient U_V (a part of the costs that is directed to reproducing the base capital); D_4 is a value of the coefficient U_T (a part of the costs that is directed to preventing migration of the pollution out of the region); D_5 is a value of the coefficient U_Q (a past of the costs that is directed to decreasing risk of the illness of the liquidators and suffered); D_6 is a value of the coefficient U_C (a part of the cost that is directed to progressing the production technology of the useful products and technology of the wastes utilization and elimination of the contamination); D_7 is a value of the function V^{+} (flow of the costs); D_5 is a value of the function V^{+} that determines a flow of the costs; D_8 is a value of the function V_2 that determines the money equivalent of the funds produced in the region.

Each parameter $D_i = D_i(t), i = \overline{1..K}$ is shown as the polynomial of the 3rd range: $D_i(t) = d_{0i} + d_{1i}\tau + d_{2i}(1-\tau)\tau + d_{3i}(\tau-1)\tau^2$, where $\tau = \frac{t - t_1}{t_K - t_1}$.

The area of the parameters of the model is given: $\vec{X}(t) \in X'$ and limits to the control influence $\vec{D} \in D'$, where $\vec{D} = (D_1, \dots, D_K)$. The value $\vec{X}_1 = \vec{X}(t_1)$ at the time moment t_1 is given. It's a task to define a value $d^*_{1i}, d^*_{2i}, d^*_{3i} (i = \overline{1..K})$ that reaches to the minimum of the derivation in contents of the mean square values of sample $\{X_i^j = X_i(t_j)\}, i = \overline{1..N}, j = \overline{1..M} (t_j \in [t_1, t_k])$ from the results given by model (1) $\{\overline{X}_i^j = \overline{X}_i(t_j)\}, i = \overline{1..N}, j = \overline{1..M} (t_j \in [t_1, t_k])$, e.g.

$$\sqrt{\sum_{\substack{i=1\\ \langle i,j\rangle \in A}}^{N} \sum_{j=1}^{M} \left(\frac{X_j^i - \overline{X}_j^i(\vec{D}^*)}{X_j^i} \right)^2} = \min_{\substack{\vec{X}(t) \in X' \\ \vec{D} \in D'}} \sqrt{\sum_{\substack{i=1\\ \langle i,j\rangle \in A}}^{N} \sum_{j=1}^{M} \left(\frac{X_j^i - \overline{X}_j^i(\vec{D})}{X_j^i} \right)^2} .$$

1.2 Results

The modified method of the accidental search is a method of statistic gradient that used for the numerical result of the identification task of the model parameters. There were used data from works (Baloga V., Kholoscha V., et al., 2006) at the preparation of the information to modeling of the technogenic and eco-safety. The data were defined more precisely after the decision of the identification task. The result of the model parameters analysis is given in the Table 1 where U_V - a part of the costs that is directed to reproducing the base capital; U_Z - a part of the costs that is directed to liquidating pollution; U_T - a part of the costs that is directed to preventing migration of the pollution out of the region; U_R - a part of the costs that is directed to restoring the resources; U_C - a part of the cost that is directed to progressing the production technology of the useful products and technology of the wastes utilization and elimination of the contamination; U_Q - a past of the costs that is directed to decreasing risk of the illness of the liquidators and suffered by the technogenic accident; V^+ - flow of the costs, mln. fixed unit; I_N - a speed of the flow generation of the pollution out of the region; C_M - cost of the unit of the power renewal of binding of the pollution; fixed unit; T_V - an average period of the working term of the power of binding resources, year; V_1 - a summary cost of the main industrial funds, mln.fixed unit.

Year	Parameter of the model										
	U_V	U_Z	U_T	U_R	U_C	U_Q	V^+	I_N	C_M	T_V	V_1
1991	0.13±0.02	0.1±0.02	0.4±0.15	0.25±0.08	0.05±0.02	0.05±0.02	-	0.01±0.002	50±15	30±10	700±60
1992	0.1±0.02	0.03±0.01	0.48±0.16	0.25±0.08	0.05±0.02	0.05±0.02	600±50	0.02±0.004	50±15	30±10	680±60
1993	0.08±0.02	0.02±0.01	0.51±0.17	0.25±0.08	0.05±0.02	0.05±0.02	600±50	0.02±0.004	50±15	30±10	640±60
1994	0.07±0.02	0.02±0.01	0.51±0.17	0.26±0.08	0.06±0.02	0.06±0.02	650±60	0.02±0.004	50±15	35±15	620±60
1995	0.08±0.02	0.05±0.02	0.48±0.16	0.27±0.08	0.06±0.02	0.06±0.02	750±60	0.01±0.002	60±20	40±15	600±50
1996	0.08±0.02	0.05±0.02	0.41±0.15	0.26±0.08	0.06±0.02	0.06±0.02	800±60	0.01±0.002	60±20	40±15	560±50
1997	0.09±0.02	0.07±0.02	0.4±0.15	0.25±0.08	0.06±0.02	0.07±0.02	890±60	0.01±0.002	60±20	50±15	530±50
1998	0.1±0.02	0.08±0.02	0.35±0.12	0.25±0.08	0.07±0.02	0.07±0.02	930±70	0.01±0.002	60±20	50±15	500±50
1999	0.1±0.02	0.09±0.02	0.32±0.10	0.25±0.08	0.07±0.02	0.08±0.02	900±70	0.009±0.002	60±20	50±15	480±50
2000	0.1±0.02	0.1±0.02	0.30±0.10	0.24±0.07	0.07±0.02	0.08±0.02	850±60	0.009±0.002	60±20	45±15	450±40
2001	0.1±0.02	0.1±0.02	0.28±0.10	0.22±0.07	0.07±0.02	0.09±0.03	800±60	0.009±0.002	60±20	45±15	450±40
2002	0.1±0.02	0.08±0.02	0.26±0.10	0.21±0.07	0.07±0.02	0.10±0.03	820±60	0.009±0.002	70±20	45±15	400±40
2003	0.1±0.02	0.07±0.02	0.24±0.08	0.19±0.06	0.08±0.02	0.12±0.03	850±60	0.009±0.002	70±20	40±15	380±40
2004	0.1±0.02	0.06±0.02	0.22±0.08	0.17±0.06	0.08±0.02	0.14±0.03	880±60	0.009±0.002	70±20	40±15	360±40
2005	0.1±0.02	0.05±0.02	0.20±0.08	0.15±0.05	0.05±0.02	0.17±0.03	890±60	0.01±0.002	70±20	40±15	320±40

Table 1. The result of the mathematical model parameters analysis for modeling of the technogenic, ecological and medical safety

The results of the modeling, redistribution of the means allow us to increase the quality of life and cost\benefit ratio on the interval from 2002 to 2010. At the same time and on the same interval we have an increasing the pollution connected with decreasing a part of the means directed to liquidating consequences of the technogenic catastrophe and increasing a part of the means directed to increasing risk of illness of the liquidators of the consequences of the technogenic catastrophe.

2. Risk Assessment of the Extreme Situations Occurrence on the Object "Ukrytie" (Shelter) in the Chernobyl Alienation Zone

2.1. The main idea of the method. Overall risk assessment of the extreme situations occurrence was researched on the base of theory of statistic functions. But some aspects of the quantitative calculations of risks need to be improved in further because of the difficult conditions of absence of the initial data and statistically represented samples. The main idea of the method is to present all parameters of the potentially dangerous object in a form of the dynamic system; to reduce the system to one equation that presents a surface of the three-dimensional area to an universal deformations of the theory casp catastrophes (I.V.Sergienko et al., 2000), (Yanenko V.M., 2003). There are sub surfaces that have stable and unstable positions of the balance. In the parametric area the distance between trajectory of the current condition of the potentially dangerous object and limit of stable and unstable position of the balance is a measure of risk of extreme situations occurrence.

2.2. Computing and modeling technology of solving the problem of quantitative risk assessment of extreme situations occurrence and assessment of nuclear safety of the object "Ukrytie". To solve the problem quantitative risk assessment of extreme situations occurrence and assessment of nuclear safety we have developed mathematics and software for blocks "Modeling" and "Forecasting" to join a technological and computing link. The link is a sort of information conveyor where are linked the problem-oriented program modules for risk assessment of extreme situations, on example of handling with radioactive waste on the potentially dangerous object "Vector" in the Chernobyl Alienation Zone. At the same time we need to improve mathematic tools to join data of monitoring using a vector map of the Alienation Zone (geo-information technology) with information about of pollution level of radionuclides 137Cs and 90Sr. The vector map of the Alienation Zone is used on the stage of data preparation for risk assessment of the technogenic and ecological accidents in the Chernobyl Alienation zone.

To solve the problem quantitative risk assessment of extreme situations occurrence let us remind of some states. According to general note, the base of safety of the objects of nuclear power composes: a) safe construction and quality of the objects of nuclear power; b) appropriate and safe maintenance of the objects of nuclear power minimizing risk of extreme situations occurrence; c) possibility of realizing the technical measures minimizing negative consequences of extreme situations on the objects of nuclear power if they still occur. The majority of experts working on the problem of the liquidation of the Chernobyl catastrophe consequences consider that the object "Shelter" doesn't answer safety characteristic and represents the greatest threat for nuclear safety among all objects of the Alienation zone (Yanenko V. et al., 2003).

The following factors make the main contribution to such an unsuccessful situation, first of all, progressive wear of building constructions of the object "Shelter", secondly, destruction processes proceeding in the fuel-containing masses and, at last, the threat of fire connected with plenty of easily inflammable materials in the object "Shelter".

Mass gush of a radioactive dust outside the object "Shelter" will take place with disastrous radiation consequences in case of extreme situation occurrence, caused by any of the above-listed reasons, namely: a) disturbance of the object "Shelter" shell (including the destruction of carrying constructions and roof), b) transition of radionuclides from the binding state to sliding dust particles, c) gush of aerosols.

The values of parameters are constantly varying, characterized as a state of the fuel-containing masses, contained in the object "Shelter", and its shell state, when because of inaccessibility to the most places for measuring and stochastic character of the transformation of nuclear power are occurred the deficit of the objective information. In these conditions the task of decreasing the degree of uncertainty with regard to the developing the joined technological and calculation link for risk assessment and ranking of extreme situations becomes of prime importance.

The international committee for radiological defense introduced the conception "RISK" and "COST" as the third principle of the radiological safety to quantitative definition of danger factors. The ecological risk is a possibility of unfavorable consequences for ecological resources any anthropogenic changes of nature objects and factors. The international committee for radiological defense formulated the principal "ALARA" that means construction and waste of sources and connected with it the practice activities must guarantee as the lowest irradiation as the most it's possible practically with regard to social and economical factors.

Traditionally the quantitative risk assessment is a multiplication of size of the event (P) and measure of possibilities its realization (q):

$$R = \sum_{i=1}^{k} P_i q_i \; .$$

The size of consequences can be presented by the number of the killed at the accident or by cost form, and a possibility can be presented by the number of the accidents at the unit of time.

In numerous works connected with various aspects of the Chernobyl catastrophe risk assessment is based on the application of probability theory methods. Thus the authors should meet with difficulties of authentic probability assessment not only because of the incompleteness of the sample, but also more often because of the complete absence of the information, owing to the impossibility of taking measurements in some premises of the object "Shelter" which are inaccessible because of a high radiation level. Besides, there are some methodological difficulties stipulated first of all by uniqueness of the event - accident on the nuclear power plant, by virtue of which it's impossible to calculate probability of the event on the basis of statistical data processing.

As an example can be a results connected with calculation of probability of the roof destruction of the object "Shelter" made by Research institute for building constructions. Their main hypothesis is that if the object "Shelter" has stood under loadings without destruction for 10 years so Shelter will stand under the same loadings and further, and its probability is 0.1.

To except such difficulties another approach to risk assessment was developed. The approach is based on methods of theory of caspoidal catastrophes (Poston T., Stuart I., 1980), (Guastello S.J., 1981). The risk is estimated on a degree of approximation of system parameters to their bifurcation values, which characterize system transition from one state (norm) to another one (catastrophe). The realization of this approach allows us not only the assessment of the risk of extreme situations occurrence, but also the receiving the quantitative characteristic of reserve possibilities of the system and its components.

The obtained calculations allow us to describe the current state of the system by ranking the set of the risks of extreme situations occurrence into their separate links, and by that to find the "weakest" link, the strengthening of which is necessary to direct main efforts.

By using this method, the probability account of the Shelter's roof collapse will be based on parameters of separate elements of building construction the most sensitive to earthquakes, tornadoes and other natural cataclysms, which are close to critical values (exhaust tower of the block "C", supports of beams B1 and B2, the southern shields - between axes B-C, the western zone of the object t "Shelter").

3. Risk Assessment and Ranking Risk Systems

Modeling and assessment of levels of risk occurrence of ecological and technogenic catastrophes for concrete potentially dangerous object - object "Shelter". The task of risk assessment of extreme situations occurrence is presented in terms of the theory of caspoidal catastrophes using the universal deformation type "swallow's tail". It's necessary to research a state of the safety of the system of the object "Shelter" to the effect of different factors: within the model - the universal deformation type "swallow's tail", which has three steady states. We have to research a balance of one from them at the defined parameters. There is defined the safety state of the system of the object "Shelter" after estimation of the balance of one from the states.

3.1. To assess the risk of extreme situations occurrence in the Alienation zone let us consider the first state (state 1) corresponds to a regular situation in the Alienation zone. The second state (state 2) is an occurrence of some local extreme situations, which do not result to a growth of radiation pollution levels outside of the Alienation zone. The third state (state 3) corresponds to extreme situations occurrence, which results to growth of radiation pollution level outside of the Alienation zone.

The biggest threat for the Chernobyl Alienation zone safety is the object "Shelter". Thus, let us base on that the threat of the Alienation zone safety disturbance is connected, on the one hand, with a state of elements of building constructions in the Shelter (subsystem A), and, on the other hand, with destruction processes, proceeding in fuel-containing masses (subsystem B). But the safety level also depends on flows of radionuclides from different sources (subsystem C) and human factor (subsystem D).

According to the universal deformation type "swallow's tail" the risk of extreme situations occurrence is described by the following equation:

$$\frac{dX}{dt} = X^5 + AX^3 + BX^2 + CX + D \ , \tag{2}$$

Where coefficient A includes the integral expert parameters assessments of type – state of support beams B1 and B2, state of the western zone, state of the southern shields between axes B-C, state of block B;

- coefficient B is estimated on the base of the following parameters – activity of radioactive waste in fuel-containing masses, activity of radioactive waste in the air, activity of the water samples in wells;
- coefficient C is estimated on the base of the following parameters – efficiency of the protective systems of the Shelter, condition of the dams and systems of flood control, condition of the system of the radioactive monitoring, condition of fire-prevention devices, a level of the technologies, a level of the capital investments;
- coefficient D is estimated on the base of the following parameters –a level of staff qualification, state of work condition and labor payment. We use a point estimation based on expert conclusion at the estimating values of the parameters.

Let the system, presenting a condition of the Shelter, be at norm (state 1). There are trajectories of change of its parameters, which pass the system at first in state 2 (local extreme situations, not causing pollution growth outside of the Alienation zone), an then in state 3 (extreme situations that result in pollution growth outside of the Alienation zone). Also there are trajectories immediately passing the system from state 1 to state 3. If the initial state of the system corresponds to state 2 or 3, the trajectories that return the system to state 1 (normalization of ecological radiation situation) can be determined. Thus, the task of risk assessment of the extreme situations occurrence can also be formulated to define a concrete stationary state in the model (2). For that can be used:

1) The values of the parameters corresponding to the current state of the system are determined.
2) The array of their bifurcation values corresponding to changes of number of stationary states is determined.
3) The 4th dimensional vector of distance $R_i, i = \overline{1,...,4}$ from an initial state of the system up to surfaces, which divide parameter areas corresponding to different number of stationary states is determined.
4) The risk value $Risk_i, i = \overline{1,...,4}$ is determined as the ratio of this vector to a vector describing appropriate distance in norm $R_i^{(N)}, i = \overline{1,...,4}$, as follow:

$$Risk_i = R_i \big/ R_i^{(N)}, i = \overline{1,...,4}. \tag{3}$$

5) The reserve values $res_i, i = \overline{1..4}$ is determined as distances from the initial state of the system up to surfaces, which divide the parameter areas corresponding to different number of stationary states.

The index of the state is calculated for every subsystem.

The state index of the subsystem "A" is calculated by:

$$I_A = \frac{1}{4}\sqrt{\sum_{i=1}^{4} a_i (X_i - X_i^{\blacklozenge})^2}\ ,$$

where I_A is an index of the subsystem "A", X_1 - estimation of the condition of beams support B1 and B2, X_2 - estimation of condition of the western zone, X_3 - estimation of condition of the southern screens between axes B-C, X_4 - estimation of the condition of the block B, $X_i^{\blacklozenge}$ - values of the appropriate parameters in a norm, $a_i\,(i=\overline{1..4})$ - norm coefficients.

The state index of the subsystem "B" is calculated by:

$$I_B = \frac{1}{3}\sqrt{\sum_{i=1}^{3} b_i (X_i - X_i^{\blacklozenge})^2}\ ,$$

where I_B is an index of the subsystem "B", X_1 - estimation of activity of the radioactive waste in the fuel-containing masses, X_2 - estimation of activity of the radioactive waste in the air, X_3 - estimation of activity of the water samples in wells, $X_i^{\blacklozenge}$ - value of the appropriate parameters in a norm, $b_i\,(i=\overline{1..3})$ - norm coefficients.

The state index of the subsystem "C" is calculated by:

$$I_C = \frac{1}{4}\sqrt{\sum_{i=1}^{4} c_i (X_i - X_i^{\blacklozenge})^2}\ ,$$

where I_C - is an index of the subsystem "C", X_1 - estimation of the condition of efficiency of the protective systems of the Shelter, X_2 - estimation of the condition of dams and systems of flood control, X_3 - estimation of the condition of the radiological monitoring systems, X_4 - estimation of the condition of fire protection devices, $X_i^{\blacklozenge}$ - value of the appropriate parameters in a norm, $c_i\,(i=\overline{1..4})$ - norm coefficients.

The state index of the subsystem "D" is calculated by:

$$I_D = \frac{1}{4}\sqrt{\sum_{i=1}^{4} d_i (X_i - X_i^{\blacklozenge})^2}\ ,$$

where I_D - is an index of the subsystem "D", X_1 - estimation of the technology level, X_2 - estimation of the investments, X_3 - estimation of the staff qualification, X_4 - estimation of the payments and condition of work, $X_i^{\blacklozenge}$ - value of the appropriate parameters in a norm, $d_i\,(i=\overline{1..4})$ - weight coefficients.

3.2. Modeling and risk assessment of extreme situations occurrence on the Shelter. Let us consider the results of the modeling and risk assessment of extreme situations occurrence with help of the method and software – the subsystem "Risk assessment of extreme situations occurrence on the Shelter. These results have been obtained at the solution of control examples for mathematics modeling and risk assessment of extreme situations occurrence in the Alienation zone. The results have a general type and can be used for the same type of potentially dangerous object. Let us use the input data from the Table 2 for solution of two examples.

The results of the modeling are presented in the Table 3, where state is a current stat of the system, risk is a summary risk of conversion in the state 3, $I_i, i = \overline{1,...,4}$, I are indexes of the states of the subsystems, $R_i, i = \overline{1,...,4}$ are risks of conversion for subsystems to the state 3, $rez_i, i = \overline{1,...,4}$ – reserve values for the subsystems.

Subsystem	Parameter	Task No.1	Task No. 2
A	Condition of beams support B1 and B2, points	8	7
	Condition of the western zone of the Shelter, points	8	6
	Condition of the southern screens between axes B-C, points	8	6
	Condition of the block B, points	8	6
B	Activity of the radioactive waste in the fuel-containing masses, points	4	3
	Activity of the radioactive waste in the air, points	4	3
	Activity of the radioactive waste in the water samples in wells, points	4	3
C	Efficiency of the protective systems of the Shelter, points	7	7
	Condition of the dams and systems of flood control, points	5	5
	Condition of the radioecological monitoring systems, points	6	6
	Condition of the fire protection devices, points	6	6
	Technology level, points	7	6
	Level of the investments, points	8	6
D	Level of staff qualification, points	8	8
	Condition of payments and condition of work, points	7	6

Table 2. Input parameters for examples No.1-2

The values of reserves $res_i, i = \overline{A,...,D}$ are determined as a distances from the current state up to the surface that divide the area of the parameters corresponding to change of the number of stationary states $res_i = \left|R^*_i - R_i\right|, i = \overline{A,...,D}$.

Current state	Risk	Systems of the parameters											
		Subsystem A			Subsystem B			Subsystem C			Subsystem D		
		IA	RA	rezA	IB	RB	rezB	IC	RC	rezC	ID	RD	rezD
1	0.004	0.1	0.0	-	0.2	0.0	-	0.2	0.0	-	0.2	0.004	0.005
2	0.008	0.2	0.0	-	0.1	0.0	-	0.2	0.0	-	0.2	0.008	0.003

Table 3. The results of the control examples o risk assessment for conversion to the state 3

As we see from the examples No.1 and 2, the main factor at the given set of input data, having influence to the extreme situations occurrence, a is state of the subsystem D (technology level, level of the investments, level of staff qualification, condition of payments and condition of work). The decreasing the protection level of the Shelter that at the same time corresponds to decreasing the technology investment levels, level of staff qualification, condition of payments and condition of work leads to double increasing risk of extreme situations occurrence from 0.004 to 0.008.

4. Research of Risk Ranking of the Various Technogenic Accidents on the Potentially Dangerous Objects and its Medical and Ecological Consequences

The models of faultness of the technological systems on the potentially dangerous objects depend on its destination and conditions of use. There are known more than ten models of faultness at the handling with the radioactive waste. The base of them is the first exponential model of distribution of duration of Mean Time Between Failures that leads to the extreme situations occurrence. That model is correct to Poisson flow of failures.

With help of developed software it was performed a risk ranking of the various technogenic operations at the disposal and conservation of the radioactive waste.

The results of risk ranking research of the various technogenic operations are presented in Table 4.

Elements of scheme of processing radioactive waste - hard radioactive waste	Rank	Elements of scheme of processing radioactive waste – liquid radioactive waste	Rank
Mechanisms of giving of hard radioactive waste	1.0	Mechanisms of receiving of liquid radioactive waste	1.0
Bunkers for substances	0.98	Mechanisms of mechanical depuration	0.98
Batchers	0.7	Filter-press	0.9
Devices for mixing	0.85	Devices of cementation	0.9
Devices of steam heating	0.84	Evaporator	0.8
Temperature press	0.79	Condenser of steam	0.7
Mechanisms of packing	0.6	Pressing mechanisms	0.82
Loaders	0.5	Mechanisms of packing in plastic	0.95
Mechanisms of packing in plastic	0.5		

Table 4. The results of risk ranking research

4.1. Research and risk assessment of personnel illness on the potentially dangerous object. We will perform the research and assessment on example of risk assessment of illness occurrence of personnel of the Chernobyl Alienation zone at the possible safety violation at the handling with the radioactive waste.
At the same time, we have to calculate of the additional irradiation that the personnel get at the handling, shuttling and transportation of the radioactive waste at the extreme situations.
The situations connected with handling and shuttling of the radioactive waste can be divided to three groups: D1 - accident-free handling and shuttling, D2 - accident leaded to partial damage of the part of containers without ground pollution; D3 - accident leaded to atmosphere and ground pollution.
General additional dose of radiation that the personnel gets at the transportation, shuttling and storage of the radioactive waste - D, is calculated by:

$$D = D1 + D2 + D3, \tag{4}$$

where D1, D2, D3 are calculated from (5), (6) and (8).

Consideration of the additional factors that define the level of the catastrophe weight from that depends the quantity of the radioactive waste from the containers at the transport catastrophe, allows us calculate the parameters k3 and k2 . Those are practically possibilities of appropriate catastrophes. To the number of such factors can be taken the followings: speed of collision, fire, angle of blow, meteorological condition, relief etc.
The calculation of risk of illness occurrence after received the additional radiation dose is obtained by (Yanenko V.M., 2003) and by additional coefficients:

$$R\ il = Kr*D, \tag{5}$$

where Kr is an additional coefficient of risk (see Table 5); D is received dose (Gr) (see formula (4)).

Tissues	Additional coefficient of risk (1e-2 1/Zv)	Weight factor
Gonads	0.40	0.25
Mammary gland	0.25	0.15
Red bone marrow	0.20	0.12
Lungs	0.20	0.12
Thyroid gland	0.05	0.03
Bone surface	0.05	0.03
Other	0.50	0.30

Table 5. Additional coefficients of risk calculation Kr of tumor with death or with inherited effects of person of any sex and age

Additional coefficients of risk Kr for some illnesses: - leukemia - 1 e^{-8} 1/Zv; death from cancer - 4 e^{-5} 1/mZv; - cancer - 0.8 e^{-5} 1/mZv; death from cardiovascular diseases - 4 e^{-5} 1/Zv; worsening of inheritance - 8 e^{-6} 1/mZv.

4.2. Modeling of distribution of the radioactive waste release in result of the accident without fire and explosion. Let us consider the task solution of mathematical modeling of distribution of the radioactive waste at the following set of input data (example No.1): type of explosion – gas substance, radionuclide Cs137, duration of the accident – 40 hours, speed of the wind – 0.5 m/s; activity of the explosion – 100 Bk/z; modeling is performed on the section of the area 3000×400 m. The screen form with results of the modeling is presented on the Figure 1. In the point of observation density of pollution is 1.2*104 Bk/sq.m, individual dose is 1.0*10-3 Zv. The risks of diseases occurrence: Leukemia - 1.0*10-8, Cancer - 8.1*10-6, Death from cancer - 4.0*10-5, Worsening of inheritance - 8.1*10-6. The risks of tumor occurrence with death results and inherited effects: Honads - 4.0*10-6, Mammary gland - 2.5*10-6, Red bone marrow - 2.0*10-6, Lungs - 2.0*10-6, Thyroid gland - 5.1*10-7, Bone surface - 5.1*10-7, other - 5.1*10-6.

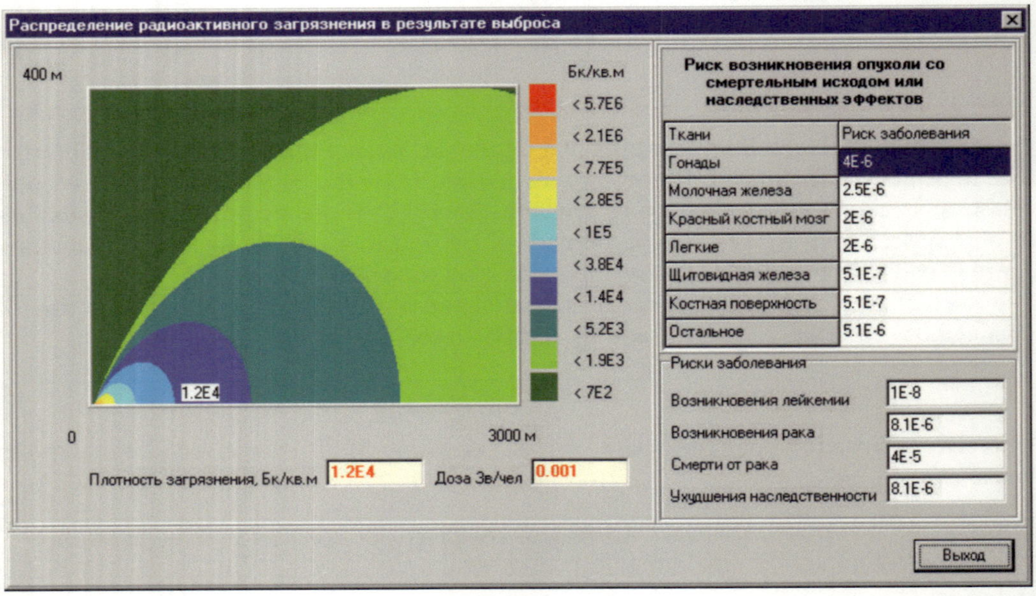

Fig. 1. Screen form with results of the modeling for task No.1

5. Medical and Cybernetics Systems

5.1. Software and information technologies allowed one to research the condition, reserves and risks of illness of the liquidators of the catastrophe under the influence of the negative factors of the Chernobyl catastrophe.

The traditional register of changes only conservative values of the parameters of cardiovascular system and system of regulation of protective functions of organism to the object doesn't give all-round estimation of possible self-healing of the subject, their reserve possibilities and risk assessment of the pathological changes of different systems of regulation of the organism.

To make the process of decision making more effectively it should be noticed the dynamic characteristics of the subject including the estimation of the irreversible changes and the estimation of reserve possibilities of the investigated object.

According to the traditional approach it's required to reduce the dynamic models of cardiovascular system and system of regulation of protective functions of organism to the feature of smooth reflection "swallow's tail" of the universal deformation in the theory of casp catastrophes. Then there are investigated the types of steady functioning of the systems and initial conditions of the system up to surface divided area of the parameters that correspond to changes of the number of stationary conditions.
Lets' base on that condition of the initial models after the reduction to the model "swallow's tail" is given by (3), where the parameters of the 4th subsystems correspond to: A - energetic subsystem, B - immune, C – myeloid and D – cardiovascular.
The task of reserve possibilities assessment and risk of the pathological changes in cardiovascular and regulation immune systems lead to estimation of characteristics of the stationary conditions (3) and others.
The results of the risk modeling are presented in the Table 6, where S is the current condition of the system, P is the summary risk of conversion to the condition of pathology, I is an index of subsystem condition, R is a risk of conversion for appropriate subsystem in the condition of pathology, Res is a value of reserve for the appropriate subsystem.

No	S	RISK	Subsystems											
			Subsystem A			Subsystem B			Subsystem C			Subsystem D		
			IA	RA	RezA	IB	RB	RezB	IC	RC	RezC	ID	RD	RezD
1	Norm	0.22	0.0	0.0	0.12	0.0	0.0	-	0.03	0.22	0.1	0.09	0.0	-
2	Norm	0.50	0.44	0.0	-	0.0	0.0	-	0.03	0.0	-	0.14	0.50	0.138

Table 6. Results of the indexes of conditions, reserves of the subsystems and risk the cardiovascular diseases

As it's shown in the control examples the main factor that defines the risk of the cardiovascular disease is the condition of the subsystem C (condition of the blood system). In that case the risk of pathology equals to 0.22. In case of another set of data the main factor is the subsystem D (cardiovascular system) with insignificant worsening of the parameters of the energetic system. The risk of the cardiovascular disease increases more then two times and equals to 0.50.
5.2. The research of the neuro-immune and endocrine regulation and system of regulation of protective functions of organism let us develop the software to restore damaged data for risk assessment of illness and for forecasting some processes (Yanenko V.M. et al., 2006).
Mathematical modeling of the neuro-immune and endocrine regulation. To provide the mathematical modeling of the condition of the system of neuro-immune and endocrine regulation the data of five patients have been chosen. The condition of immune system of the patients is characterized by indexes in the Table 7. The indexes of peripheral blood are presented in the Table 8. The results of hormone research are presented in the Table 9.

Parameter	Number of the patient				
	1	2	3	4	5
Lymphocytes, %	30	60	42	40	48
T-lymphocytes, %	50	80	60	80	70
T-active lymphcytes, %	30	50	30	50	50
T-helpers, %	35.3	50	28	50	55
T-suppressors	20	35	12	30	30
Coefficient helper/suppressor	1.8	1.7	2.3	1.7	1.9
B-lymphocytes, %	20	85	35	45	45
Ig, g/l	2	4	2.5	4	4
Ig, g/l	2	2.5	1	2.5	2.5
Ig, g/l	10	17.5	15	15	10
Ig, g/l	0.05	1.75	0.05	1.0	0.8

Table 7. The indexes characterizing the condition of immune system of the patients

Parameter	Number of the patient				
	1	2	3	4	5
WBC – leucocytes, 109/l	6	9.4	8.2	7	7
RBC – erythrocytes, 1012/l	4	6	5	6	5
HGB – hemoglobin, g/l	120	175	140	110	115
PLT – platelet, 109/l	247	550	350	500	450
LYM – lymphocytes, %	35	70	42	60	50
MO – monocytes, %	5	5	5	4	4
Eosinophils, %	2	4	4	3	4
Stab, %	3.6	5	5	5	4
Segmentonuclears, %	51	60	60	55	55
SOE, mm/hour	10	12	12	12	14

Table 8. The indexes of peripheral blood of the patients

Parameter	Number of the patient				
	1	2	3	4	5
TTG, med/l	2	20	1	1.25	1.5
T-4, nmol/l	90	35	242	85	245
T-3, nmol/l	1.8	0.8	5.7	6	2
FT – 4, pmol/l	15	12	21	15	21
FT - 3, pmol/l	4	6	7	7	5
r – 3, pmol/l	0.65	0.5	0.5	0.4	0.6
KT, pg/mol	240	300	350	300	300
TSG, mkg/mol	18	20	22	20	22
AKTG, pg/mol	50	65	40	40	35
Cortisol, nmol/l	290	300	340	320	330

Table 9. The indexes of the endocrine system condition of the patients

The numerical experiments were performed. The screen form with predicted dynamics and with appropriate dynamics obtained in the result of the task of optimal control is presented for the patient No.2 on the Figure 2 . The screen form with graphics of control influences (activators of oxidative phosphorylation (U1), activators of calcium transportation (U2), level of iodine (U3)) obtained in result of the optimal control is presented for the patient No.2 on the Figure 3. The screen form with risk assessment of pathological changes is presented for the patient No.2on the Figure 4.
The results of risk assessment are presented for the patient No.2 in the Table 10.

Pathology	Risk predicted	Risk obtained in result of task of optimal control
Hypothyroidism	1.0	0.41
Hyperthyroidism	0.02	0.09
T3- thyrotoxicosis	0.03	0.11
T4- thyrotoxicosis	0.04	0.14

Table 10. The results of risk assessment of pathological changes for the patient No.2

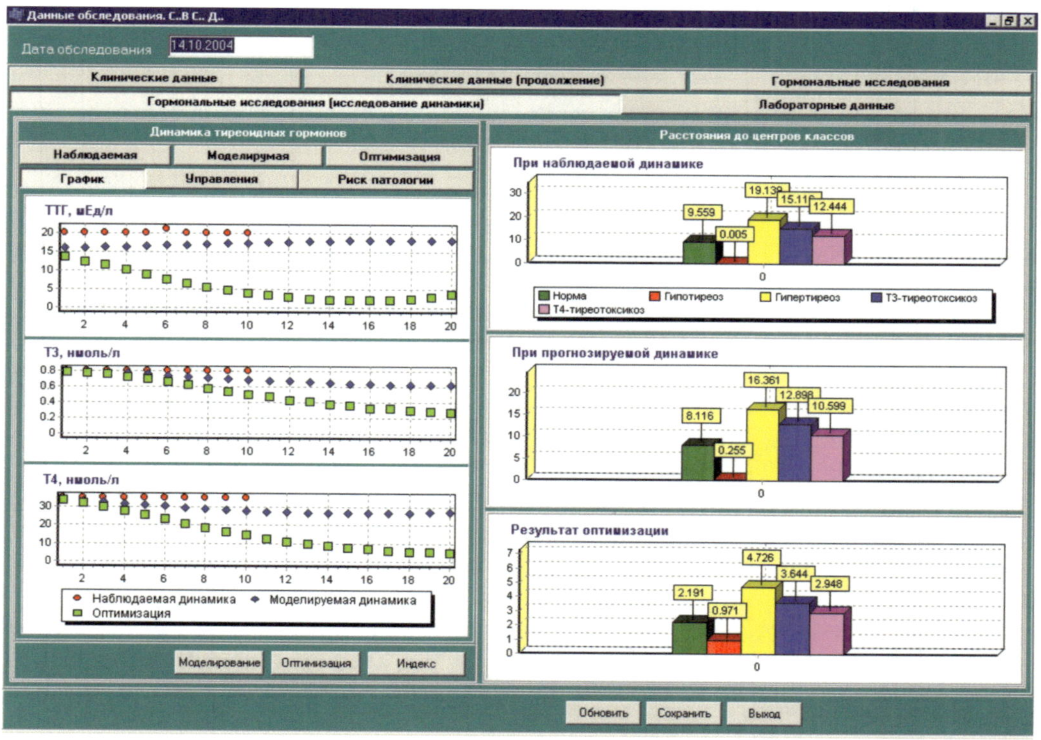

Fig. 2. The form "Research Data", subsection "Hormonal research (research of dynamics)", page "Graphic". .for the patient No.2

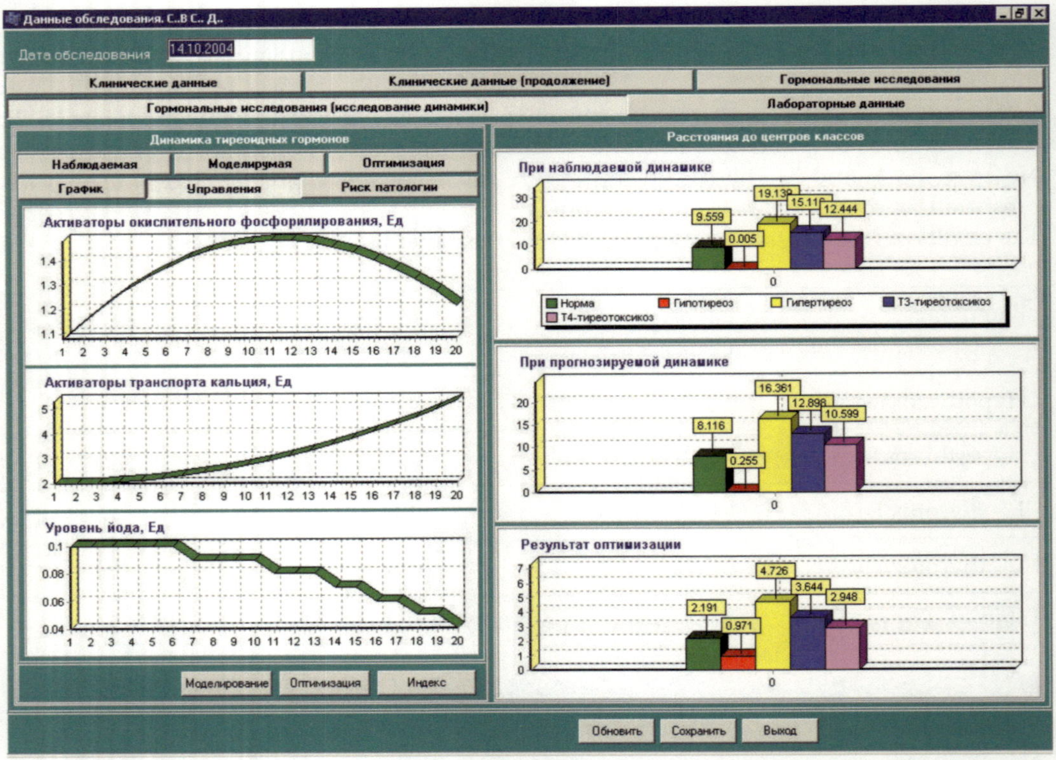

Fig. 3. Screen form with graphics of the control influences obtained in result of task of optimal control ofr the patient No.2

Thus, the condition of the patient No.2 is characterized as hypothyroidism. The risk of hypothyroidism equals to 1.0. In result of task of optimal control the risk of hypothyroidism decreased to 0.41.

5.3. Information software. There is developed a software product C/BR-RAW-ChAZ-2.0 (volume 40.4. Mb) - "System for database administration" described the 10th km of the Chernobyl Alienation zone, subsystems "Risk assessment and rating», «Modeling and forecasting dynamics of cost/benefit ratio from consequences of possible accidents and impact of radiation at the hand ling with radioactive waste of the Alienation zone», scientific and technical documentation (volume 17.3 Mb).

5.4. Information and program-technical providing with "Medical decision making for endocrinologist" (volume 7.0 Mb), "Medical decision making for cardiologist" (volume 7.0 Mb) support: administrating database of the patients, forecasting the influence of post-Chernobyl thyroid and cardiovascular pathologies to evolution of appropriate human organism systems, assessments of pathological changes in thyroid gland and in cardiovascular systems caused by the Chernobyl catastrophe's factors. These software products also provide the work with database using technology File-Server.

6. Gratitude. This work was started in 1979. We express our gratitude for all colleagues for creative and assiduous work.

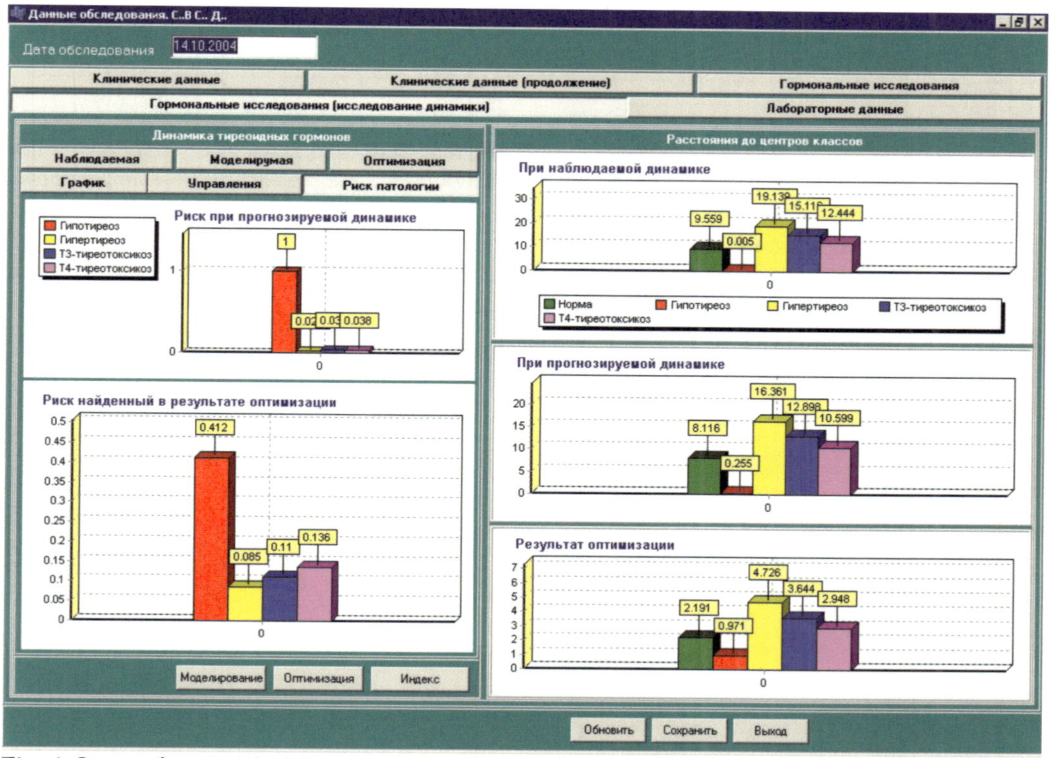

Fig. 4. Screen form with risk assessment of the pathological changes for the patient No.2

6. Literature

Lyashenko I., Yanenko N. (1999). The Usage of cost-benefit ratio in risk assessment of region contaminated by Chernobyl accident, *Proceedings of the 9th Annual Conferece: „Risk Analysis: Facing the New Millennium*, pp. 745-747, ISBN 90-407-1954-3, Rotterdam - The Netherlands, October, 10-13, 1999.

Yanenko V.M., Rykhtovsky V.O., Yanenko N.V. (2006). The ecological, medical and cybernetics technologies: modeling, catastrophes risk assessment, cost/benefit analysis at the prevention of the possible ecological and medical accidents. *Proceedings of the International Conference: Twenty Years after Chornobyl Accident. Future Outlook*, pp. 116-121, ISBN 966-96638-2-2, Kiev - Ukraine, April 24-26, 2006.

Baloga V., Kholoscha V. et al., (2006). National report of Ukraine. In: *Twenty Years after Chornobyl Accident. Future Outlook*, 224 p., Atike (Ed.), ISBN 966-326-166-8, Kiev - Ukraine.

Sergienko I.V. et al. (2000). Computer technology for waste management for Chernobyl remediation, *Environmental Management and Health - An International Journal*, Volume 11, Number 5, 2000, pp.410-421, ISSN 0956-6163.

Yanenko V.M. (2003). Risk ranging of ecological and medical violations caused by influence of the Chernobyl catastrophe factors, *Proceedings of the International conference: „Anthropogenically changed environment of Ukraine: risks for the population's health and ecological systems"*, pp. 77 – 98 (in ukr), Kiev – Ukraine, March 25 – 29, 2003. Special issue of the journal «Ecological bulletin» - Kiev, Chernobylinterinform, 2003

Yanenko V. et al. (2003). The Dynamics of Special Natural Investigations of the Chernobyl Alienation Zone during 1998-2002, *Proccedings of «XXVI International workshop "Modeling of developing systems: analysis and control of ecological technogenic and telecommunication risks. Problems of rehabilitation and recreation of the Chernobul Alienation zone (MRS-XXVI - 2003). To the memory of Viktor Glushkov – LXXX"*, pp. 44-53, ISBN 966-02-2985-2, Kiev - Ukraine, September 11 - 17 2003.

Guastello S.J. (1981). The organizational security subsystem: some potentially catastrophic events. *Behavioral Science,* Vol. 26, 1981, pp. 48-58.

Poston T., Stewart I. (1980). Theory of catastrophe and its applications, 607 p., Mir (Ed.), Moscow –Russia (in rus).

16

Habitation and Noise

Vesna Zlatanovic-Tomasevic M. Sc. Arch.
College of Structural Engineering and Geodesy, Serbia

1. Introduction

Noise is a physical phenomenon, which spreads along all traffic arteries, seizing more settlements every day, and especially expressive in large cities where it endangers the population, not only with its influence on hearing aparatus, but on the whole human organism.

The largest effects in preventing this phenomenon are accomplished by proper planning of the cities and by respecting all normatives and standards in order to prevent the noise.

While planning the new settlements, all sources of the noise must be taken into consideration; the intensity, the frequency and lasting of the noise must be estimated and appropriate measures must be proposed.

When urban tissue reconstruction is done, controlling the noise is much complex problem and challenge, because it is often necessary to preserve the existing urban grid and the buildings of the special cultural and historical significance. Considering that one of the solutions-widening the streets, is often impossible or very expensive, problem is solved by technical intervention in or outside the buildings, by traffic restriction, by transforming some parts of the city into pedestrian zones or by constructing the underground objects.

The principles and the elements of environment management are:

- protection of the environment politics
- planning
- introduction (introduction and carrying out)
- measuring and evaluation (checking and correcting measures)
- reexamination and improving
- permanent improvement

2. Noise as Environmental Problem

The noise in the settlements, mainly from the traffic, is one of the important environmental polluters. Depending on lasting and intensity, it leads to permanent organism damage, primarily effects the hearing apparatus, neurological-vegetative system, decreases the power of concentration and working ability.

It is interesting that even the ancient Romans had some regulations for noise protection, by which the blacksmiths and coppersmiths were banned to work in some periods of the day (in the afternoon and during the night).

In Yugoslavia, the first studies of communal noise from medicinal aspect were began in 1939. (S. Ramzin) and those measurements were conducted by subjective method, while the first measurements by objective method were taken during 1953. (Dr. R. Felix). The first medicinal technical research of harmful affects of communal noise were organized by Federal Institute of Public Health, Clinic of Medical School in Belgrade, Military-Medicinal academy and School of Electrical Engineering..
Beside the problem of communal noise, the problem of noise inside the buildings appeared, which is primarily connected to the change of construction methods, from traditional to industrial.
The noise is every undesirable sound. To be called the noise, it has to be strong enough to be heard. There is the difference between the noise from the **natural sources** (like thunder, wind, water, etc.), which is not relevant for urban environment, and the noise from **man made sources**, in working environment and in the rest of environment.
The noise is made at the place where the source is: on the street, in the apartments, in the backyard, etc. The noise can be various in types: **continual**-constant intensity (easier to bear); **discontinued** (more difficult to bear); **isolated sounds, murmurs**. It depends for example on traffic frequency and street width.
To alleviate the influence of noise on people's health, the noise protection measures are taken: the **preventive** and **sanitation** measures.

- **preventive** measures are those which are in the plan and project of the building, and where suggested means do not anticipate the possibility of noise creation
- **sanitation** measures are partial and restricted to sanitation of individual buildings

The measures which are taken in noise prevention are **technical** - on the way of the sounds; **physiological** - on organism protection and **legal** - regulations for allowed level of noise, and the measures which are taken in the buildings are **constructively-technical** which assume the wall construction, thickness, window and door solutions, etc.

3. Traffic Noise in Banjica Housing Project

The city of Belgrade covers the area of about 3221 km^2 (metropolitan area), where the urban aglomeration is about 765 km^2. Residential area covers the surface of 137 km^2 and traffic about 32 km^2. Belgrade has approximately 167 000 buildings. Banjica borough spreads on the surface of 85ha and has a population of 3500.
The urban noise is measured in Belgrade on 18 spots and the results show the high level of noise. The measurements in Banjica housing project are taken in 1999. on 9 measuring spots. The highest level of noise for this location as residential area are: during the day 55dB (A), and during the night 45dB (A). The measurements were done at the traffic arteries on the edge and in the heart of the settlement during the day with statistic analyzer (built in memory, type 4426 BRUEL):

- 2m away from the road at the height of 1,5m
- in the range of 36-100 dB (A)
- frequency range of 20-20 000Hz
- with 2 000 samples taken in the intervals of 0,2 sec
- all results are the statistic processing of the noise which lasted 6.6 min

The results showed the exceed of the maximum level of noise of 24,0 dB (A).

Measuring spot	MEASURED EQUIVALENT LEVEL OF NOISE		ALLOWED EQUIVALENT LEVEL OF NOISE		DIVERGENCE	
/	24.12.82.	01.03.99.	24.12.82.	01.03.99.	24.12.82.	01.03.99.
	dBA		dBA		dBA	
1	70,2	73,1	55	55	+ 15,2	+ 18,1
2	71,1	79,0	55	55	+ 16,1	+ 24,0
3	66,0	76,5	55	55	+ 11,0	+21,5
4	61,3	69,0	55	55	+ 6,3	+ 14,0
5	48,2	64,9	55	55	- 6,8	+ 9,9
6	61,3	59,8	55	55	+ 6,3	+ 4,8
7	68,9	76,3	55	55	+ 13,9	+ 21,3
8	63,3	60,7	55	55	+ 8,3	+ 5,7
9	53,6	53,9	55	55	-1,4	-1,1

Table 1. Overstepping of average values, or + 6,3 dBA of 16,1dBA, or +5,7 dBA of 24,0 dBA

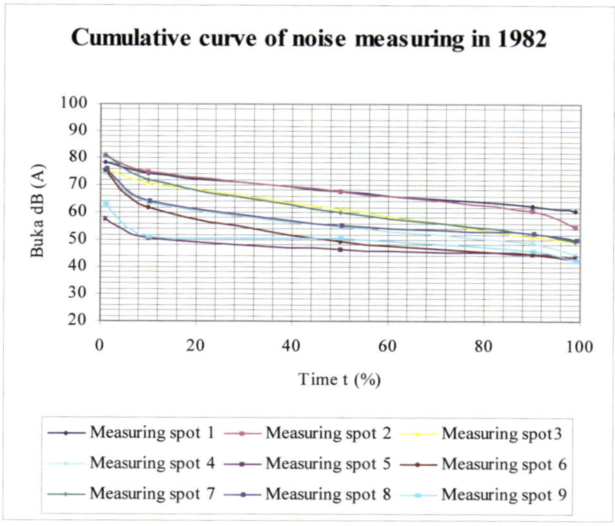

Fig. 1. Cumulative curve of noise measuring in 1982

On the base of the presented measurements, it is necessary to take some measures, primarily through the change in means of transportation (from individual to public transportation) and to minimize the need for traveling. Aiming that, it is essential to take the following measures:
preventive:

- long term strategy in organization of public transportation
- protection of the locations for future public transport
- setting the strategic frame for public approaches
- to define the environment criterion for new traffic installations
- constructing the housing projects beside already existing lines of transportation
- to build housing projects where infrastructure exists
- to revise the locations for construction which are accessible only by vehicle
- to prevent the construction on the locations with difficult approach

- to build the business centers near the housing projects

sanitation:

- traffic regulations (regulating traffic regime in the noisiest streets, modifying cargo transport to lighter vehicles, limiting transport hours, etc.)
- protection through sound barriers
- protection through vegetation
- sanitation of the most endangered buildings

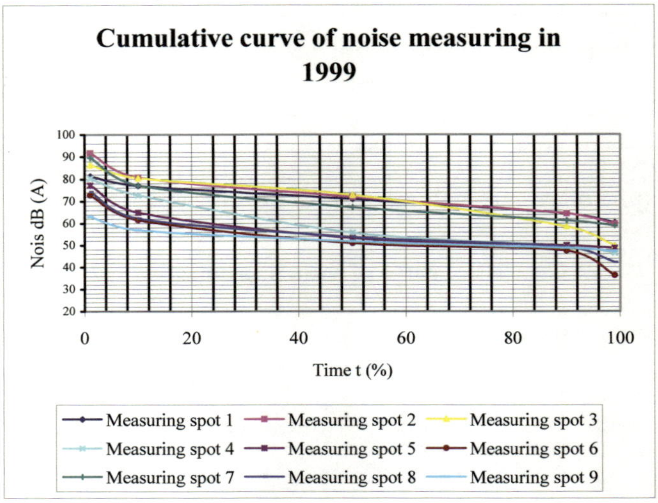

Fig. 2. Cumulative curve of noise measuring in 1999

4. Conclusion

Having insight into planned, technical documentation and by insight into the existing situation on terrain, it is established that Banjica housing project is a quiet city residential district.

The measurements, which were carried out inside the housing project, show some deviation on certain measuring places from noise level allowed by regulations. That calls for taking some measures of protection from the urban noise.

If the average level of noise in Banjica housing project is taken into consideration, and if the calculations are done on that base, it could be said that the results are satisfying.

5. Literature

V.Zlatanovic-Tomasevic, Habitation and noise, Belgrade 1977.

V.Zlatanovic-Tomasevic, Analysis of Banjica housing project from the aspect of noise protection, *Masters Thesis*, Belgrade 1982.

V.Zlatanovic-Tomasevic, Managing the noise as an ecological risk on the example of Banjica housing project, *Sustainability and city*, Faculty of Architecture, University of Belgrade, 1999.

V.Zlatanovic-Tomasevic, Environment and noise, Summer school of urbanism, Sipovo 1999.

V.Zlatanovic-Tomasevic, Managing the noise as an environmental problem, International Symposium on planning and urbanism, Vrnjacka Banja 2001.